Benedetta Chizzolini
Between Cure and Control

Vigilanzkulturen /
Cultures of Vigilance

Herausgegeben vom / Edited by
Sonderforschungsbereich 1369
Ludwig-Maximilians-Universität München

Editorial Board
Erdmute Alber, Peter Burschel, Thomas Duve,
Rivke Jaffe, Isabel Karremann, Christian Kiening and
Nicole Reinhardt

Band / Volume 16

Benedetta Chizzolini

Between Cure and Control

Doctors, Convicts, and Slaves in Tuscan and Papal Galleys (16th–18th Centuries)

DE GRUYTER

Funded by the Deutsche Forschungsgemeinschaft (DFG, German Research Foundation) –Project-ID 394775490 – SFB 1369

Co-funded by the European Union (ERC, FemSMed-101097386). Views and opinions expressed herein, however, are those of the author only and do not necessarily reflect those of the European Union or the European Research Council. Neither the European Union nor the granting authority can be held responsible for them.

This book is a revised version of a doctoral dissertation written at the University of Munich (Ludwig-Maximilians-Universität München) and the University of Padua (Università degli Studi di Padova) and defended in November 2024.

ISBN 978-3-11-165289-4
e-ISBN (PDF) 978-3-11-165413-3
ISBN (EPUB) 978-3-11-165511-6
ISSN 2749-8913
DOI https://doi.org/10.1515/9783111654133

This work is licensed under the Creative Commons Attribution 4.0 International License. For details go to https://creativecommons.org/licenses/by/4.0/.

Creative Commons license terms for re-use do not apply to any content (such as graphs, figures, photos, excerpts, etc.) that is not part of the Open Access publication. These may require obtaining further permission from the rights holder. The obligation to research and clear permission lies solely with the party re-using the material.

Library of Congress Control Number: 2025940760

Bibliographic information published by the Deutsche Nationalbibliothek
The Deutsche Nationalbibliothek lists this publication in the Deutsche Nationalbibliografie; detailed bibliographic data are available on the Internet at http://dnb.dnb.de.

© Copyright 2025 the author(s), published by Walter de Gruyter GmbH, Berlin/Boston, Genthiner Straße 13, 10785 Berlin
The book is published open access at www.degruyterbrill.com.
Cover illustration: *Sleeping rowers*, watercolor by Ignazio Fabroni – sketchbook, Firenze, BNCF, Rossi Cassigoli 199, c. 120r.

www.degruyterbrill.com
Questions about General Product Safety Regulation:
productsafety@degruyterbrill.com

A mia nonna Irma, che mi ha insegnato che una donna vale tanto quanto un uomo.
A mio nonno Massimo, che mi ha insegnato la gioia nelle piccole cose.
A mio nonno Antonio, che mi ha insegnato la curiosità, e a sognare in grande.
A mia nonna Gabriella, che oltre ad essere stata la mia prima lettrice appassionata, mi ha insegnato l'amore incondizionato.

To my grandmother Irma, who taught me that a woman is worth just as much as any man.
To my grandfather Massimo, who helped me to find joy in the smallest of things.
To my grandfather Antonio, who nurtured my curiosity and urged me to dream big.
To my grandmother Gabriella, who not alone was my first passionate reader, but taught me unconditional love.

Acknowledgments

This study is a revised and shortened version of my PhD thesis, which I defended in November 2024 at the History Department of the Ludwig Maximilian University in Munich, with a co-tutelle agreement with the Università degli Studi di Padova. The three years of my doctorate have been the most intense of my life—both professionally and personally. I have lived in six cities, worked at five interdisciplinary and prestigious research centers, attended dozens of conferences, lost count of the number of journeys and modes of transportation I've taken. En route, I have had the privilege of meeting many colleagues, some of whom were more or less directly involved in the creation of this thesis, and many of whom I am now proud to call "friends." A few words of gratitude are certainly in order.

First of all, I would like to thank my PhD's supervisors, Arndt Brendecke, Mariacarla Gadebusch Bondio, and Paola Molino, for accompanying me along this path, for their invaluable advice and unwavering support, and, above all, for always believing in me and my potential. Thank you for your guidance not just academically, but emotionally as well. I would also like to thank Lucio Biasiori, Andrea Caracausi, and Maria Conforti, who, although not directly involved in the supervision of the thesis, constantly gave me their support and affection, and who always read my drafts with pleasure.

I would also like to thank Andrea Addobbati, Paolo Calcagno, and Luca Lo Basso, without whose valuable research advice I would have been lost, especially during the early stages of this project.

Thanks to Fernanda Alfieri, Giulia Bonazza, Lucia Dacome, Silvia De Renzi, Christian De Vito, Vincenzo Lagioia, Chiara Lucrezio Monticelli, Alessandro Pastore, and Andrea Zappia for the discussions and the exchanges.

A *remerciement* also to the archivists at the State Archives of Rome, Florence and Livorno, the Vatican Archives, the Vicariate of Rome and the Historical Archives of the Jesuits, as well as the staff at the Biblioteca Corsiniana, who have always been extremely helpful, and whose assistance has helped me to unravel the "wonderful world" of miscellany.

In this regard, I would like to thank Roberto Benedetti and Cesare Santus, whose advice and support were essential for the research on Civitavecchia and Livorno, respectively.

Special thanks are due to the institutions that generously awarded me research fellowships during the final year of my PhD: the Rolf und Ursula Schneider-Stiftung at the Herzog August Bibliothek –Forschungs-und Studienstätte für europäische Kulturgeschichte (Wolfenbüttel, Germany); the Max Weber Stiftung at the Deutsches Historisches Institut in Rome, Italy; and the Eva Schler Fellowship at the Medici Archive Project, Florence, Italy.

Before that, of course, I would like to express my gratitude to CRC 1369 Cultures of Vigilance at LMU, where I had the honor of working as an assistant researcher during the first two years of my Ph.D., thus starting my academic career in the best possible way.

A special word of thanks goes to Alina Enzesberger, without whose help and support I would probably still be lost in the complexities of German bureaucracy.

I am deeply grateful to Tamar Herzig, Principal Investigator of the ERC project *Female Slavery in Mediterranean Catholic Europe* (1500–1800), of which I am currently working as a Post-Doc. fellow, for her advice during the final stages of revising this work.

Finally, my sincere thanks to Martina Heger and John Barrett, whose hard work and efficiency were instrumental in the publication of this book.

Contents

Acknowledgments —— VII

Abbreviations —— XI

Remarks —— XIII

Introduction —— 1

Chapter 1
Theorizing Medical Vigilance between the 16th and 18th Centuries —— 15
1. Vigilance as a defining virtue of the *optimus medicus* —— 15
2. *Optimus medicus* as *medicus cautus*: the evolution of ethical and deontological inquiry —— 27
3. The complex nature of the doctor-patient relationship —— 42
4. The social and political utility of medical expertise —— 49
5. Medicine and jurisprudence: a complicated relationship —— 55
6. A new self-awareness: the publication of the *Methodus Testificandi* —— 60
7. A question of loyalty: the physician's dilemma —— 67

Chapter 2 Early Modern Galleys as Spaces of Vigilance: Livorno and Civitavecchia from the 16th to the 18th Century —— 69
1. Vigilance aboard galleys: punishing, controlling, and correcting deviance —— 69
2. Early modern galleys and rowers from the 16th to the 18th century —— 77
3. Juridical discourse on the galley sentence —— 85
4. Cultural discourse concerning the galleys —— 94
5. Galleys and vigilance practices in early modern Tuscany: the case of Livorno —— 99
6. The Papal States and the galleys of Civitavecchia —— 113

Chapter 3
Medicine and the Early Modern Galleys —— 125
1. The profile of a galley doctor —— 126
2. Health and manpower at sea —— 135
3. The rower's physical examination —— 168
4. Controlling the crew: crime, punishment, and medical expertise —— 178

Chapter 4
On the Sin Abhorred by Beasts: Sodomy aboard Early Modern Galleys —— 191
1 Criminals and immoral individuals: Cultural representations of early modern *galeotti* —— 191
2 Sodomy and its condemnation in the early modern period —— 196
3 Moral order and sodomy: sexual transgressions among early modern *galeotti* —— 205
4 Criminal prosecution of sodomy in 18th-century Civitavecchia —— 224
5 Theoretical gaps in early modern medicine: the case of sodomy —— 235

Conclusion —— 243

Bibliography —— 251

Index of Persons —— 273

Index of Subjects —— 275

Abbreviations

ASF	Archivio di Stato di Firenze, Mediceo del Principato (MP)
ASL	Archivio di Stato di Livorno
AAV	Archivio Apostolico Vaticano
ARSI	Archivum Romanum Societatis Iesu
ASR	Archivio di Stato di Roma
ASVR	Archivio Storico del Vicariato di Roma
BAV	Biblioteca Apostolica Vaticana
BCR	Biblioteca Corsiniana Roma
BNCF	Biblioteca Nazionale Centrale Firenze
BNCR	Biblioteca Nazionale Centrale Roma
HAB	Herzog August Bibliothek, Wolfenbüttel

Remarks

I have opted to use Italian abbreviations for the archival sources:
b., bb.= *busta, buste* (envelope/s)
f., ff. = *folio, folii* (sheet/s)

For the Archivio di Stato Firenze sources:
filza, filze = (volume/s),
c., cc. = *carta, carte* (sheet/s)

Unless otherwise indicated, all the Latin passages have been translated by the author.

The transcription of the sources—including the Arab names of the enslaved individuals—reflects the original manuscripts; hence any spelling and grammatical errors have been retained.

In the Italian context, galley rowers are generally referred to as *galeotti*. Initially, the term *galeotto* referred exclusively to convicts, but by the mid–16th century it had come to include convicts and slaves serving aboard Mediterranean galleys. While the literal translation into English would be "galley slaves," this translation, I believe, can be misleading, as it risks obscuring the presence of convicts on board, and suggesting that the galleys were manned solely by enslaved oarsmen.

Given that this research focuses on both convicts and enslaved seamen, I will use *galeotti* (unless otherwise specified) to refer to the general category of galley crew—whether convicted or enslaved—reflecting the original Italian usage. The English term "galley slaves" will be reserved exclusively for enslaved individuals.

Although imperfect, I will use the terms "slaves" or "Turks," in keeping with the early modern Italian sources, to refer to galley slaves. I am fully aware of the complex implications of the term *slave*, especially in light of the Anglophone preference for person-centered language such as "enslaved individuals," which aims to foreground personhood over legal status. In this study, however, I have chosen to retain the term *slave* to remain linguistically close to the original Italian sources. Whenever the term *slave* appears, it refers specifically to galley slaves.

The term "Turk" was commonly used in early modern sources to refer to not only subjects of the Ottoman Empire, but more broadly to Muslims and non-Christian Africans.

When referring specifically to convicts, I will use "convicts," "forced rowers," or the Italian term *forzati*.

Finally, the term *doctor* will be used as a general designation for medical practitioners, including physicians and surgeons. While the term may appear anachronistic at first glance, it is frequently found in archival sources (as *dottore*), even in reference to educated surgeons, such as those aboard the galleys.

Introduction

> Into whatsoever houses I enter, I will enter to help the sick, and I will abstain from all intentional wrong-doing and harm.[1]

With these words, the Hippocratic Oath has, since antiquity, affirmed the physician's fundamental duty to protect the well-being and recovery of the patient, regardless of their social status. While attentiveness to symptoms was already central in ancient medical traditions, it was during the Middle Ages that the figure of the *optimus medicus*—the vigilant, discerning physician—was more fully conceptualized.[2] However, it was only in the early modern period that this attentiveness acquired new political, cultural, and institutional meanings. The emergence of physicians as authoritative experts on the human body marked a pivotal transition in the doctor-patient relationship.[3] Medical vigilance took on an increasingly complex character: it could be protective, but also became invasive, coercive, and disciplinary. The physician's gaze, once associated primarily with care, began to intersect with broader concerns of control, suspicion, and the maintainance of social order. In short, during the early modern period, the ethical commitment of physicians was frequently subordinated to the necessities of social control when a patient's behavior suggested they might pose a threat to the wider community.

As theorized in the 20th century by Michel Foucault, what appears to have been put into practice was a genuine "disciplinary" project in which the actions of doctors were often considered central, if not decisive.[4] Rather than expanding upon Foucault's analysis of the disciplinary character of medicine, this book takes his insights as a point of departure. It seeks instead to examine the inherently complex nature of the doctor-patient relationship throughout the early modern period—a relationship marked by shifting dynamics of cooperation and conflict, trust and suspicion, care and control.

In *Naissance de la clinique: une archéologie du regard médical* (1963), Foucault traces the transformation of medical knowledge in the transition from the 18th to the 19th century, arguing that the practice of care became inseparable from the exercise of power through the clinical gaze. In this epistemological shift, the patient ceases to be a speaking subject and instead becomes an object of medical

---

1 Quoted in MacKinney, Medical Ethics, p. 31.
2 Münster, Deontologia medica; Linden, Gabriele Zerbi; Schleiner, *Medical Ethics*; MacKinney, Medical Ethics, pp. 2–5; Minois, *Il prete e il medico* (translation); Arrizabalaga, Medical Ideals,
3 For the Italian context see Pomata, *La promessa*; Malatesta, *Fiducia.*
4 Foucault, *Naissance de la clinique; L'archeologie du savoir; Surveiller et punir.*

∂ Open Access. © 2025 the author(s), published by De Gruyter. [CC BY] This work is licensed under the Creative Commons Attribution 4.0 International License. https://doi.org/10.1515/9783111654133-004

observation—classified, interpreted, and treated as a "case." Care thus becomes not merely a therapeutic intervention, but also a "function of knowledge:" it is possible to treat only what it is possible to know. This gaze is inherently disciplinary: it inspects, categorizes, and normalizes, embedding medical care within broader structures of surveillance and control. The patient, within this framework, is reduced to an object, subject to the power of interpretation and intervention.[5]

The argument develops further in *L'archeologie du savoir* (1969), where Foucault shifts from examining the content of knowledge to analyzing the conditions of its possibility: discursive formations, institutional constraints, and "regimes of truth." Within this framework, "care" as a scientific and institutional practice can only exist within specific epistemic configurations that determine its legitimacy. Discipline, in this context, operates not only on bodies but also through discourse: it governs who is authorized to speak, in what terms, and on what subjects. In doing so, it challenges the presumed neutrality of scientific knowledge, revealing its embeddedness within historical and political structures of authority.[6] Finally, in *Surveiller et Punir: Naissance de la Prison* (1975), Foucault addresses discipline as a "technology of power" that molds bodies and behaviors. Institutions such as hospitals, asylums, and prisons become sites where care and punishment overlap due to the internalization of surveillance, rather than through direct coercion. In this context, care becomes a mechanism of normalization: the objective is not solely to heal, but to realign individuals with normative expectations of conduct and thought.[7] Across these three seminal works, the relationship between discipline and cure is revealed as profoundly ambiguous. Cure functions as a vehicle for disciplinary power—expressed through observation, classification, and correction—while discipline frequently takes the form of care, framing control as protection and normalization as healing.

To develop my analysis of the doctor-patient relationship, I deliberately distance myself from Foucault's disciplinary perspective, which tends to reduce the interaction between doctor and patient to a mere exercise of power and control.[8] In doing so, I will focus on a figure often overlooked in historical scholarship: the galley doctor. Physicians and surgeons aboard galleys were tasked not only with safeguarding the health of the crew, but also with upholding discipline among convicts and slaves—groups deemed inherently suspect due to their behavior, so-

---

5 Foucault, *Naissance de la Clinique.*
6 Foucualt, *Archeologie du savoir.*
7 Foucault, *Surveiller et punir.*
8 As a summary on the critiques of Foucault's perspectives on medicine, see Jones/Porter, *Reassessing Foucault.*

cial origins, or religion. Central to this study are the judicial proceedings brought against galley rowers accused of serious offenses, such as rioting, murder, feigned illness, and engaging in sodomitical acts. These cases illustrate the ways in which medical expertise was mobilized in contexts where health, discipline, and morality converged. This dynamic is further highlighted by the setting of the galleys—secluded, tightly regulated spaces where medical practice became deeply intertwined with enforcing authority and maintaining social order.

The narrative centers on several key questions: How did doctors' ability to observe and diagnose contribute to the health and well-being of the crew, and support order in such a tightly regulated environment? How did the medical gaze function as both a tool for understanding and treating illness, and a means of exerting authority in broader political and economic contexts? This book investigates how the boundaries between care, control, and punishment often overlapped and explores how the medical profession shaped perceptions of criminality and morality in the early modern Mediterranean world.

The decision to study the *galeotti* lies in the marginal nature of these figures. Their social status as convicts and slaves led physicians and society at large to treat them with suspicion and without scruple, as they were considered an inherently dangerous social group. Medical scrutiny was most intense when they were accused of violating the moral or military codes that governed life aboard the galleys. Medical practitioners were required to be vigilant primarily to keep the crew healthy, with the practical aim of ensuring successful navigation. However, this medical vigilance was also necessary to maintain discipline among rowers and to prevent criminal behavior. In this sense, accusations of feigning illness or practicing sodomy are exemplary cases. Any suspicion of simulating illness, on one hand, and of sexual deviance,[9] on the other, called for a scrupulous examination of the rowers' bodies by surgeons, who were considered the only professionals capable of establishing obvious truths through a precise physical examination. Yet, when it came to fights and murders, the severity of the crime—the determination of the aggressor's guilt—was invariably established on the basis of a meticulous medical examination of the wounds or corpses.

This research also underscores the designation of doctors as "experts of the body."[10] In this context, special attention will be given to the objectification of the rowers' bodies through the medical gaze, which was employed for both military and economic purposes. Furthermore, it is crucial to recognize that the phys-

---

9 Although rooted in 19th-century connotations, the term "deviance" will occasionally be employed throughout this work to denote any behavior or individual that diverges from established norms.

10 de Ceglia, *Body of Evidence*.

ical examination of galley slaves and forced rowers also provided a crucial opportunity for advancing medical knowledge. The confined and controlled environment of the galleys allowed for more precise studies of diseases and even facilitated experiments on rowers' bodies, who, during their service, were stripped of their humanity and reduced to the status of mere beasts.

The geopolitical focus of this book is the Italian peninsula, with particular attention to the ports of Livorno and Civitavecchia. As the principal naval bases of the Tuscan and Papal fleets, these ports functioned not only as hubs of commercial and cultural exchange but also as operational centers for galleys engaged in corsairing activities against the "Turks."[11] Their urban landscapes were shaped by a diverse population that included merchants, condemned prisoners, and galley slaves. Both the Grand Duchy of Tuscany and the Papal States played prominent roles in the Mediterranean corsairing wars, aimed at undermining the economic and military power of Muslim adversaries, while simultaneously serving as a means of acquiring captives for ransom or religious conversion.[12] Although other kingdoms, such as France and Spain, were actively engaged in pirating activities in the Mediterranean, following patterns quite similar to those observed in the Italian peninsula, the picture that emerges from this analysis must be regarded as specific to the peninsula itself. This is particularly evident in the medical dimension, which appears to be notably more developed compared to that of other political realities involved in corsairing. Equally distinctive is the influence of the Roman Church, whose role in the organization and discipline of naval crews represents a distinctively Italian feature.

While these two ports differ in multiple and significant ways, the commonalities—particularly in terms of crew management—are equally noteworthy. I chose to analyze Livorno and Civitavecchia because, in both contexts, it was the Capuchin Friars who were ultimately responsible for the daily oversight, spiritual welfare, and management of the crews. Consequently, comparable symptoms of control and management of convicts and galley slaves emerged, grounded in principles and implemented by the same actors.[13]

---

[11] The corsairing wars encompassed a series of military engagements, naval confrontations, and territorial disputes, particularly in the eastern Mediterranean. Their primary objective was to weaken the military and economic interests of opposing forces, often through the capture of prisoners of war. These captives were frequently enslaved and employed in public labor, with the potential for ransom or sale to private individuals.

[12] To cite some titles: Benedetti, Servi introvabili; Bono, *Schiavi*; Fiume, *Schiavitù mediterranee*; Santus, *"Il turco"*.

[13] The presence of Capuchin Friars will be later attested in other contexts, such as in Genoa, but only from the 18th century onward.

Furthermore, Livorno and Civitavecchia represent two complementary contexts in terms of crew composition. As Lo Basso observed, the Papal fleet relied more heavily on convicts to complete its crews, while the fleets of knightly orders such as the Knights of St Stephen in Tuscany depended more on enslaved rowers, captured during corsairing activities.[14] Both fleets, however, followed the same model, with crews composed predominantly of forced oarsmen—slaves and convicts—unlike ports such as Venice, which relied more on free oarsmen, and Genoa, which aimed to maintain a certain balance between all three categories of oarsmen (free, convicted, and enslaved).[15] A comparative approach between these two realities provides a coherent view of how crews were managed, shedding light on the role of the doctor aboard fleets, which were primarily deployed to fight the "Turks," a context in which the influence of the Catholic Church was extremely strong.

The research presented here likewise faced the challenge of chronological gaps in archival sources. The relatively small number of documents preserved from the late 16th century can be attributed to the fact that this period was one of institutional transition and consolidation. During this era, both state chanceleries and navies had yet to be systematically organized. Despite the scarcity of sixteenth-century sources, I have opted to begin my research with the Battle of Lepanto (1571), which contemporaries regarded as a pivotal juncture in managing naval crews, particularly in terms of health and sanitation practices. It is important to clarify that, by selecting the Battle of Lepanto as the starting point of my research, I do not intend to imply that this event marked a sharp rupture, or ushered in entirely new practices and strategies. Rather, the healthcare and vigilance practices under examination have much earlier origins, although they gained greater visibility and momentum in the Italian navies from 1571 onward.

The galley thus emerges as an exceptional space—physically and socially—within the early modern world. The *galeotti* formed a marginal social category, not only because they were convicts and enslaved individuals, but also because they lived in a liminal space separated both physically from the rest of society —in the ports, on the galleys, at sea—and socially, as they were subjected to a form of autonomous legislation. Galley slaves and convicts were thus a marginal social category in relation to the dominant group of free and Christian citizens. They represented an "otherness"[16] deliberately kept at a distance and separated from the rest of society—due to their criminal and immoral behavior, in the

---

14 Lo Basso, *Uomini da remo*, p. 21f.
15 Ibid.
16 See the definition of *alterità* which I find to be clear and convincing in Di Nepi, *I confini della salvezza*, p. 25.

case of convicts; or their adherence to a non-Catholic faith, in the case of galley slaves.

Even though separated from society at large, galley rowers were still not completely isolated from it. This was especially true of slaves who worked in the port's taverns and workshops during non-shipping periods. Indeed, in Civitavecchia, and even more so in Livorno, opportunities for encounters and cultural exchanges between Muslim slaves and Catholic citizens formed an integral part of everyday life, and a certain degree of integration into Christian society was even possible. As Giovanna Fiume has suggested, it almost seems as though slavery, despite its aberrant nature, had the merit of assigning the foreigner a place—albeit temporarily—by regulating his admittance and presence in a society alien to him.[17] Even amid prevailing antagonism toward the Ottoman presence, the "Turk"—while remaining the primary enemy of Christianity and the object of prejudice, stereotyping, and discrimination—could become familiar.[18]

Convicted rowers, on the other hand, had fewer opportunities to interact with the rest of society, as they were only used for forced labor on land. It seems plausible that this greater freedom afforded to enslaved rowers can be attributed to the fact that they likely did not need to be forcibly isolated; their skin color, physical features, language, and even their attire were sufficient to identify them as different, alien individuals.

Furthermore, galleys and their associated shore facilities constituted real "spaces for vigilance," where various strategies for controlling and disciplining the *galeotti* were implemented at different levels, with medical vigilance being just one of many. Particularly after the Battle of Lepanto in 1571, one can observe a systematic effort to regulate and re-educate the behavior and morals of the *galeotti* aboard the early modern Italian galleys. This process aimed to create a crew of obedient and God-fearing oarsmen who would ensure successful sea voyages, in accordance with a disciplinary and confessional framework.[19]

To study the galley context in all its specificity affords us insights into the complexities surrounding the role of doctors in early modern times, as well as the function and meaning of "vigilance" within medical practice. Indeed, doctors' obligations toward galley convicts and slaves were ultimately identical to those toward ordinary patients. However, the status of the former as forced rowers necessitated heightened attention from medical practitioners—a sustained vigilance.

---

17 Fiume, *Schiavitù Mediterranee*, p. 33; Bono, *Schiavi*.
18 Bono, *Schiavi*; Ricci, *I turchi*; Valensi, *Ces étrangers familiers*.
19 Lavenia, *Dio in uniforme*. I will return to the historiographical issue of disciplinarity later.

In approaching the figure of the galley doctor, it is essential to first offer a brief historiographical overview of the limited studies that exist at the intersection of medical history and maritime history. Existing works primarily focus on epidemics—most notably plague outbreaks—and the institutional responses they triggered, such as quarantine.[20] They also engage with themes from the Age of Discovery, notably the notion of *unification microbienne du monde*.[21] By contrast, studies probing the role of physicians prior to the 18th century are almost entirely lacking.

The British context has been a notable exception, with several works addressing what could be termed "naval medicine" in the early modern period. However, these works focus on military vessels and tend to overlook the Mediterranean world.[22] While Anglo-Saxon historiography deserves credit for addressing a subject largely neglected elsewhere in Europe, it often overstates the pioneering role of English naval medicine—largely due to the absence of broader comparative research. Although a few recent studies have begun to examine other European contexts, these too remain largely centered on colonial settings. Their focus is typically limited to the professional figure of the Atlantic slave trade ship's surgeon, the conventional themes of maritime hygiene, or the advancement of medical knowledge through contact with hitherto unknown diseases.[23] Moreover, scant attention has been paid to physicians who—though fewer in number than surgeons—were nonetheless a constant presence in the Mediterranean fleets.

In contrast, this book contributes to a growing—albeit nascent—body of scholarship on medicine aboard Mediterranean fleets. While recent research has begun to address the cultural and social dimensions of maritime medicine, especially in relation to slavery and race, these works remain overwhelmingly focused on the Atlantic world.[24] The Mediterranean—particularly its western basin—has only recently begun to attract renewed scholarly attention.[25] Despite this

---

20 Biraben, *Les hommes face à la peste*; Harrison, *Contagion*; Inì, *Lazzaretti*, pp. 83–123; Hengerer/Demichel, *Vigilance and the Plague*; Ziegler, *Coastal Policing*.
21 Le Roy Ladurie, L'unification microbienne du monde, pp. 627–696.
22 See, for instance, Tröhler, *To Improve Evidence*; Hudson, *Military and Naval Medicine*; Caputo, Sickness, Agency, pp. 749–769.
23 Bruijn, *Ship's Surgeons*; Koslofsky/Zaugg, *A German Barber-Surgeon*; Linte, *Médicine des colonies*.
24 Smallwood, *Saltwater Slavery*; Mustakeem, *Slavery at Sea*; Murphy, Re-writing Race. See the outgoing project *Medicine and the Making of Race, 1440–1720*, led by P.I. Dr. Hannah Murphy at the King's College, London.
25 For the Mediterranean context, although related to the Medieval period, see: Ferragud, Role of Doctors, pp. 143–169; Barker, *That Most Precious Merchandise*, pp. 92–120. These texts are the

burgeoning interest, the role of medical practitioners at sea in the Atlantic and in the western Mediterranean remains an urgent research objective. For instance, the database *Slave Voyages*, which documents transatlantic and intra-American slave trade, notably omits terms such as "medicine," "doctor," and "surgeon" from its glossary.[26] Moreover, although some projects explore the relationship between medicine and slaves aboard ships in the early modern period, there is a complete lack of in-depth studies on the intersection between medicine and convicts—notably the medical care for rowers—within Mediterranean navies.

Up to this point, I have used the term *medical vigilance*—but what exactly does it mean? The Latin verb *vigilare* denotes an action that is essentially "horizontal,"[27] meaning it requires active participation from individuals who must remain "vigilant"—responding to any noteworthy stimulus. The term, now widely used in the humanities, was first introduced into the medical field by the British neurologist Henry Head (1861–1940) in the first half of the 20th century. In its original medical context, it referred to a function of the nervous system, where responsiveness to stimuli was essential for proper operation. Over time, the term has expanded into other domains, but it continues to refer to a specific form of heightened human attention—one activated in relation to particular tasks and duties. One could argue that writing a history of vigilance is, in effect, akin to writing a history of human attention.[28]

In historical research, the term *vigilance* has been proposed as an alternative to the widely recognized concept of *surveillance*—deeply linked to ideas of *population control* by a central institution.[29] To *surveil* presupposes the presence of an authority that—from a commanding position—ensures that its subjects conform to established norms, both legislative and behavioral. The primary aim of surveillance is to detect any deviation—legal, social, or cultural—and to punish infrac-

---

points of reference for the historiography of the medical examination of the slave's body to determine their selling price in the Mediterranean. For the early modern Italian context see the ongoing project *Healing Slaves: Medicine and Slavery in Early Modern Italy* led by Professor Lucia Dacome at the IHPST, University of Toronto.

26 The database is sponsored by the National Endowment for the Humanities, and was carried out originally at Emory Center for Digital Scholarship, the University of California at Irvine, and the University of California at Santa Cruz. The Hutchins Center of Harvard University also provided support. https://www.slavevoyages.org/about/about#

27 A horizontal action refers to an initiative, intervention, or measure that takes place between actors operating on the same social level.

28 Brendecke, Attention and Vigilance, p. 18.

29 For a historiographic overview of the subject, see Lyon, *Surveillance Studies*; *The Culture of Surveillance*. Central to these studies by Foucault are: *Surveiller et punir*, and *Naissance de la clinique*. For the Italian context, see Prodi/Penuti, *Disciplina dell'anima*.

tions through what has been described as "vertical discipline." By this, I refer to that process of a "fundamental and lasting transformation of societal attitudes and social behavior," a transformation championed by political authorities through the enforcement of normative laws.[30]

The concept of social discipline [*Sozialdisziplinierung*] developed by Gerhard Oestreich in 1968,[31] offers a lens through which to understand this wider historical phenomenon aimed at producing a regulated and orderly society. This disciplinary process fundamentally alters social attitudes and fosters the internalization of externally imposed norms.[32] According to this view, discipline plays a key role in the transformations associated with the emergence of the modern European nation-state.

In contrast to the teleological theory of the formation of the modern state based on the concept of "social discipline," historians such as Wolfgang Reinhard and Heinz Schilling have instead proposed viewing the state as a historical phenomenon—its emergence contingent, even almost accidental. According to this perspective, European "confessionalism" was the determining factor in shaping the modern European state. Beginning with the Council of Trent (1545–1563), the reform of the Catholic Church during Counter-Reformation can be seen as the ecclesiastical authority's response to the secular state's growing demand for modernization.

Just as the political authorities sought to discipline the body, the ecclesiastical authorities began to discipline the soul, through a process which from 1981 onward became known as "confessionalization." This project aimed at internalizing models, ideals, and behaviors, imposed from above, rooted in the values and teachings of a particular confessional doctrine and morality. Since 1983, confessionalization has been interpreted as closely intertwined with the concept of "social discipline"—the two being increasingly seen as two sides of the same coin.

Confessionalization, in this view, facilitated the broader disciplinary project by laying the foundations for modern rationality: it established clear theological orthodoxy in opposition to superstition, disseminated and imposed new ideological and moral norms, and bureaucratized religious practice. A striking example is baptism, which came to function not only as a religious rite but also as a social

---

30 Härter, Disciplinamento sociale, p. 636.
31 The concept was first elaborated in a lecture given in Hamburg in 1968, and published in 1969 under the title *Geist und Gestalt des frühmodernen Staates. Ausgewählte Aufsätze*.
32 Härter, Disciplinamento sociale, p. 637.

and political mechanism for recognizing and institutionalizing membership in the Catholic community—and, by extension, in the broader social order.³³

Over recent years, the concepts of surveillance, social discipline, and confessionalization have come under sharp criticism for their excessively rigid character. Critics contend that these models not only fail to adequately capture the evolving dynamics of early modern social relationships, but also tend to exclude any form of subjective agency. This oversight limits individuals' capacity to adapt to historical contingencies and to negotiate their own "agency."³⁴

In historical research, the term *vigilance* has been theorized as an alternative to the more familiar concept of *surveillance*.³⁵ Vigilance studies aim to shift the focus away from institutional mechanisms toward the participatory role of private citizens, who voluntarily report what they have seen and heard. The underlying premise is that, for a security system to be truly effective, it must enlist the active cooperation of individuals working alongside institutions in pursuit of broader goals, such as maintaining public order. Authorities sought to stimulate this private attentiveness in order to safeguard public security—essentially by "integrating human attention into social tasks."³⁶ Vigilance arises whenever people are asked to focus on specific phenomena and to react to them, or report them to the authorities, when necessary.

Similarly, in medicine, a "culture of vigilance" has been theorized and ultimately practiced as an integral part of the medical profession. The ideal of the "vigilant doctor," characterized by the exercise of the virtue of prudence [*prudentia*] and caution [*cautela*] has been a subject of historiographical attention since the early 20th century.³⁷ However, in-depth studies exploring the various meanings and applications of medical vigilance remain limited. The deontological

---

33 Reinhard, Konfession und Konfessionalisierung, pp. 165–189; Schilling, *Konfessionskonflikt*. On the idea of "confessionalization" as a result of the Counter-Reformation in Italy, see, e.g., Prodi, *Il sovrano pontefice*; Prodi/Penuti, *Disciplina dell'anima*; Prosperi, *Tribunali della Coscienza*; Alfieri, L'età della disciplina cristiana.
34 Brendecke/Molino, *Cultures*; Brendecke, Warum Vigilanzkulturen?, pp. 10–17. On the limits of the concept of "disciplining," see also Schiera, Prodi, "Disciplinamento", pp. 349–351; Blockmans/Holenstein/Schläppi, *Empowering Interactions*. On the concept of "agency" see Thompson, *English working class*; Johnson, *On Agency*; Sewell, *Agency*.
35 For a historiographic overview of the subject, see Lyon, *Surveillance Studies*; *The Culture of Surveillance*. Central to the studies by Foucault are *Surveiller et punir*, and *Naissance de la clinique*. For the Italian context, see Prodi/Penuti, *Disciplina dell'anima*.
36 Brendecke/Molino, *Cultures*, p. 11f.
37 See, for example: MacKinney, *Medical Ethics*; Wear/Geyer-Kordesch/French, *Doctors and Ethics*; Schleiner, *Medical Ethics*; Linden, *Gabriele Zerbi*; Gadebusch Bondio/Förg/Kaiser, *Zerbi, Über die Kautelen der Ärzte*.

ideal of the *medicus cautus* [cautious doctor] or *medicus prudens* [prudent doctor], in fact, encompasses different forms of "vigilance," depending on the situation at hand.

After an initial focus on this early modern ethical-deontological reflection, the vigilant attitude of doctors came to be predominantly described in negative terms throughout the 20th century. It has been viewed as a precursor of a model of vertical social control, rooted in concepts of discipline and surveillance, which reached its apogee toward the end of the early modern period with the rise of "clinical medicine."[38] Medicine's intrinsically disciplinary nature, as explored in Foucault's seminal studies, has garnered significant attention—particularly for its critique of a whole set of social norms perceived as overly intrusive in the individual's private sphere.

Despite the widespread acceptance of these theories, many historians of medicine have been skeptical about Foucault's approach, which they considered too limited to fully define the complexity of relations within the medical world during the "classical age." Instead, these historians then began to focus more dispassionately on the relationship between medicine and power—or, one could say, between medicine and discipline—concentrating on the practical political roles that doctors played rather than the theoretical implications of their work.[39] Early modern physicians were acutely aware of their central political function, so much so they even wrote and published treatises known as *medico-political* works, which discussed political issues related to public utility, such as its role in criminal proceedings.[40] In fact, the role of physicians and surgeons as experts in court [*periti*], and their contributions to the establishment of forensic medicine as an official science, beginning in the 16th century, has been a recognized field of research since the 1980s.[41]

Despite an abundance of academic literature on the topic, some crucial questions concerning doctors' fundamentally vigilant character remain unanswered—notably about their ambivalent role in general, and their stance toward the inter-

---

38 See Foucault, *Folie et déraison*; *Naissance de la clinique*.
39 See Pastore, *Le regole dei corpi*; Arrizabalaga, Medical Ideals; Mandressi, Medicus politicus. For the political function of doctors during epidemics, see, in particular, the work by Cohn, *Cultures of Plague*.
40 Ibid.
41 For the naissance of forensic medicine, as well as for the strict relationship between medicine and political power in early modern Italy, see the works by Pastore: *Il medico in tribunale*; *Le regole dei corpi*. For a history of forensic medicine in general, see Clark/Crawford, *Legal Medicine*; Watson, *Forensic*; de Ceglia (ed), *The Body of Evidence*. The work of Fischer-Homberger *Medizin vor Gericht—Gerichtsmedizin von der Renaissance bis zur Aufklärung*, published in 1983, marked the beginning of this shift in the direction of research.

ests of their patients in particular. Drawing on a wide array of archival sources—medical records, criminal proceedings, institutional correspondence—this study seeks not only to reconstruct the everyday medical practice aboard the galleys, but also to interrogate the ways these documents represent, interpret, and sometimes distort the lived realities of those they describe.

Aware of the inherent limitations of such documents—often fragmentary and also frequently presenting a singular, typically institutional, perspective—this book approaches the available archival sources not as mirrors of the past, but rather as mediated representations of daily life aboard the galleys, filtered through the naval authorities in their interactions with slaves and convicts. Particular attention is given to what Natalie Zemon Davis has termed the "fictional" qualities of the sources—how narratives are shaped, situations described, and words chosen and employed. On one hand, the archival narrative structure must be interrogated to uncover meaning beneath its surface. On the other, it must be understood as both a conscious and unconscious interpretation of reality, offering valuable insights into the social and cultural context in which these narratives were constructed.[42] This approach not only gives voice to figures traditionally excluded from or rendered silent in historiography due to their apparent invisibility, but also exposes the thematic silences embedded within archival narratives themselves.

This book is divided into two distinct sections. The first delves into the theorization of medical vigilance during the early modern period, exploring how physicians and surgeons were regarded as the experts responsible for observing and interpreting the bodies under their care. The second part turns to archival research, focusing on the galleys as "spaces of vigilance" and re-education for rowers, and examines the specific role played by medical staff in these settings. These two sections are linked by the figure of the doctor, whose vigilance was central not only to healing but also to maintaining discipline over the rowers' behavior and moral conduct.

Chapter 1 introduces the historical context of early modern medicine, highlighting how vigilance became a fundamental competency for medical practitioners. It also explores two particular types of medical literature that were influential during the period. The first are the medical treatises written between the 15th and 17th centuries that outlined the characteristics of the ideal physician [*optimus medicus*]—both as a learned expert and a paragon of ethical conduct. The second are the *Methodus Testificandi* [Methods of Testifying], treatises published during

---

[42] Zemon Davis, *Fiction in the Archives*, pp. 2–5.

the 16th and 17th centuries that offered guidelines for doctors summoned to provide expert testimony at legal proceedings.[43]

Chapter 2 serves as a bridge between the theoretical and the practical realms. It lays the groundwork for understanding the structure of galley slavery and penal conviction during the early modern period, with a particular attention to the Grand Duchy of Tuscany and the Papal States. The chapter reveals how the galleys functioned as "spaces of vigilance," in which the medical role formed part of a broader system of discipline and control.

Chapter 3 shifts focus to the complex and ambivalent relationship between doctors and patients within this historical context. It examines the duties of doctors in detail, illustrating how their work was not only crucial for maintaining the crew's physical health, but also instrumental in upholding order and discipline within the galleys. Doctors were frequently called upon to provide expert opinions in cases of violence or disturbances, reinforcing their role as enforcers of authority. The chapter also delves into the personal and professional profiles of the medical practitioners involved, while considering the galley hospitals—spaces where healing and confinement coexisted, and where medical authority often intersected, and at times even clashed, with religious authority.

Finally, Chapter 4 focuses on a particularly sensitive and controversial issue aboard the vessels: sodomy. Through an analysis of legal proceedings involving convicts and galley slaves, this chapter investigates the role of medical experts in these cases, shedding light on the complex interplay between medicine, morality, and the law during the early modern period. It considers the doctor's role as a legal expert and explores how physicians and surgeons became central figures in determining the moral and criminal status of the individuals they examined. This chapter also underscores the pivotal role of medical vigilance in maintaining order and correcting perceived moral transgressions among crew members.

This book offers a historically grounded exploration of medical practice aboard Tuscan and Papal galleys, shedding light on how the intertwined functions of care, discipline, and social control shaped early modern maritime life. By focusing on the figure of the galley doctor—a practitioner situated at the intersection of healing and authority—it uncovers the lived experiences of marginalized individuals whose bodies became sites of both medical attention and institutional regulation. Through this lens, the galleys emerge not merely as instruments of war or punishment, but as spaces where the boundaries between health, morality, and legality were continuously negotiated. Drawing upon a rich array of archival sources, this study contributes to a broader historiography that seeks to understand

---

43 Pastore, *Medico in tribunale*.

how medicine functioned within the mechanisms of early modern governance, and how the ideals of vigilance, expertise, and salvation impacted the daily realities of galley rowers. In doing so, it restores visibility to those often omitted from the history of medicine and opens new avenues for exploring the entangled relationship between bodily care and coercive authority across the early modern Mediterranean world.

Chapter 1
Theorizing Medical Vigilance between the 16th and 18th Centuries

1 Vigilance as a defining virtue of the *optimus medicus*

Since ancient times, a "culture of vigilance" has been both theorized and practiced as an integral component of medical theory and practice. Physicians have long been expected to demonstrate vigilance, particularly in relation to the signs and symptoms of illness observed or reported by patients. During the early modern period, a more formal ethical and deontological reflection on the concept of vigilance (*vigilantia, prudentia, cautela* and related concepts) emerged, reinforcing its significance within medical discourse and clinical practice. A clearly defined ethical profile became a crucial tool for asserting the professionalism and social status of physicians and surgeons, grounded in the assumption that a competent doctor was, in every sense, a vigilant one: the *optimus medicus*.

The notion that the ideal physician was one who made the fewest mistakes was already established in antiquity, notably in the teachings of Hippocrates and Galen.[1] Fallibility has always been an intrinsic feature of both medicine and human nature in general. The awareness of navigating a field marked by a higher degree of uncertainty than other disciplines undergirded the development of a series of reflections within the medical world on how to minimize errors. The emergence of a *methodus* would not only have recognized the legitimacy of medicine's claim to the status of science, but also have guaranteed a technical superiority, manifesting as a distinct professional identity and exclusivity. Early modern physicians understood that in order to secure a monopoly in the health market and to assert a central position within society, they needed to equip themselves with every possible means to avoid making mistakes. Despite the extreme degree of uncertainty inherent in the humoral theory, its logical structure and repetitiveness still allowed doctors to operate with a certain degree of predictability.

---

[1] Gadebusch Bondio, *Avoidable Mistakes*, p. 1.

1.1 Medicine between theory and practice

The teachings of Hippocrates and Galen conveyed the idea that man should be considered a microcosm integrated into the cosmological macrocosm, and thus subject to the four cardinal physical laws of heat, cold, dryness, and humidity. The human body was thought to consist of a certain "temperament," the result of different combinations of the four elements—fire, air, earth and water—and four visible fluids known as the "humors": blood, phlegm, black bile, and yellow bile. Each humor was believed to correspond to one of the four principal organs of the human body—heart, brain, liver, and spleen.[2] Equilibrium between these four humors was considered indispensable for maintaining health. Bile and phlegm were identified as the primary humors responsible for disease. In addition to their natural presence in the body, their production was linked to specific seasons. Winter colds were associated with phlegm, while summer dysentery and vomiting were linked to bile. These two bodily fluids only became visible when expelled during a bout of illness and their appearance was almost universally considered dangerous. Blood, on the other hand, occupied a more ambiguous status. While it was associated with life, the fact that it was naturally expelled from the body at particular times—such as during nosebleeds or menstruation—was considered suspicious. Indeed, the human body was expected to regulate itself to some extent to keep these humors in balance and to expel them when in excess. The body's occasional need to rid itself of blood reinforced the idea that it could be the cause of disease.[3]

Scholars recognized that significant differences existed between individual temperaments and humor balances, often influenced by climate and habitat. People were identified as "sanguine"—corresponding to air on a macrocosmic level, being hot and humid; "choleric"—corresponding to fire, being hot and dry; "phlegmatic"—corresponding to water, being cold and humid; or "melancholic"—corresponding to earth, being cold and dry. Age and sex were also seen as determining factors in an individual's humoral balance, and personal temperament not only shaped one's personality but also one's susceptibility to certain diseases. For example, women were considered colder and wetter than men, leading to the belief that their blood was thicker.[4]

---

[2] Siraisi, *Medieval Renaissance Medicine*, pp. 104–106.
[3] Nutton, Medicine in the Greek World, p. 24f.
[4] Ibid., p. 102. The idea that women were seen as anatomically and physiologically imperfect versions of men seems to find support in Galen in *On the Use of Parts*, as argued, among others, by Laqueur, *Making Sex*, p. 124f. However, many historians have criticized this version, arguing that a strong dichotomous model of gender differences has existed since antiquity. For an

Disease was not perceived as an independent entity that could afflict the human body, but rather as an excess of one of these four humors, resulting in a state of "dyscracy." Healing, consequently, was only possible by restoring equilibrium, either through a particular "diet"[5] or through the expulsion of the corrupted humor from the body, primarily by the evacuation of blood.[6]

The conception of the human body as divided into an inner and an outer dimension, the former being of greater importance and concern due to its association with the soul and the exercise of the intellectual faculties, formed the basis of a theoretical distinction within the "official" medical world into three hierarchical categories of professionals.[7] At the top of this pyramid stood physicians—those doctors who attended university or another recognized medical school, completed their studies, passed their final examination, and obtained the doctorate. Below them came pharmacists, followed by surgeons, who, in theory, were the only professionals permitted to use their hands and were restricted to dealing with external diseases. Conversely, internal diseases were the sole domain of physicians, who were the only ones who could prescribe medicines to be ingested.

This tripartite image of medieval and early modern medicine, however, represents a conjectural framework, predicated on the hierarchical distinction within universities between theoretical and practical medicine. The reality was clearly more complex. This was particularly true on the Italian peninsula, where, between the 12th and 15th centuries, a category of "learned" surgeons emerged—those who had had studied at university and obtained a degree in surgery. Given that they were recognized as belonging to a professional category with established academic skills and credentials, their relations with physicians were certainly more equal.[8]

The traditional image of the absolute distinction and antagonism between physicians and surgeons—or rather of the perceived inferiority of the latter—is further challenged by the fact that, by the early 17th century, hospitals were increasingly configured as spaces where surgeons exercised autonomy and initiative, often independent of the authority of physicians.[9] Moreover, given that a basic knowledge of the principles of anatomy and physiology was indispensable

---

effective summary of the criticisms of the concept of the one-sex body paradigm, see Park, "One sex" Body, pp. 150–175, and, in particular, the bibliography cited on p. 153f.
5 A diet not only in the sense of a nutritional regime, but in the broader sense of a set of rules for life—nourishment, physical activity, rest, and more—designed to maintain good health.
6 Brockliss/Jones, *Medical World*, pp. 108–114.
7 Pomata, *La promessa*, p. 134f.
8 Siraisi, *Medieval Renaissance Medicine*, pp. 175–180.
9 Conforti/De Renzi, Sapere anatomico, p. 444.

for the successful practice of surgery, a number of surgical manuals began to be published from the 16th century onward, most written in the vernacular in order to make them accessible to a wider audience. Drawing more from everyday experience than from classical medical texts, these treatises elevated the role of the surgeon over that of the physician, indirectly accusing the latter of blindly accepting a bookish medical culture.[10]

It is important to note that, although the statutes established by various medical colleges prohibited doctors from using knives, scissors, or any other surgical instruments, they still needed a basic understanding of surgery and anatomy to effectively supervise the work of surgeons. In theory, surgeons could not practice without the permission and instruction of physicians. In his treatise *De Optimo Medico* (1551), Antonio Siccus used the analogy of a ship, as well as that of an army in battle: for a captain, a king, or an emperor to succeed in their endeavors, they had to command from above, directing subordinates conscientiously and diligently.[11] Furthermore, many physicians were also qualified as surgeons, thus becoming *doctor utriusque medicinae*—doctors of both branches of medicine.[12] In contrast, the differentiation between theoretical and practical medicine was more pronounced outside the Italian peninsula, especially in northern Europe, where surgery was theoretically excluded from university curricula until the 18th century. In such contexts, the low esteem in which surgeons were held was exacerbated by the presence of such hybrid figures as the "barber-surgeon"—simple craftsmen who, though adept with a razor-blade often lacked formal training in medicine.[13]

This insistence on the superiority of the *theorica* over practice can be explained by the deeper concern about the fundamental nature of medicine—whether it was *ars* or *scientia*. This question delved deeper than it might at first appear; it touched upon the discipline's status, its social utility and, crucially, its relationship to truth. Avicenna, in the incipit of his *Canon* (1025), defined medicine as "the science by which we learn the various conditions of the body; in health, when not in health; the means by which health is likely to be lost; and, when lost, is likely to be restored. In other words, it is the art whereby health

---

10 Ibid., p. 454.
11 Siccus, *De optimo medico*, p. 8v : "medicus [...] similis est gubernatori navis, qui remigare sciat, & mlum scandere_ nec vero solu scit, sed ipse etiam haec faciat, & alia similia, quae ad nautas pertinet. Similiter etiam reges, & imperatores exercituum non paucos invenies, scire ea, & facere, quae militum sunt munera."
12 Robison, *Healers in the Making*, pp. 18–20; Savoia, Early Modern Italian Surgeon, p. 32; Stolberg, *Learned Physicians*, pp. 73–75.
13 Brockliss/Jones, *The Medical*, pp. 93–96.

is concerned and the art by which it is restored after being lost."[14] By the 13th century, however, it was widely accepted in the medical sphere that only specific aspects of medicine could provide the certainty required for *scientia*, in the Aristotelian sense of the term.[15] The particularity and extreme variety of patients and diseases, and the consequent difficulty of postulating universal criteria and treatments, led to the recognition that medicine was a *scientia coniecturalis*—conjectural science.[16] As Ian MacLean has rightly noted, recognizing the conjectural nature of medicine did not necessarily mean that the precepts of doctrine were as uncertain as its practice. Indeed, medical theory was not—and never has been—entirely abstract or self-referential: theoretical knowledge always aimed at concrete cases just as logical skills were intended to solve practical problems.[17]

1.2 A methodological approach to treatment

Therapeutic activity primarily involved the careful monitoring of the state of equilibrium—or disequilibrium—between the humors, with the ultimate goal of maintaining their balance. The Hippocratic-inspired diagnostic approach was essentially based on examining the patient in all their complexity and individuality, adopting what has been described as a "patient-oriented rather than nosographic-oriented form of inquiry."[18] In other words, before deciding on the optimal treatment for a patient, it was essential to conduct a thorough analysis not only of the disease's trajectory but also of the patient's habitual lifestyle. Therapeutic prescription in humoral medicine unfolded in three stages: anamnesis—collating information about a patient's medical history; diagnosis—identifying the disease; and prognosis—predicting the disease's course and outcome.[19] In a medical system where both diagnosis and treatment options were often fraught with uncertainty, the importance of a correct prognosis had consequences not only for the success of the therapy proposed but, more significantly, for the physi-

---

14 Ibid., p. 78.
15 Gilly, *Theodor Zwinger*, p. 139 f. For Aristotle, the sciences were divided into theoretical (*physica, mathematica, prima philosophica*), productive (arts and techniques), and practical (*ethica, oeconomica, politica*). In the strict sense of the word, "science" only referred to the theoretical sciences, grounded in the deductive method, and not the practical and productive sciences, which are based on contingent objects.
16 Agrimi/Crisciani, *Edocere Medicos*, pp. 139–150; Siraisi, Bresadola, Segni evidenti, p. 733 f.
17 Maclean, *Logic, Signs and Nature*, pp. 68–73.
18 Nicolson, The Art of Diagnosis, p. 802.
19 Siraisi, *Medieval Renaissance Medicine*, pp. 123–136.

cian's credibility. The two principal diagnostic methods employed were urine and pulse tests, as explained by Galen in his treatise *De pulsis et urinis*.[20] The examination of urine was considered particularly useful. Being naturally excreted by the body, the study of its color, quantity, and presence of sediment would allow the physician to observe, by means of visible characteristics, what was otherwise invisible within the body: the humors and their fluctuations.[21]

Despite attempts to establish medicine with a proper methodology, similar to other sciences like logic—aiming to generate unquestionable and definitive results—the inherently unique nature of each disease as a disorder of humors that always differed from one individual to the next and, by extension, of the administrated therapy, kept medicine in the early modern period in a state of persistent uncertainty, making prognosis difficult to predict. Furthermore, it was believed that the optimal therapy was rooted in the principle of opposites—an excess of cold and wet humors should be countered with warm and dry remedies—and that the patient's own input was essential for the treatment's success. Indeed, it was up to the patient to know their own temperament and clinical state better than anyone else, enabling the physician to prescribe the most appropriate remedy for their specific case.[22] Moreover, the inability to clearly define the difference between cause, symptom, and disease only aggravated this state of confusion, thus amplifying the perplexity surrounding the field. While medicine continued to claim its scientific nature, it lacked the degree of certainty typical of the theoretical sciences.[23] To avoid errors during these critical moments in therapy, physicians had to rely on their own experience, both theoretical—rooted in the teachings of the ancient authors, and practical, predicated on sensory and personal experience.[24]

According to the Roman protophysician Paolo Zacchia (1584–1659), a doctor could commit three types of error: ignorance, negligence, and willful misconduct [*ignorantia, negligentia, et dolus*]. From a purely legal standpoint, willful misconduct was the only type of error punishable by law, given that it entailed an intentional criminal act. For Zacchia, however, the most serious errors a medical

---

20 Ibid., p. 134.
21 Moulinier Brogi, *L'uroscopie*, pp. 7–9.
22 Brockliss/Jones, *Medical World*, p. 299.
23 Maclean, *Logic, Signs and Nature*, p. 261.
24 The term "experience" referred to any kind of knowledge the physician acquired during his career and which contributed to the increase of his knowledge and skills. The ideal product of the Renaissance educational system was thus personified in the figure of the "rational doctor." He had to be equipped with adequate knowledge, based on his own experience, derived both from the study of books and from sensory knowledge of the particular.

practitioner could commit were the "sins" of *ignorantia*, and even more so, of *negligentia*. Both, in part, stemmed from poor or non-existent vigilance on the part of the medical practitioner. Indeed, physicians could sin not only *in committendo* [through positive action], but also *in omittendo* [through omission]. While ignorance could be partly excused due to the vastness of medical knowledge—or a physician's lack of experience when young—it became unforgivable when it led to negligence. In such cases medical practitioners failed to recognize the limits of their knowledge or abilities; or chose not to act appropriately due to recklessness or carelessness.[25] To protect themselves from accusations of negligence, physicians needed to align themselves with an ideal model of physician—one they could aspire to and embody in practice. Drawing on medieval precedents, a series of treatises were published from the late 15th century onward, stipulating the behavior and practices that medical practitioners were expected to follow.

Let us begin by examining the knowledge considered essential for any physician to be deemed "good." Their education should commence at an early age, when the mind is more malleable. Theory—defined as the possession of a general understanding of basic physiological principles—was seen as the starting point for medical practice. Without an in-depth understanding of the human body and its workings, it would, practically speaking, be impossible to treat a patient. Therefore, the basic requirement for a physician to avoid errors was a thorough theoretical knowledge of medicine, knowledge which could be acquired both through a rigorous study of the subject, within a defined academic path, and through the mastery of a *methodus:* a codified and regulated procedure designed to acquire correct theoretical knowledge [*scientia*] that could then be applied in practice.[26]

Theoretical knowledge was anchored in the ancient authors' teachings, beginning, of course, with the works of Hippocrates and Galen. Any young physician had to be well-versed in the rules of natural philosophy. As Galen had observed, a good physician was, first and foremost, a philosopher. Alongside philosophy, other sciences were seen as complementary to medicine. According to influential physicians such as Pietro d'Abano (1250–1316), a physician should have mastered the *artes liberales*, especially dialectics. Ever since the Middle Ages, the liberal arts had constituted the two foundational groups of academic teaching: the *trivium* which included grammar, rhetoric, and dialectics, and the *quadrivium*, which encompassed arithmetic, geometry, music, and astronomy.[27] As disciplines, astrology

---

25 Marchisello, La responsabilità del medico, pp. 221–248.
26 See Zwinger's reflections on the Hippocratic method in Gilly, *Theodor Zwinger*, pp. 54, 119.
27 Gadebusch Bondio, *Avoidable Mistakes*, p. 7.

and music were particularly relevant to medical practice. Both astrology and medicine were considered conjectural arts reliant upon predicting future events through signs incomprehensible to non-specialists.[28] Astrology, in particular, was believed to play a key role in a doctor's prognosis, notably in relation to the theory of critical days, a branch of medical astrology. This field involved the application of astrological techniques for medical purposes, such as calculating the position of the planets on any given day in order to decide whether a particular medical procedure should be carried out. It was also believed that the position of the planets at the time of birth could influence a person's temperament.[29] The study of music, too, was viewed as decisive, as it was thought to aid in properly interpreting and analyzing the rhythm of the human pulse.[30]

Some physicians, however, believed this educational system was flawed, as it was overly grounded in accumulating abstract concepts that had little practical value for therapeutic practice. This was the case, for example, with the Turin physician Leonardo Botallo (1519–1558), who, in his *De medici et de aegri munere* [the duty of the doctor and the patient] (1565), did not reject the centrality of the liberal arts, but rather argued that they were essentially barren when studied in isolation. Specifically, Botallo accused his colleagues of overly concentrating on the definition of things, losing sight of their essence and thus depriving themselves of the possibility of acquiring genuine knowledge about them.[31]

For Botallo, a thorough knowledge of diseases and the human anatomy was of paramount importance. The practice of human dissection for academic purposes was first documented in 1315, when the anatomist Mondino de' Liuzzi conducted a dissection on the body of a criminal in Bologna. Over time, this practice spread to other European centers and gradually became an official part of the university curriculum.[32] At the University of Florence, for example, the 1388 statute of the *Studium* mandated that two dissections be conducted each year—one on a male corpse and the other on a female. Based on the assumption that "no one can be a good or fully trained doctor unless he is familiar with the anatomy of the human body,"[33] participation in both dissections was compulsory for graduation.[34]

---

28 Siraisi, *Medieval Renaissance Medicine*, p. 69.
29 Ibid., p. 135f.
30 Siccus, *De optimo medico*, p. 7r.
31 Botallo, *De medici munere*, p. 62.
32 Nutton, Medieval Western, p. 177.
33 Park, Criminal and Saintly Body, p. 14.
34 Park, *Doctors*, p. 59f.

The burgeoning significance of autopsies—specifically the *anatomia medica*—can be traced back to attempts to unite the two branches of medicine (*theorica* and *practica*) in order to overcome the probabilistic nature of medical knowledge.[35] Academic dissections had an exemplary purpose: to provide students with visual aids that would help them better grasp the teachings in their texts.[36] Dissecting a corpse had the merit of eliminating the obstacle of fragmented medical knowledge, thereby allowing universal conclusions to be formulated on the basis of specific results, in keeping with an inductive approach to investigation.[37]

Similarly, a physician was expected to possess a thorough understanding of both the natural and supernatural causes of disease, as well as be familiar with the "six non-naturals" [*sex res non naturales*]. These factors, while belonging to the realm of nature, did not directly relate to the body's constitution, but nonetheless influenced how it functioned. These "six non-naturals" encompassed air, food and drink, sleep and wakefulness, movement and rest, evacuation and satiety, and finally the *passiones animi*, which included sexual activity.[38]

Furthermore, physicians were required to have a solid grasp of both surgery and pharmacology in order to effectively supervise their subordinates. A good physician needed a method rooted in Aristotelian logic—one that would enable him to avoid errors in judgment. The idea of logic as fundamental to the medical profession had been firmly established since the Middle Ages. Pietro d'Abano, for instance, believed logic had to take precedence over philosophy and must adhere to the principles laid down by Aristotle.[39] Only through the rigorous application of logical rules could medicine be transformed from a conjectural science into an exact one, enabling physicians to identify signs of disease without the risk of erroneous diagnosis or prognosis. This involved adapting the variability of individual cases to the universal framework of logic. The ultimate goal was to equip medicine, once and for all, with the necessary means to become an exact science, through the use of analytical logic and systematic quantification.[40]

As Hippocrates taught, "experience"—in the broadest sense, encompassing both particular cases derived from practice and universal cases arising from theory—played a central role in training doctors and in medical practice.[41] Despite

---

[35] Liboni, Humanist Post-Mortem, p. 23.
[36] Siraisi, *Medieval Renaissance Medicine*, p. 89.
[37] Siraisi, Bresadola, Segni evidenti, p. 723.
[38] Maclean, *Logic, Signs and Nature*, p. 252.
[39] Siraisi, *Medieval Renaissance Medicine*, p. 67.
[40] Ibid., p. 164 f.
[41] Gilly, *Theodor Zwinger*, pp. 52–57.

the highly theoretical nature of the curriculum, the works selected for study in university classrooms were primarily aimed at equipping future physicians with the skills and knowledge necessary to treat their patients in the most practical and effective manner. Being both *ars* and *scientia*, medicine "was at once a system of explanation and a set of techniques; the acquisition of medical expertise was both an intellectual enterprise and a process of gaining skills."[42] As such, a period of apprenticeship with a more experienced practitioner was required, along with attending autopsies and anatomy lessons at least once a year, as these were considered essential for acquiring knowledge about the "hidden parts" of the human organism. Thus, theory and practice—which traditional medical historiography had for centuries presented as dramatically opposed—were, even within the academic realm, far more interconnected than previously thought. This is not to deny that, within universities, theoretical teaching took precedence over practical training.

With the growing focus on practice and experience in the medical field, a range of literary genres with explicitly practical objectives and character proliferated from the 16th century onward. This renewed attention to practical experience can be traced primarily to the humanist physicians' direct engagement with the works of Hippocrates. In particular, Books I and III of the *Epidemics* served as a key reference model for employing anecdotal narration [*per historiae*] as a guide for medical treatment. In the *Epidemics*, the narrative unfolds by means of presenting specific cases—through a description of how symptoms of the disease progressed in specific individuals, as observed firsthand by the physician from Kos.[43]

From the latter half of the 16th century, and increasingly in the 17th century, *per historiae* descriptions became a highly successful literary genre throughout the European medical world. By discussing individual medical cases, physicians —especially surgeons—were able to identify paradigmatic cases on which to optimally base their therapies.[44] In 17th-century medical treatises, the term *historia* explicitly referred to knowledge derived primarily from direct observation, often accompanied by bibliographical references, yet never subordinated to them.[45] The exemplary nature of history as *magistra vitae* [teacher of life] was thus adopted within the medical practice, with the narrative *per historiae* solidifying as an expression of *sensata cognitio*—practical, non-abstract knowledge rooted in sensory perception and direct observation.[46] To this end, collaboration among

---

42 Nutton, Medicine in the Greek World, p. 49.
43 Pomata, Praxis Historialis, p. 112.
44 Ibid.
45 Pomata,/Siraisi, *Historia*, p. 28.
46 Pomata, Praxis Historialis, p. 106.

experts was of paramount importance, especially in the form of the *consilium*—a collective consultation or advisory process in which physicians would come together to provide guidance in uncertain situations. As Chiara Crisciani noted, the act of "giving *consilia*" was not merely a competence exercised by the physician; it was, in many ways, central to defining their identity. The *consilium* served as an expert assessment of *what to do* in situations of uncertainty.[47] In the face of doubt, physicians had to skillfully navigate the challenges confronting them, carefully considering the nuances of each case, where variability was the norm. As a metaphor from the early modern period aptly put it: "The profession of doctor is akin to the profession of seafarer." Like navigators at sea, medical professionals had to be prepared for all eventualities, never allowing themselves be taken off guard by complications.[48]

Medical advice encompassed three spheres of action: the physician's direct instructions to the patient; collective decisions made by multiple healers about a single patient; and offering "opinions" to a colleague who explicitly sought help with a particularly difficult case. In the latter context, the specific case afforded an opportunity for broader reflection, allowing physicians to hypothesize from the illness at hand. This marked a departure from the traditional approach in early modern medicine, which had, for centuries, aimed to apply general and universal concepts to particular contexts. While the *consilia* were still designed with practical intent, they increasingly took on a distinctly theoretical and doctrinaire character. This shift was further influenced by the fact that those doctors offering advice often did so at a distance: the patient was not directly in front of them. Instead, they would read a letter from a colleague who had carefully summarized the clinical case and sought counsel. As a result, direct engagement with the patient—a hallmark of earlier medical case studies—gradually diminished. The *consilia* evolved in such a way as to assume the appearance, scope, and argumentative structure of the scholastic *tractatus*, eventually becoming an integral part of the academic education system.[49]

---

47 Crisciani, Consilia, p. 260.
48 Da Villanova, *Explicatio super Canonem*, quoted in Crisciani, Consilia, p. 260: "Officium medici est simile officio nautae." The metaphor is also found in Siccus, *De optimo medico*, p. 15r: "Nec vero male mihi videntur sentire, qui medicos impitos comparant mali navirum gubernatoribus: sicut enim minime percipiuntur, istorum errore tranquillorum pelago et pacato, in magni vero & adversis tempestibus perspicui sunt, ac tunc cuiuis liquet per imperitiam gubernatoris naugragium contingere, ita imperitorum medicorum errores, in levibus morbis latent vulgarem quemquae & inexercitatum, quem non eveniunt ex illis magne offesiones."
49 Crisciani, Consilia, pp. 266–268.

In the medical field, the literary genre of *Observationes*, which emerged in the second half of the 16th century, epitomized the focus on experience and individual cases. Initially, *Observationes* served as a means of self-promotion for physicians, highlighting their previous successes to demonstrate professional competence. However, the *Observationes* genre was not confined to medicine. As early as the 1630s, compilations of *Observationes Legales* or *Forenses* began to be published in the legal realm, featuring collections of solutions to hypothetical court cases based on jurists' common opinions. Reports of authentic cases, discussed and resolved in court, were also collected during this time.[50] In parallel, *Observationes* in medicine evolved as collections of real-life cases, with a focus on therapeutic success. Unlike the medieval *consilia*—which sought to abstract diseases and formulate general theories and doctrines—these cases were presented within their specific contexts, thereby offering future therapeutic models. Over time, however, the emphasis shifted to a more descriptive knowledge of the disease, derived from direct, first-hand observation.[51] Ultimately, *Observationes* were intended for publication and, above all, for circulation among colleagues. The goal was to contribute —by dint of the detailed description of therapeutic cases—to the creation of a shared and effective body of knowledge within the community of medical scholars and professionals—the *respublica medicorum*.[52]

This focus on the particular nature of therapeutic practice reminds us that therapy involves more than just the interaction between the doctor and the disease, or between colleagues. In fact, there is a third, and perhaps more crucial, component: the patient. The importance of the doctor-patient relationship, emphasized since ancient medicine, and notably reflected in the Hippocratic Oath,[53] is central to effective treatment. For treatment to succeed, the physician needed not only a deep understanding of the disease and the human body, but also the skill to conduct tests to detect and interpret signs of disease practically and correctly. Equally essential was gaining the patient's trust—an element always deemed vital for the success of any treatment. With the twin objectives of avoiding errors and establishing clear standards of practice, early modern physicians had to rethink the nature of their chosen profession, carefully crafting a model of practice to guide their interactions with patients at the bedside.

---

50 Pomata, Sharing Cases, p. 201f.
51 Ibid., p. 205.
52 Ibid., p. 197.
53 Rigato, Medico divino e razionale, p. 40.

2 *Optimus medicus* as *medicus cautus:* the evolution of ethical and deontological inquiry

To explore how a physician could embody the ideal of the attentive physician [*medicus cautus*] in practice—and thus how this ideal could be realized—treatises were published from the late 16ᵗʰ century that sought to offer behavioral models for physicians to follow in practice. The *cautelae* genre was seen as most fitting for this purpose, bridging the gap between the ideal model of a physician and its practical application.[54] As the medical profession consolidated in the early modern period and as medical knowledge was increasingly scrutinized, there arose a growing need to codify ethical and deontological standards, with the ultimate goal of safeguarding medicine's social standing.[55] In this context, the ethical profile of a conscientious physician began to reflect not just technical expertise, but also an evolving framework that sought to position medicine more strategically within society. Adherence to these standards allowed good physicians to distinguish themselves not only by their skill and knowledge but also by their ethical conduct from their adversaries.

2.1 A competitive marketplace

In early modern society—where reputation was closely linked to fame and honor—a bad reputation could result in outright exclusion from political and social life. The importance of a good reputation in medicine was particularly relevant, in a field with an uncertain and irregular clientele, where no guaranteed remuneration existed for any services rendered. Patients often selected the healer to whom they would turn based on purely "word-of-mouth" recommendations. As Gianna Polmata highlighted, a patient's judgment of a healer's ability, in practical terms, carried more weight than that of professional colleagues. Recognition from the authorities "above"—in the form of a degree or a license to practice[56]—was

---

54 Linden, Gabriele Zerbi, p. 20 f.
55 Münster, Deontologia medica, pp. 60–83.
56 The need to bring order to the vast and varied world of healers led the political authorities to make several attempts to regulate them, both politically and socially. Institutions such as guilds and medical colleges were created to exercise greater control over this profession, which, despite numerous criticisms, was recognized as essential to the upkeep of a resource that the authorities considered increasingly important—public health. Following the example of the medieval *arti*—secular associations of individuals practicing the same profession—guilds of physicians were established. One of their key functions was to keep special registers to identify those members

insufficient without what might be called "legitimation from below."[57] Physicians, however, faced widespread skepticism toward the medical profession, largely driven by the inherent uncertainties of medical practice and theory. This general suspicion and mistrust of medicine's capabilities were compounded by many notorious polemics against medical professionals, leading to accusations to which they had to respond in order to save their profession's social standing and good repute.

The primary accusation levelled at doctors—next to incompetence—was that of greed. It was widely believed during the early modern period that physicians and apothecaries intentionally mislead their patients with incorrect or inappropriate treatments, aiming to prolong the course of the disease and thereby enrich themselves.[58] This accusation became particularly intense during times of plague, with physicians and apothecaries sometimes accused of deliberately infecting the population to boost their earnings. Cardinal Sforza Pallavicino's chronicle of the 1656 outbreak of plague in Rome recalls how, at the onset of the epidemic, numerous rumors circulated suggesting that doctors—in league with the city's magistrates—were intentionally spreading the disease. The contagion was described as a fully-fledged "artifice of secret politics," a tool of political control set in motion by the authorities in cahoots with the medical sect. It involved machinations of covert politics, with doctors and political leaders colluding to control the population.[59]

Accusations that doctors endangered their patients due to ignorance, arrogance, and recklessness, clearly greater carried weight when voiced by fellow medical professionals. This criticism is particularly evident in *De medici et de aegri munere* by the Turin anatomist Leonardo Botallo, who explicitly criticized academic medicine for being overly reliant on book learning. He advocated a more pragmatic approach, arguing that for the good of the patient, medical knowledge had to be rooted in direct knowledge of the subject rather than in

---

who were registered, and therefore, were authorized to practice and enjoy certain rights and obligations. The guild itself was responsible for examining its members and granting them a license to practice, as well as ensuring that they met the professional and moral standards required for the profession. In Italian cities such as Venice, as early as 1316, guild-like organizations were set up in which only medical professors and graduate physicians could be admitted as members: these were the "colleges of physicians." These colleges presented themselves as genuine judicial bodies, responsible for examining and licensing all those who wished to practice medicine in a given territory. See Gelfand, Medical Profession, p. 1124; Gentilcore, I Protomedicati; Andretta, *Roma medica*.

57 Pomata, *La promessa*.
58 Siraisi, *Medieval Renaissance Medicine*, p. 85.
59 Pallavicino, *Descrizione del contagio*, p. 30.

abstract definitions.⁶⁰ Botallo contended that the conjectural nature of medicine rendered certainty impossible and that claiming to have all the solutions based solely on book knowledge was a clear sign of ignorance and hubris.⁶¹

During the early modern period, lacking a good reputation posed an enormous obstacle for any physician seeking to practice medicine. Medicine was viewed as a commodity in a competitive "medical marketplace," structured according to similar rules employed in commercial exchanges,⁶² where supply often far outstripped demand. Unlike the traditional tripartite model of medicine that assumed the monopoly of academic medicine, many healers—both official and unofficial—were active throughout this time.⁶³ Patients were free to choose their healer, often basing their decision on the expectation that their needs would be met effectively and swiftly. To achieve this, patients did not hesitate to turn to all sectors of medical practice, whether licensed or not. As Gianna Polmata noted, the hope of finding relief from their illness drove patients to engage in what she called "therapeutic experimentalism," trying every available option. In the patient's eyes, the two universes of medicine—the official and the unofficial—were not separate entities but rather part of a unified world from which they could select different cures.⁶⁴ Furthermore, the recurrent failures of official medicine—often demonstrating its impotence—made the use of pseudo-medicine and other non-institutionalized channels more acceptable to the general population. For official medicine to retain patients and convince them of the superior quality of care it offered, it needed to overcome mistrust while establishing an outright monopoly on the practice of healing.

Throughout the early modern age, physicians had to compete with a wide array of healers. It should not be overlooked that at that time there was a marked presence of "folk" healers, practicing a form of medicine rooted in empirical knowledge of the animal and plant kingdoms, passed down orally over generations. More often than not the boundaries between learned and popular medicine were blurred. Even physicians did not hesitate to use remedies of popular origin, particularly those validated by tradition as effective. At the same time, numerous concepts from academic medicine were made accessible and translated into vernacular knowledge.⁶⁵ Many physicians published treatises called *secreta* in everyday Italian as spoken by a layman, explaining how to obtain effective remedies

---

60 Botallo, *De medici munere*.
61 Gadebusch Bondio, *Avoidable Mistakes*, p. 11f.
62 Fissell, *The Medical Marketplace*, p. 533.
63 Robison, *Healers in the Making*, p. 8.
64 Pomata, *La promessa*, p. 249f.
65 Wear, Medicine Early Modern, p. 238.

for everyday ailments such as headaches or constipation—using a few simple ingredients that could typically be found in any well-stocked kitchen.[66] Paradoxically, the sheer simplicity of the treatments on offer was the key factor leading to a preference of domestic medicine over learned medicine. The use of technical language, particularly in Latin, only further alienated the largely illiterate population and increased the distrust patients harbored toward university-trained physicians. Folk medicine frequently appeared the optimal solution for patients seeking a relationship with the healer grounded, as far as possible, in a sense of "equality."

Alongside domestic medicine—unrecognized by the authorities' official licensing system—existed a distinct realm of practitioners linked to healing practices collectively known as *medici*, including barbers, midwives and others.[67] In *Quaestiones medico-legales* (1621-1635), Zacchia dedicates the entire first section of the sixth book to errors committed by doctors punishable by law [*De medicorum erroribus a lege punibilis*].[68] While eight of the thirteen *quaestiones* address errors committed by physicians, the remaining five consider those by other health practitioners, including, in order of importance, surgeons; pharmacists and apothecaries; empirics (i.e., quacks) and chemists; midwives; and finally, nurses and assistants. Zacchia, whose goal was to cover the subject as exhaustively as possible, adheres to the medical world's traditional tripartite hierarchy, yet also acknowledges those figures who, despite lacking official recognition—either due to the absence of a specialized corporation or formal academic training—were socially recognized as health professionals. Although indirectly recognized, Zacchia is particularly hostile to these healers who, despite possessing rudimentary curative skills, "abuse the title of doctor"[69] and were accused of doing more harm than good to their patients because of their extreme ignorance and lack of formal training. "Empirics" were harshly criticized for making mistakes every time they sought to treat a patient, as they acted without the slightest understanding of the disease they were supposed to treat or its causes, thus committing a "mortal sin."[70] The overriding reproach against them was their lack of a rationale.

Zacchia acknowledged that theoretical teaching was essentially incomplete without practical experience, for not only was practice without theory insufficient for the medical profession but was perilously so. Consequently, empirics were un-

---

66 Park, *Doctors*, p. 48 f.
67 Ibid., p. 136.
68 Zacchia, *Quaestiones*, pp. 371–406.
69 Ibid., p. 371: "omnes enim hi abusive Medici nomine veniunt."
70 Ivi p. 402: "toties peccato mortali subijciuntur, quia medentur non cognito morbo, nec causa inquisita."

equivocally defined as "enemies of men." The animosity toward them was so pronounced that the Roman protophysician did not hesitate to liken them to those figures regarded with even greater hostility by the academic medical world: charlatans.[71] The Portuguese physician Rodrigo de Castro (1550–1627) critically viewed empirics, devoting an entire chapter in his influential treatise *Medicus Politicus* (1614) on how to distinguish between true and false physicians [*De veri et falsi medici agnitione*], thereby warning his readers—particularly the easily impressionable women [*mulierculae*]—about the deceptive nature and inherent danger of those remedies offered by these "pseudo-medics."[72] While the practice of buying remedies from peddlers dates back to antiquity, as David Gentilcore noted in his study *Medical Charlatanism* (2006), early modern charlatanism emerged as a distinct socio-historical phenomenon. The very fact that charlatans were required to hold a license from the mid-16th century onward exemplifies the reality and novelty of the development. Also known as "empirics," charlatans were considered by the Italian *protomedicati* as a legitimate medical category with specific functions and characteristics, practicing medicine without any official qualifications—lacking formal training in medicine, enrollment in college, or membership to a guild.[73] To obtain a license, the charlatans had to convince the authorities of the novelty of their remedies, which had to align with pharmacopoeia and official medical theories, and prove effective by curing a large number of people. Obviously, the remedy had to be external, as ingestible or "internal" medicines were reserved for physicians.[74] According to Gentilcore, several factors contributed to the rise of early modern charlatanism. First, the progressive medicalization of Italian society in the 15th and 16th centuries, and the consequent organization of health professionals within faculties and colleges, transforming medical treatment into a service for which payment was required.[75] Alongside this, increased literacy, the rise of a mainly mercantile economy and the decline of the traditional craft economy all played a role. Many charlatans, being literate, could not only access medical and pharmacopoeia texts, but also engage on a more equal footing with medical and institutional authorities.[76]

---

[71] Ibid.: "sed empirici, a ratione ita sunt alieni, ut ei sint inimici manifesti."
[72] De Castro, *Medicus Politicus*, pp. 200–205.
[73] Brockliss/Jones, *Medical World*, p. 230.
[74] Gentilcore, *Negoziare rimedi*, p. 76f.
[75] Gentilcore, *Medical Charlatanism*, pp. 91–98.
[76] Ibid., pp. 99–106.

2.2 Ethical and deontological inquiry

To provide a positive code of practice, the authors of treatises on the *optimus medicus* felt it necessary to present a negative model by identifying and exposing the deceit and fraudulent practices committed by bad doctors. In his prologue to *Opus perutile de cautelis medicorum* (1495), the Paduan professor Gabriele Zerbi—while also defining the concept of *cautela*—argued that his primary goal was to provide his medical colleagues with the necessary means to avoid making errors. Zerbi offered physicians a code of conduct for use in all practical situations where errors could occur: "to make as few mistakes as possible [*peccare*], to err as seldom as possible [*errare*], and accordingly not be liable, or hardly ever, for legal prosecution [*delinquere*]."[77] Even more explicit was Giovanni Antonio Sicco, who, with his treatise *De Optimo Medico* (1551), strove to offer young aspiring physicians a model of excellence, contrasting it with the typical defects of his contemporaries. The ideal profile of the *optimus medicus* was personified by Sicco's professor at the University of Padua, Vittore Trincavelli, to whom his treatise was dedicated, as can be seen from the work's first page.[78] Unlike other contemporary physicians—accused of being arrogant and reckless, or timid and reticent [*multi enim praecipites sunt, ac audaciores; non pauci vero timidi & segnes*]—Trincavelli always acted diligently and prudently thus avoiding serious mistakes and invariably finding a concrete and most appropriate solution for each particular case.[79] Traditional medicine's strongly polemical nature is evident in Botallo's *De medici et de aegri munere*, where physicians who made mistakes were labelled "bad" and "false" practitioners, and often compared to actors on stage. In his treatise *De optimo medico*, published in Messina in 1637, the physician Pietro Castelli reformulated a traditional analogy: "such [bad] physicians resemble characters introduced in tragedies, for just as they share the appearance, attire, and persona of those they portray, yet are not truly those individuals, so too, many doctors exist only in reputation and in name, yet very few are so in reality."[80]

The desire to correct errors committed by contemporary physicians was expressed through the writing and publication of numerous treatises, not just in the form of *cautelae*. A notable example is the work by Sicilian protophysician

---

77 Gadebusch Bondio, *Avoidable Mistakes*, p. 5.
78 Siccus, *De optimo medico*.
79 Ibid., pp. 2r–4 l.
80 Castelli, *De Optimo medico*, p. 15: "Simillimi enim huiusmodi Medici sunt personis, quae tragoediis introducuntur; quemadmodum enim illi figuram quidem, et habitum, ac personam eorum quos referunt habent, illi ipsi autem vere non sunt: sic et Medici fama quidem et nomine multi, re autem, et opere valde pauci."

Giovanni Filippo Ingrassia: *Iatropologia liber quo multa adversus barbaros medicos disputantur* (1547).[81] While not a *cautela*, the *Iatropologia* represented a strongly polemical attack on Sicilian medical practitioners, whom Ingrassia considered barbaric [*barbaros medicos*] due to their outdated and completely ineffective, if not downright dangerous, medical practices. He also accused his peers of excessive presumptuousness and arrogance. Aimed at warning younger physicians against repeating the mistakes of their elders, Ingrassia's work tackled methodological issues, notably the need to unify medical theory and practice and the importance of subordinating medicine to philosophy, as Galen had taught. Recognizing the primarily philosophical nature of medicine, Ingrassia argued, would restore its erstwhile prestige, distinguishing and defending it from accusations of being a merely lucrative exercise.[82]

Ethical reflection in medicine has ancient origins, with references to the ideal qualities in a doctor's training, qualifications, and behavior dating back to ancient Greece. A key reference is unquestionably the Hippocratic Oath, which examines the healer's behavior in their relationship with their patient and their relatives—the *adstantes* as they were called, those surrounding the patient during illness. During the Middle Ages, these ideals were revived and adapted to align with Christian moral values.[83]

The earliest known work that grapples with such issues is *Arsenio's Letter to Nepoziano*, a manuscript likely written between the 5th and 8th centuries as a series of paternal recommendations to a son, a doctor. While obviously fictional, it allows for a broader ethical reflection on the medical profession. The text can be divided into four parts. In the first, the author dwells on the doctor's essential attributes: in addition to being sober and modest, he ought, above all, to be vigilant, compassionate, and skillful in all circumstances.[84] Following a similar model, a whole series of prescriptions was written between the 9th and 14th centuries on the qualities of a good physician, detailing the subjects he should master along with practical instructions—on procedures like checking someone's pulse and examining urine—as well as on how medical practitioners should interact with patients. These texts also highlight medical etiquette, stressing moderation. As in the earlier texts, the "good" doctor had to be sober and modest, eager to learn, humble and benevolent, able to control himself and avoid all excesses, especially those arising from the "passions of the soul." His approach should be discreet, primarily

---

81 Ingrassia, *Liber adversus barbaros*.
82 Gadebusch Bondio, *Avoidable Mistakes*, p. 8f.
83 MacKinney, *Medical Ethics*, pp. 2–5.
84 Minois, *Il prete e il medico*, (translation) p. 46.

aimed at not frightening the patient while fostering trust and cooperation.[85] From the 14th century onward, this genre of writing became more practical, with greater attention paid to how physicians should present themselves at the bedside. While medieval medical writing primarily sought to secure respect and obedience for the profession, a wider range of motivations emerged, with the patient becoming a co-protagonist in therapeutic practice. Acknowledging the predominately conflictual nature of the doctor-patient relationship, the physician had to find ways to earn his patient's trust and ensure his obedience for therapeutic purposes.[86]

Alberto de' Zancari's (1280–1348) *Libellus de cautelis seu documentis medicorum habendis*, likely written between 1301 and 1325, is an example of such a treatise. Concerned with preserving the reputation of his peers, Zancari stressed the need to avoid rushed or superficial diagnoses. Only a thorough study of the patient's medical history and signs of disease could enable a physician to correctly pronounce on the disease's progression and its future evolution.[87] This marked the beginning of medical examinations as detailed investigations, where physicians were called upon to examine both the symptoms and the patient in order to identify the true causes of illness and prescribe the most appropriate treatment.[88]

Another notable treatise, *De cautelis medicorum*, attributed to the Catalan physician Arnaldo da Villanova (1235–1311), insisted that honesty and meticulousness were essential qualities for a physician's competence and professionalism. This treatise is divided into four parts, offering general reflections on prudent conduct, rules for examining patients, health regimes, and a section entirely devoted to precautions regarding urine analysis [*cautelae circa urinas*].[89]

The content of these medieval texts does not substantially differ from that found in treatises on the characteristics of the *optimus medicus* published during the early modern period. The key differences are essentially twofold: first, the later treatises marked a growing departure from Christian sources, embracing a more secular and materialistic approach to medical practice; and second, they introduced a clearer formulation of the medical profession's deontological framework. While ethical reflection was already present in classical and medieval medicine, it was limited to the expression of ideals for conduct that medical practitioners were expected to keep in mind and endeavor to follow in practice. In contrast, the growing professionalization and regulation of the medical class

---

[85] MacKinney, Medical Ethics, p. 18.
[86] Duranti, Confidentia, p. 64f.
[87] Linden, Gabriele Zerbi, pp. 31–33.
[88] MacKinney, Medical Ethics, p. 24.
[89] Linden, Gabriele Zerbi, p. 32.

in the sixteenth and seventeenth centuries transformed these ideals into an actual code of practice, essential for full recognition as a member of the medical profession. Consequently, these codes of conduct—which were essentially ethical in nature—assumed a deontological character. Adherence to certain moral standards became a reflection of proven technical superiority, which had the merit of upholding the reputation and dignity of doctors and their profession.

2.3 "With diligent attention": the medical examination

The moment when a doctor's vigilance should have been at its sharpest was during the medical examination. Loren MacKinney discovered five manuscripts, dating from the 10th century to the 15th century, outlining how doctors were expected to examine patients.[90] These works emphasized that each patient was unique and therefore required a tailored approach. The primary objective of the examination was to "know everything" about the patient and their state of health. The process began by questioning the patient about the nature and intensity of their symptoms, followed by an assessment of their pulse and urine. The doctor would then inquire about the smallest details of the patient's illness—whether the pain was severe, whether they had difficulty sleeping or breathing, and additional complaints. After a thorough analysis of any symptoms revealed by the physical examination and the patient's responses to questions, the skilled doctor would be able to identify the causes of the illness and prescribe an appropriate treatment.[91] Since the Middle Ages, the acts of inspecting and inquiring have remained at the heart of the physician's role during the medical examination. This approach stemmed from the recognition that, in order to formulate a correct prognosis, doctors had to approach the patient's bedside in what MacLean called a "Sherlock Holmesian" manner—with a sharp eye for detail.[92] Therapy could succeed only if the doctor applied an investigative method, uncovering every possible clue to extract the truth. The level of attention required, then, had to be exceptionally high, turning the act into a genuine appraisal.

Though not exceptionally successful at the time, Zerbi's work surely remains the best example of this type of treatise. Zerbi, an anatomist and professor of medicine and logic, initially lectured at the University of Bologna (1475–1483) and later at the University of Padua, where he taught theoretical medicine from 1495 to

---

[90] MacKinney, Medical Ethics, p. 23.
[91] Ibid., p. 24 f.
[92] Maclean, Logic, Signs and Nature, p. 98.

1505. In the intervening years, he settled in Rome, where he lectured medicine at the university and served as Papal *Archiatra*—the pope's personal physician—at the courts of Pope Sixtus IV (1471–1484) and Pope Innocent VIII (1484–1492). Although historians have traditionally viewed Zerbi's work as the earliest and most comprehensive systematic work on medical ethics written during the early modern period, recent scholarship has shown that Zerbi was, in fact, reformulating long-standing ideas, referring not only to authoritative physicians and philosophers such as Galen and Hippocrates, but also to Pietro d'Abano and Alberto de' Zancari.[93] Katharine Park further emphasized Zerbi's great indebtedness to his Florentine colleague Niccolò Falcucci, author of the well-known *Sermones Medicinales* (also known as *Practica*), a compendium of contemporary medical knowledge on various diseases, their causes, symptoms, and treatments.[94] The primary goal of this work was to systematize medical knowledge and render it more accessible to a wider public. In the first *sermo*, Falcucci outlined the figure of the ideal doctor, for whom personal and ethical virtues were considered pivotal attributes in earning the patient's trust and, by extension, ensuring the treatment's success.[95] As Linden has argued, Zerbi's outstanding innovation lay in revitalizing the medieval *cautelae* genre by dedicating an entire work to ethical and deontological reflections—rather than limiting such material to a chapter or section within a general medical textbook. In doing so, he acknowledged their pivotal importance to the medical profession.[96]

Zerbi devoted the entire fourth chapter of his book to the subject of the medical examination and the physician's conduct toward patients during this process, titled *De modo se habendi medici erga patientes et maxime erga egrotantes* [On the manner in which the physician should behave toward the patient, and especially toward the sick]. For a prognosis to be successful, Zerbi advised that physicians maintain vigilance before, during, and after the examination. Under normal circumstances, the doctor was expected to visit the patient twice daily—morning and evening. In exceptionally grave cases, however, the doctor was not to leave the patient's bedside at all.[97] While overseeing the course of the treatment, the doctor was not to act alone but in close collaboration with the patient's family and friends. Their role was particularly important during the anamnesis phase, which involved collating information about the patient's medical history, temperament, symptoms, and the trajectory of the illness. Given the critical importance

---

[93] Gadebusch Bondio, *Avoidable Mistakes*, p. 3.
[94] Park, *Doctors*, pp. 110–112.
[95] Duranti, Confidentia, p. 67.
[96] Linden, Zerbi, p. 34.
[97] Zerbi, *De cautelis medicorum*, p. 92.

of gathering information to ensure a proper understanding of the symptoms of the disease, it was necessary to obtain as much detail as possible. Interviewing those who knew the patient and had cared for them during their illness was therefore indispensable. As Zerbi wrote: "Listening to the patient and to those around them; studying all the changes that have affected the patient in their life—and those that will affect them in the future, not only the major ones but also the smallest details—in order to consciously avoid all deceptions and mistakes."[98]

Once the physician entered the patient's room, he was expected to carefully observe the surroundings, looking for any fruits, herbs, or prepared palliatives from which he could deduce the patient's illness.[99] This preliminary phase was followed by the examination proper, which included close inspection of the patient's face—pallor, for instance, being considered a good indicator of the patient's overall health—followed by an evaluation of the pulse, palpation of various body parts, and, finally, a urine examination. After completing the examination, the doctor was to gather the patient's friends and family to inform them of the real causes and nature of the illness, and possibly provide a prognosis. Given the sensitivity of this final step, Zerbi advised that the physician proceed with the utmost caution, as a premature or inaccurate prognosis could not only harm the patient but also the doctor's professional reputation.[100] With this investigative phase concluded, the curative part of treatment—prescribing the necessary therapy—could begin.

2.4 Controlling gestures and words: medical etiquette

Another important precaution that doctors had to observe during the medical examination concerned their own conduct. As Zerbi noted, the physician was expected to present himself as praiseworthy to the public not only in his behavior and manners, but also in his attire. According to what might be termed a "medical eti-

---

98 Ibid., p. 104f. "Audiens ab infirmo, et ab astantibus omnia que possunt, investigans de omnibus mutationibus egro supervenientibus, et que supervenerunt, et ita in posterum que supervenient in processu vite non solum magnis, sed etiam qua<n>tumcunque parvis, ut certior factus deceptiones, et fallacias quascunque evitet."
99 Ibid., p. 98. "Utatur preterea medicus altera cautela dum locum ubi residet infirmus ingressus est si forte viderit fructus, herbas, aut fomenta aliqua parata per que coniecturari possit super egri infirmitate."
100 Ibid., p. 110.

quette," the physician had to be extraordinarily attentive, diligent, and courteous toward both the patient and any bystanders.[101]

As previously noted, the way the doctor presented himself to the patient was considered important, especially in the Middle Ages, when moderation was viewed as a defining quality of a good physician. The doctor's sobriety had to be evident in every aspect of his presence: in his gestures, speech, and appearance. From the medieval period onward, medical authors consistently emphasized the need for a highly careful and measured approach to the patient—one that would inspire confidence, foster trust and thereby contribute to the success of the proposed treatment.[102]

In this context, Sicco's moral portrait of the excellent physician is exemplary. He begins by emphasizing the doctor's training, which, according to tradition, should begin at an early age and be grounded primarily in the study of dialectics. This discipline was intended to continue throughout the doctor's life, instilling in him a habit of self-questioning and guarding against arrogance. The good medical practitioner was to be modest, sincere, and prudent, devoid of vices—especially those of gluttony, wine, and luxury. Finally, as Zerbi noted, a physician had to be God-fearing, recognizing that nothing could occur without the intercession of divine will.[103]

In addition to general rules, a distinct code of conduct was expected at the patient's bedside. Above all, the doctor was to remain calm and display good manners. He was expected to be friendly and courteous with the patient and his family, without being overly talkative. To ensure clarity and avoid arousing suspicion, doctors were advised to minimize the use of Latin and refrain from excessive courtly formalities. Beyond his behavior, the physician was also to be mindful of his appearance—neither negligent nor overly fastidious. Particular attention was paid to personal hygiene: doctors were expected to wash and apply perfume before seeing patients, though they were cautioned not to overdo the scent. These various expectations can be summarized by the maxim: *refrain from excess and follow moderation.*[104] The purpose of such medical etiquette was not only to distinguish the true doctor from his rivals, but also to win favor with the sick and, in turn, secure their compliance with treatment.

The physician's self-vigilance is similarly highlighted in the treatise of the Portuguese *converso* Rodrigo de Castro, written in the early 17th century as he left Lisbon

---

[101] Ibid., pp. 102 f.
[102] MacKinney, Medical Ethics, p. 18.
[103] Siccus, *De optimo medico*, p. 6r: "Verus autem, bonusque medicus Deum semper in animo habet, unum illim contemplatur, illi omnia refert accepta in sanandis morbis."
[104] Ibid., p. 10r: "sed ab omni excessu declinare, & mediocritatem sequi."

for Hamburg.[105] In the third book of his *Medicus Politicus*, de Castro argued that the physician's "prudence" should be twofold: the first type, defined as *militarem* [military] or *oeconomicam* [economic], was to be applied to others—particularly to patients and assistants—and resembled the prudence of generals with their troops or masters with their household servants. The second type, described as *eremiticam* [hermit-like], was to be turned inward, toward the physician himself. Once again, the aim was to provide doctors with a code of conduct by which to live prudently, cultivating a careful and self-restrained disposition.

De Castro also portrayed the good physician as a man of moderation, stressing that abstinence from vices and excesses was not only essential to maintain focus on one's work, but also for establishing the physician as a moral exemplar within society. Doctors were particularly advised to abstain from lust—a vice that, according to de Castro, could drive a man insane and reduce even the most capable to a beastly state *"quae nomine reddit belvinos ac bruti similes, & animum stupidum & ad sapientiam inertem"*—rendering the mind dull and inert in relation to wisdom. Above all, however, they were urged to refrain from melancholy.[106]

In line with contemporary polemical literature against the medical class, de Castro also identified avarice, pride, and envy as the most widespread vices among his fellow physicians. So pervasive were these flaws that physicians had to remain particularly vigilant against them. He reserved particular criticism for overly proud doctors who, in claiming the success of their treatments, failed to recognize that the therapeutic success and healing ultimately came from God acting through man.[107] Building on what his predecessors had written, de Castro further emphasized that modesty should never be mistaken for mediocrity.[108]

The impression the doctor aimed to convey was that of a competent yet humble professional; affable, but never inappropriate. Central to this perception was the physician's outward appearance. Although, as de Castro reminds us, "clothes don't make the man,"[109] it remained important to present oneself in a way that created a favorable impression. Once again, the guiding principle was "moderation": a doctor had to be polished, though not excessively so, and impeccably hygienic. Equally important was how doctors approached their patients. Their gestures were never to be rash or abrupt; instead, their conduct had to be adapted to the temperament of each individual. In this way, a doctor needed not only to be a reassuring presence, but also to project authority. This made the performative

---

105 Arrizabalaga, Medical Ideals, p. 108 f.
106 De Castro, *Medicus Politicus*, pp. 110–112.
107 Ibid., pp. 113–121.
108 Ibid.
109 Ibid., p. 124: "vestis monachum non facit."

dimension of medical practice crucial—carefully staging diagnostic gestures, such as pulse-taking and urine inspections.[110] Particular care was required when treating patients prone to melancholy, as the wrong words or mannerisms could deepen their despair and accelerate the progression of the illness.[111] Prognoses, especially in cases marked by uncertainty, had to be delivered with the utmost caution. While doctors were encouraged to instill hope, they were strictly warned against guaranteeing recovery—for the final outcome, ultimately, rested with Divine Will.[112]

The treatises dedicated to describing an excellent doctor were fully integrated into the contemporary cultural climate. In fact, medical etiquette appears to have absorbed and, in some ways, adopted the principles of all those court etiquettes published from the 16th century onward, the prototype for which is undoubtedly Baldesar Castiglione's *Il libro del Cortegiano*. Published in Venice in 1528 from the press of Aldo Manuzio, *Il Cortegiano* set out to provide a manual of conduct for those living at court. As in the medical treatises, Castiglione's analysis integrates both ethical and aesthetic considerations, ultimately culminating in the ideal of moderation. As Cicero declared in *De officiis* [*mediocritatem illam tenebit, quae est inter nimium et parum*], one should pursue the ideal of moderation. Castiglione, in turn, applied this principle to the figure of the courtier who "through study and effort [should] treat and largely correct [their] natural defects."[113] This was the art known as *sprezzatura* [effortless grace], the antithesis of *affettazione* [mannerism].

Whereas *affettazione*, Castiglione noted, referred to an overtly forced and contrived behavior aimed at earning approval from others, *sprezzatura* can be defined as a form of "nonchalance." It was not a natural and spontaneous attitude, but rather one so controlled and refined as to appear so. For the courtier, *sprezzatura* came to represent a *regola generalissima*—a very general rule—that required continuous exercise and relentless self-surveillance. Such meticulous control over the smallest gestures and movements had the merit of perfecting them through practice, thus concealing the learning involved and lending the impression that everything was done with ease.[114] To this end, the courtesan had to be endowed with the virtues of caution and prudence, equipping themselves with

---

110 Malatesta, *Fiducia, fiducie*, p. 13.
111 De Castro, *Medicus Politicus*, p. 126 f.
112 Ibid., p. 135 f.
113 Castiglione, *Cortegiano*, p. 40 f: "e con studio e fatica liminare e correggere in gran parte i difetti naturali."
114 Ibid., p. 59.

true inner discipline in a bid to govern their outward appearance.[115] They had to resort to the art of *sprezzatura* to achieve grace, and, consequently, harmony. Similarly, a doctor was expected to maintain constant vigilance over himself, striving to uphold all those ethical-deontological rules of behavior—internalizing them and transforming them into genuine habit.

Alongside descriptions of the ideal physician, some deontological reflections on the professional identity of the "good surgeon" were also published by graduate surgeons. This is evident, for example, in Giovanni Andrea Della Croce's *Cirugia universale e perfetta* (1583), and Giovanni Battista Cortesi's *In Universam Chirurgiam Absolutam Institutio* (1633). The attributes of the ideal surgeon scarcely differ from those of the ideal physician, and are similarly rooted in the teachings of medical authorities such as Celsus (c. 143–37 BC) and Galen.[116] Like the good physician, the good surgeon had to be vigilant, temperate, and God-fearing. He also needed to be healthy, as no one would trust their wellbeing to an unhealthy person. Furthermore, he had to be sober in judgment and modest in conduct. As with physicians, the ideal surgeon was expected to be well-versed in theory and practice, especially in how to conduct himself at the patient's bedside. The key difference was that a surgeon needed to have "learned hands"—confident in what he was doing and never trembling,[117] while being fully equipped with all the tools necessary for exercising his profession. He should not move too briskly, make fewer incisions than are necessary, and not be put off by his patients' cries. Irrespective of the circumstances, the surgeon had to remain calm, with his senses fully alert, paying close attention to any visual inputs.

The surgeon's sense of moderation had to be expressed aesthetically: never overly groomed, but not unkempt either. A surgeon had to be familiar with his patients and yet maintain a certain distance, for otherwise his instructions would not be taken seriously. His words had to be clear and concise. He should neither diagnose nor operate without first being thoroughly aware of the ailment's true nature, nor should he promise recovery or make predictions of death in doubtful situations.[118] Surgeons embraced the Galenic definition of surgery as a "manual art," thereby stripping the practice of its negative connotation. As Dalla Croce wrote, surgery—being performed by hand—was not only the most intricate branch of medicine but also the most reliable, as it directly engaged with the physical body: a tangible, material entity. It also represented the oldest tradition within

---

115 Ibid., p.128.
116 Savoia, Early Modern Italian Surgeon, pp. 33–35.
117 Celsus, *De Medicina*, p. 296 f.
118 Dalla Croce, *Cirugia Universale*, p. 6v (HAB: M: MK4° 6).

medical practice.[119] Given that surgery required the utmost precision, a good surgeon had to be experienced, thus he could not be too young. Nor could he be too old, for that matter, for age—generally a sign of experience and authority—often brought with it a weakening of the senses.[120]

3 The complex nature of the doctor-patient relationship

The doctor-patient relationship throughout the early modern period was much more intricate than generally assumed. A simplistic, and quite superficial interpretation, would suggest that the doctor-patient dynamic, from the time of Hippocratic medicine until the 1970s, took the form of "medical paternalism." This view posits an unequal and highly asymmetrical relationship, a genuine power imbalance which favored the doctor—who held technical-scientific knowledge—to the detriment of the patient. While a key component in any therapeutic process, the patient was completely passive, and was expected to accept the doctor's decision without expressing an opinion.[121] As we have seen, however, the situation was much more complicated. Winning the patient's trust was not just an essential goal, but also a strategic one, crucial not only for the success of the therapy undertaken, but also for navigating the highly competitive medical market.[122] While trust was considered an essential element in any therapeutic relationship between doctor and patient, it was, nonetheless, a historically constructed and asymmetrical concept. The resulting imbalance in this relationship stems from the fact that trust presupposes, above all, that the patient is in a state of discomfort or distress, necessitating their reliance on the doctor's care. Trust, therefore, was a hetero-driven sentiment, a concept which, in medieval legal terms, referred to a binding relationship that was inherently asymmetrical.[123]

The indispensability of the patient's trust in healing has been theorized since ancient times. In the aforementioned *sermones* by Niccolò Falcucci, the lack of a patient's trust was identified as one of the greatest obstacles to successful treatment.[124] Zerbi and Sicco, likewise, stress the paramount importance of trust and goodwill for effective therapy. The patient's cooperation was seen as a funda-

---

119 Ibid., p. 1r.
120 Celsus, *De Medicina*, p. 297.
121 De, Paternalism in Medicine; Kaba, Soorakumarian, Doctor patient relationship; Sisk, Frankel, Isaacson, Truth-telling; Viafora/Furlan/Tusino, *Questioni di vita*.
122 In this respect see Pomata, *La promessa*.
123 Duranti, Confidentia, p. 61.
124 Malatesta, *Fiducia*, p. 14.

mental resource. Zerbi asserts that: "the confidence a patient places in [his] doctor does more for his recovery than all of his [the doctor's] instruments."[125] Sicco, too, describes medicine as an art based on three elements: the doctor, the patient, and the disease. To explain how these three elements interrelate, he draws an analogy to war: as in a battle, the patient's body becomes the battleground, the disease the enemy, and the doctor the commander—yet even the most skilled commander is powerless without the cooperation of his ally, namely the patient. The patient was never merely perceived as a passive party, but was positioned as a key player. Botallo, echoing this perspective, offers a vivid analogy: "the disease is the enemy, the sick person is the ally, therapies are the weapons and strongholds, the doctor is the artificer."[126] In this view, not only was cooperation vital—but so too was obedience. The patient's role was not merely to trust but also to follow the physician's instructions, as their collaboration was essential for therapeutic success.

The doctor-patient relationship exhibits an inherent asymmetry, though this did not imply that the doctor's opinion takes precedence over the patient's wishes and consent. Quite the opposite, in fact. Rather, their relationship was complex for a number of reasons. While honesty and transparency were expected between doctor and patient, and it was understood that the patient should never lie about their true state of health, there are instances when the doctor was exempt from the obligation of fully telling the truth. Situations did arise when the doctor's omission—not telling the truth—was considered therapeutic. Mindful of the suggestive power of the imagination, Sicco warns doctors to be prudent with their words, so as not to upset or annoy patients.[127] Zerbi similarly cautions doctors never to rush to answer their patients' questions, but to carefully consider them first before speaking.[128] Finally, while patients should never lie, doctors were well aware of the risk of being deceived by the patients themselves. When faced with a patient's dishonesty, vigilance no longer served a therapeutic purpose but instead transformed into a defensive mechanism.

---

[125] Zerbi, *De Cautelis*, p. 64. "Confidentia autem quam habet infirmus de medico plus valet ad sanitatem, quam medicus cum omnibus suis instrumentis."
[126] Botallo, *De medici munere*, p. 108. "La malattia è il nemico, il malato è l'alleato, le terapie sono le armi e le roccaforti, il medico è l'artefice."
[127] Siccus, *De optimo medico*, p. 12v.
[128] Zerbi, *De Cautelis*, p. 116.

3.1 The physician's right to lie

As noted, physicians were expected to be vigilant about how they expressed themselves verbally, especially when delivering a dire prognosis, or addressing patients who were easily impressionable or prone to despair. In such cases, doctors needed to exercise self-censorship, carefully choosing their words and delivery. The potential for omitting and concealing information from patients became a subject of ethical reflection within the medical world, encapsulated by the question: "Is it permissible for a doctor to lie?"

In the *Medicus Politicus*, Rodrigo de Castro devotes the whole chapter IX of Book III to this issue of whether is it permissible for a doctor to deceive a patient for the sake of their health [*Liceatne Medico Aegrum Fallere Valetudinis Gratia*].[129] This theoretical question, equally explored by other authors, takes on greater complexity here in view of de Castro's religious status. De Castro was in fact a Jewish convert who, after arriving in Germany, made it increasingly clear that his conversion to Catholicism was a professional necessity rather than a true personal conviction.[130] The Lusitanian doctor's position was that for therapeutic purposes, the doctor was authorized—even on moral grounds—to lie to the patient. De Castro based his position on a passage from Plato's *Republic* in which the Greek philosopher argued that, if done for a good cause, a lie by doctors and jurists should not be considered intrinsically evil, but rather as a form of medicine [φαρμακός]. The idea that physicians could deceive their patients has been theorized and accepted in medical circles since ancient times, as evidenced in Hippocrates' treatise *In Epidemiis*, and Galen's commentaries on that work. Ancient medical practitioners accepted the use of a lie to induce a placebo effect in their attempts to have patients believe that they had been given one substance instead of another. For example, in *Epidemiis* (VI, 5–7) there is an episode in which a sick man complains of an earache. To cure him, the doctor pretended to remove a foreign body from the patient's ear, which was nothing more than a wool ball he had concealed in his hand and which he quickly threw into the fire after its supposed removal.

While Galen was skeptical about physicians' use of deception, he distinguished between two types of patients: courageous ones—whom the physician should always inform about their current state of health—and timid ones, for whose recovery it was necessary to awaken hope, and for whom deception was not only acceptable but even recommended.[131] De Castro also observed that

---

[129] De Castro, *Medicus Politicus*, pp. 142–146.
[130] Arrizabalaga, Medical Ideals, p. 22 f.
[131] Gadebusch Bondio, Verità e menzogna, p. 73.

patients were often distrustful of doctors, trying to catch every little word they uttered, carefully scrutinizing their faces for the slightest clue that might reveal something about their true state of health. This required physicians to be alert with their words, trying not to reveal what they knew. This became essential when dealing with anxious patients, where the doctor had to dissimulate and adopt an impassive and inscrutable expression in their presence, maintaining neutrality and objectivity, for therapeutic reasons. De Castro recalled that Celsus and Damascenus (c. 650–750) supported lying to keep the patient's hopes alive and prevent despair, but he always advised informing the family of the true state of the patient's health. The legitimacy of lying is further reinforced by a biblical example from the First Book of Samuel, in which David escapes from Achish, king of Gath, through cunning and by feigning madness.[132]

However, de Castro also notes—alongside the views of those thinkers who accepted lying if it was for the patient's good—the opposing views of those who totally condemned lying, particularly those positions formulated by Aristotle in the *Nicomachean Ethics*, where he argued that simulation was contrary to the divine order of nature, and thus inherently evil. To defend his stance, de Castro distinguished between three types of lies. The first, he noted, consisted not so much of a false statement as an omission of the truth,[133] referencing the subtle distinction that had existed in early modern times between *dissimulatio* [pretending not to be what one is] and *simulatio* [pretending to be what one is not .][134] The second, de Castro emphasized, was the abominable difference between a harmful lie [*mendacium nocivum*] and a benevolent lie [*mendacium officiosum*].[135] While the former should always be avoided and condemned, the latter could have positive effects depending on the context, as with a placebo. He further supported the recourse to a lie for a good purpose by invoking the concept of *dolus* which, following Winfried Schleiner's suggestion, can be translated as "ruse," or "cunning."[136] Returning to the war analogy, a *dolus* against an enemy took on a positive

---

[132] Samuel 21:13–15. "13 And he changed his behavior before them and pretended to be insane in their hands and made marks on the doors of the gate and let his spittle run down his beard.

14 Then Achish said to his servants, "Behold, you see the man is mad. Why then have you brought him to me?

15 Do I lack madmen, that you have brought this fellow to behave as a madman in my presence? Shall this fellow come into my house?"

[133] De Castro, *Medicus Politicus*, p. 142 f.

[134] Zagorin, *Ways of Lying*, pp. 2–3. See Zacchia, *Quaestiones medico-legales*, p. 169: "simulator enim id quod non est, vel alio modo quam sit, dissimulatur vero id quod est."

[135] De Castro, *Medicus politicus*, p. 144.

[136] Literally, *dolus* means "fraud, deceit."

value in view of its outcome.[137] In my view, this military analogy is particularly thought-provoking because it sheds further light on just how multifaceted the doctor-patient relationship was. If disease was considered the enemy, it's important to recognize that it was not seen as separate from the patient. As a result, any deception used to combat illness—though intended to benefit the patient—was still directed at them. This approach was justified through the metaphor *dolus ad bonum* [good deceit], likening it to strategic deceit in warfare.

While the doctor's simulation was accepted and even encouraged in certain circumstances, the patient's condemnation of deception was unequivocal. Despite patients' total rejection of doctors lying—especially in view of the therapeutic success achieved—physicians were, in reality, aware that patients might resort to deception for a variety of reasons. For instance, patients might lie about their condition out of shame or fear—typical examples being women lying about their virginity, or men feigning to suffer from colic to avoid attending court—or they might fabricate details simply to test the doctor's competence. Zerbi argued that physicians, much like athletes, must consistently exhibit determination and maintain an alert mind [*animo semper prompto*] to avoid falling victim to deception. Caution, in this context, is not simply a matter of habitual response, but rather involves proactive, vigilant action. This concept underscores the necessity of constant vigilance, ensuring both the avoidance of dangers and the safeguarding of daily security amidst unforeseen challenges in professional practice.[138] The likelihood that a patient might resort to deception during a medical examination was considered particularly high, not only during the oral questioning about their symptoms, but also during other sensitive moments, such as urine examinations.

As previously noted, the urine examination was a central yet much-contested element of medical therapy practice. While deemed the only bodily excretion capable of furnishing detailed information about the internal state of the "humors," the urine test itself was fraught with ambiguity and easily prone to error. The emphasis on doctors being vigilant during urine analysis was primarily justified by the complex nature of the test itself, which could be influenced by a formidable array of external contingencies that might distort its interpretation. Doctors were advised, inter alia, to use containers of a specific color, to analyze the contents only under certain lighting conditions, and to do so at a set time of day.[139]

---

137 Schleiner, *Medical Ethics*, p. 12.
138 Zerbi, *De cautelis*, p. 44: "Describitur autem cautela per actum, et non per habitus cum dicitur invitatio cum diligenti attentione ut denotetur medium ad pericula assicura ac free quotidie capiti suo emergentia in actu operativo vitanda debere esse animo semper pronto, atque intento non minus qual athlete."
139 Moulinier Brogi, *L'uroscopie*, pp. 84–88.

It should also be noted that many medical practitioners were highly critical of such methods and of those who claimed to offer safe and accurate diagnoses based solely on urine without physically examining the patient. In fact, the examination was often carried out at a distance, an approach that was often met with hostility and regarded as inadequate. Beyond this purely technical concern, a second kind of fear demanded the physician's utmost caution: the likelihood of being deliberately deceived by the patient. Several publications were devoted to the precautions doctors needed to take when examining urine—the *cautelae urinarum*. The earliest examples of this type of work date back to the late Middle Ages, attributed to the Catalan doctor Arnaldo de Villanova. Its enduring authority is evident later centuries, as it cited in several posthumous treatises, including de Castro's *Medicus Politicus*.[140] The most noteworthy aspect of Villanova's *cautelae* is that he begins his reflections with a warning to doctors of the need to equip themselves with the means to guard themselves against those determined to deceive them. He framed the patient-doctor interaction as a veritable contest, urging physicians to carefully scrutinize and analyze the patient's demeanor, gestures, and words.

The second precaution advised the doctor to observe the patient's face during the urine examination. If the patient intended to deceive, de Villanova noted, they might soon burst out laughing, or their face might change color. Other precautions involved gathering information about the patient whose urine was being collected and their symptoms.[141] The doctor's line of questioning assumed an inquisitorial tone, as they not only had to pay attention to assess the symptoms, but also ensure the truthfulness of the patient's responses. The physician had to be meticulous in their practice: nothing about the patient should go unnoticed, and even the slightest nod of the head could conceal a deeper meaning. Urine could be tampered with in multiple ways. Numerous reports describe patients substituting someone else's urine, or adding ingredients such as wine or vinegar in a bid to alter its color, consistency, and odor.[142] These manipulations could lead to misdiagnoses, such diagnosing pregnancy in male patients. The doctor's constant vigilance was crucial to avoid deception and safeguard their reputation.

The ramifications of patient fraud could be more severe, even politically and socially. This is evident when considering another fraudulent practice that required careful attention: feigning illness. Simulating diseases [*De morborum simulatione*] was a significant concern for both the medical and legal professions

---

140 De Castro, *Medicus Politicus*, p. 150.
141 Da Villanova, De Cautelis Medicorum, p. 751.
142 De Castro, *Medicus Politicus*, p. 153f.

of the time, so much so that between the 16th and 17th centuries a number of treatises were published on how to detect false illnesses. The methods for simulating disease were numerous and inventive—for instance, holding a bar of soap in one's mouth to simulate foaming—as documented in numerous medical treatises, from Galen to Hippocrates. These included techniques for inducing ulcers, dropsy, madness, and more, along with ways to alter skin color, raise body temperature, and speed up or slow down pulse rate.[143] By the late 16th century, feigning illness had become the subject of a branch of forensic medicine; it was viewed as equivalent to fraud and, therefore, punishable by law. These treatises outlined how to induce fever, feign pallor and facial deformities, and falsify urine, and even produce lesions similar to those displayed by lepers, often using herbs such as Thapsia mixed with ointments and then applied to the supposedly injured areas. To simulate madness, mandrake roots were boiled in wine, which, if swallowed, caused one to lose one's mind for a whole day. Of particular interest were the methods used to simulate tumors, such as vigorously rubbing dried powder made from wasp and hornet decoctions, known as *raspiolae pulvis*, onto the body near a source of heat.[144]

The topic of disease simulation is pivotal because it allows us to further explore the multifaceted nature of the doctor-patient relationship, and to focus attention on a second type of medical vigilance—the "negative" kind, seen as a tool with which to investigate individuals suspected of criminal actions or behavior. Under normal circumstances, vigilance against patient deception was considered necessary to preserve the doctor's reputation and prevent his competence from being ridiculed. However, in extraordinary contexts, such as criminal proceedings, it took on another meaning. Here, the doctor's role was not only to safeguard the reputation of the medical profession, but also to determine whether an individual—who was allegedly guilty of wrongdoing—ought to be prosecuted. Despite the multiple ways in which illness could be faked, physicians agreed that the motives for simulating disease were invariably the same: "namely out of fear, out of shame, or out of profit."[145] In these contexts, the doctor's expertise became crucial, for only they could uncover the truth on account of their knowledge and

---

[143] Pastore, *Le regole dei corpi*, pp. 73–77.
[144] The methods for simulating diseases were essentially always the same, given that the authors, when they did not refer to each other, relied on readings from works such as those of Galen. See, for example, Selvatico, *De ijs qui morbum simulant*, pp. 7–10; Pastore, *Le regole dei corpi*, pp. 63–73; Maladies vraies et simulées, pp. 11–26.
[145] Selvatico, *De ijs qui morbum simulant*, p. 6. " ad timorem scilicet, vel ad verecundiam, vel ad lucrum."

skills. Thus, simulating diseases offers a new lens to examine medical vigilance, particularly within the more confined context of the criminal court.

4 The social and political utility of medical expertise

In the pre-modern-era, judges commonly relied on expert opinion [*consilium sapientis*] in their decision-making[146]—a practice rooted legal system's demand for evidence, which underpinned the entire criminal process. Circumstantial evidence and mere suspicion were considered insufficient: what was needed were certainties, actual evidence such as the testimony of at least one witness, but more importantly, a confession by the alleged offender.[147] Throughout the early modern era, in contrast to the Anglo-American system based on English common law—where testimony from non-direct witnesses was not considered admissible—the Italian system followed the inquisitorial model typical of Roman and canon law, whereby evidence was decisive.[148] As Paolo Zacchia, the Roman physician considered the father of modern forensic medicine, wrote in support of jurisprudence, medicine offered its services in the belief that: "The truth is always a safeguard of justice; therefore, it is always a good thing, and never evil, never harmful, never a defender of crime, but always useful and always praiseworthy; and therefore the truth must be sought, lies banished."[149]

The self-generated claims that medicine could offer those certainties and absolute truths in court may seem to contradict what has been argued in the previous pages about the inherent uncertainty of medical knowledge, especially regarding the humoral theory. As previously explained, Galenic medicine struggled to establish itself as a *scientia* across Europe due to the unpredictable results it produced when applying general and universal theories to highly individualized cases. Despite numerous efforts to confer upon it greater epistemological authority, academic medicine had to accept its status as *scientia coniecturalis*—an inferior science compared with others that could offer infallible and unquestionable truths through the use of a theoretical, logical *methodus*. The recognition of med-

---

[146] For more information on this issue, see Ascheri, Consilium sapientis, pp. 533–579.
[147] Di Renzo Villata, Paolo Zacchia, pp. 17–24.
[148] Watson, *Forensic Medicine*, p. 9. On the difference between the Continental and Anglo-American systems of collecting evidence, see Clark/Crawford, *Legal Medicine*; Crawford, Medicine and the Law.
[149] Zacchia, *Quaestiones medico-legales*, p. 570: "veritatis tutela semper pro iustitia est, ergo semper bona, numquam mala, numquam perniciosa, numquam crimini patrocinans, sed semper utilis, semperque laudabilis; et idcirco semper veritas quaerenda, mendacia repellenda."

icine as a *scientia*, rather than merely an *ars*, gained its first decisive impulse with the rise of what might anachronistically be described as "forensic medicine."

Although expert advice to assist a judge on technical matters was widespread practice in areas where Roman canonical procedure was used, doubts arose as to the binding nature of such an opinion, particularly medical ones. Jurists, while recognizing the need for doctors to intervene in certain circumstances, questioned the reliability of medical expertise, given that it was based on suppositions and not infallible and proven truths—ignoring the fact that jurists themselves rarely dealt with *proba plena* [conclusive proof], but only with traces and clues. Yet physicians asserted their authority by insisting on the particular and individual nature of medical knowledge, making their contributions indispensable in legal proceedings. While their inability to produce universal knowledge was considered a major deficiency, it became an asset in the courtroom. Given the diversity of the human physique, only a physician's expert eye, accustomed to dealing with the specifics of human physiology on a daily basis, could offer true knowledge.[150] To again quote Zacchia: "The nature, however, of particular individuals, which must be taken into account when rendering a judgment, can never be fully grasped by a judge, who must rely on a physician's expertise with extensive experience."[151]

Attempts to reconcile the specifics of individual cases with a broader framework of general causes capable of providing true explanations led to the creation of *consilia* in the Middle Ages. From the 16th century onward, procedural advice became the primary means for affirming the "scientific" nature of medicine, thereby asserting the importance of its practitioners for society. The rise of forensic medicine was symptomatic of a broader process of "medicalization" in early modern society—a growing expansion of medicine's authority.[152] It is noteworthy that the aspect of medicine that most elevated its prestige was applied in situations where the therapeutic goal—typically the key objective of medical practice—was entirely absent. Instead, the focus was on ascertaining truth through the study of signs of illness.

For medicine's indispensable role to be recognized, the doctor's vigilance during legal proceedings had to be of the highest caliber. Above all, he had to closely monitor symptoms, whether palpable or reported by the defendant, in order to transform these "conjectural signs" into "obvious signs" and thereby construct a

---

150 De Renzi, La natura in tribunale, p. 808 f.
151 Zacchia, *Quaestiones*, p. 341: "natura tamen particularium individuorum, cuius respectus habendus in sententia ferenda, numquam ut medicus callere poterit, qui etiam in hoc casu requiritur, ut si magnae experientiae ."
152 de Ceglia, *Corpses, Evidence*, p. 16 f; Watson, *Forensic Medicine*, p. 43 f.

truth, even in the absence of absolute certainty. This was undoubtedly most evident in cases involving the substantiation of pain, such as a headache, which could only be verified through the patient's own testimony. Since pain is a completely subjective experience that does not manifest through any objective signs, the only way to assess its presence was through the patient's description of how they were feeling. In light of the fact that a patient might lie about their state of health, the doctor needed specific diagnostic techniques to interpret these signs accurately and determine whether the reported pain was real or not.[153]

4.1 Autopsies

Available records show that physicians were appointed to public functions since the 13th century, when they were called upon to assess the severity of wounds or to perform autopsies. Although post-mortems were a traditional task dating back to ancient Greece, the practice was only institutionalized toward the end of the Middle Ages.[154] Physicians searched for "obvious signs" inside the body that could reveal the truth when external indications were lacking. In his commentary on Hippocrates' *Prognosticon*, the physician Girolamo Cardano 1501–1576 expressed confidence in the infallible ability of autopsies to determine the cause of death. He distinguished three kinds of medical signs: diagnostic, prognostic, and cadaveric dissection [*dissectio cadaveris*]. Diagnosis and prognosis were fraught with uncertainties, and cadaveric dissection was seen as the only method capable of producing empirical evidence.[155]

While the practice of opening corpses has been substantiated as early as the mid-12th century—primarily as a funerary practice for the elites—the first known autopsy commissioned by civic authorities occurred in 1286. As historian Katharine Park notes, the decision to dissect a corpse was driven by the goal of uncovering the cause of the high rate of sudden mortality affecting both men and chickens in the cities of Cremona, Piacenza, and Reggio. To this end, a doctor from Cremona examined the carcasses of several chickens and the corpse of a recently deceased man, discovering an *aposteme*—an abscess or pustular swelling—in the hearts of both. This finding prompted a Venetian doctor to issue a bulletin advising against the consumption of chicken and eggs. Originally a public health investigation, the

---

153 De Renzi, Witnesses of the Body.
154 Crawford, Medicine and the Law, p. 1621; Watson, *Forensic Medicine*, pp. 27–35.
155 Siraisi, Bresadola, Segni evidenti, p. 719.

autopsy in cases of death from unknown causes quickly evolved into a full-fledged forensic practice in Bologna. Doctors were called upon only to give an opinion based on an external examination of the body in cases of violent injuries or suspected murder. By the 14th century, the practice of performing autopsies on victims of unexplained deaths became increasingly common. The earliest known judicial autopsy occurred in 1302, when two physicians and two surgeons were tasked with examining the body of Azzolino degli Onesti, who was suspected of having been poisoned. Their findings, however, contradicted the poisoning hypothesis.[156]

While the practice of requesting an autopsy in cases of suspected murder has much older origins, scholars generally agree that it was only with the promulgation of the *Constitutio Criminalis Carolina* by Emperor Charles V (1519–1555) in 1532 that autopsy was finally enshrined as an integral part of criminal law practice.[157] Although the *Constitutio* was specifically designed for the territories of the Holy Roman Empire, it was to become the model for all territories where Roman law was in force.[158] Local precedents, such as those in Venice, required autopsies in all homicide cases, as well as in cases of sudden death from unknown causes.[159] For example, in Lorenzo Priori's *Prattica criminale* (1663), the chapter dedicated to "vision of corpses" emphasizes the necessity of adequately observing and describing the cadaver. Given that judges and notaries often lacked the expertise to determine whether a wound should be considered fatal, expert opinions from professionals "intelligent in the art" were not only indispensable, but also not be questioned.[160]

The criminal statutes in force in numerous cities generally stipulated that in murder cases, the criminal notary should be dispatched to the crime scene, accompanied by a doctor—and in more complex cases, by two doctors. Typically, one was a physician while the other was a surgeon, thereby combining both aspects of medicine—*theorica* and *practica*—to produce a report as accurately as possible.[161] The surgeon was the only authorized figure permitted to operate manually and directly on the body, whether dead or alive. However, in theory, such

---

156 Park, Criminal and Saintly Body, p. 5.
157 Watson, *Forensic Medicine*, p. 21 f. Out of 219 articles, only four (Art. 35, 36, 147, 149) explicitly call for the intervention of medical professsionas or midwives in cases of suspected abortion, infanticide, wounds, or uncertain death. A further twelve articles merely imply that a medical assessment was necessary.
158 Wear, Medicine Early Modern, p. 237.
159 Ruggiero, The Physicians and the State, pp. 156–166.
160 Priori, *Prattica criminale*, p. 17 f: "descriver diligentemente tutte le ferite con le sue qualità."
161 Pastore, *Il medico in tribunale*, p. 86.

operations could only take place under the watchful eye of the physician whose task was to guide the surgeon's hands by means of verbal instructions.

Dissecting a corpse was viewed as paramount in cases of sudden death, especially when poisoning was suspected yet no external signs were visible. The threat of poisoning was both a pervasive and disturbing everyday phenomenon, particularly for members of the wealthier classes. Notable examples include the suspected poisoning of Popes Julius II (1503–1513), Leo X (1513–1521), and Clement VII (1523–1534). Similarly, the scandal of the Palermo poisoners Teofania d'Adamo and Giulia Tofàna (possibly her daughter), who concocted the *aquetta* or—*acqua Tofàna*,[162] a lethal poison primarily used by women who wanted rid of their husbands, led to over 600 deaths.[163]

Suspected poisoning cases challenged physicians, putting their skill and knowledge to the test. They had to contend with widespread lack of understanding about poisons, many of which were of mineral origin, and played only a marginal role in official galenic pharmacology, which focused primarily on plant or animal substances.[164] Documentation shows that autopsies were routinely performed in cases of "sudden deaths," particularly at the onset of epidemics, dating back to the Black Death of the 14th century.[165]

With the advent of forensic autopsies, early modern physicians transitioned from being experts in examining the living to becoming examiners of the dead. Giovanni Maria Lancisi highlighted this shift in 1707 in his *De subitaneis mortibus*: "Nothing teaches us more clearly than the dissection of cadavers, which brings to light the hidden causes—unknown to us—to the light of the sun."[166]

4.2 Controlling social deviance: the obligation to denounce

In a bid to control and reduce violence in urban areas, political authorities soon began requiring doctors to report any violent injuries they encountered during their examinations. One of the earliest instances of this practice is the Venice decree of 30 April 1281, which mandated that doctors report all injuries caused by acts of violence. For non-serious injuries, doctors were obliged to report them within a maximum of two days to the *Cinque Savi e Anziani alla Pace*—

---

162 See Feci, *L'acquetta di Giulia*.
163 Pastore, *Il medico in tribunale*, p. 363 f.
164 Pomata, *La promessa*, p. 146 f.
165 Park, Criminal and Saintly Body, p. 8.
166 Cited by de Ceglia, *Corpses*, p. 7: "Nihil est, quod nos doceat apertius, quam ipsam cadaverum sectionem, quae occultas nescis causas ad solis lucem evidenter exponit."

a magistracy responsible for overseeing non-fatal disputes among commoners. Injuries deemed potentially fatal [*in periculo mortis*] had to be reported immediately to the *Signori di Notte*.[167]

In Tuscany, the obligation for doctors and surgeons to report injuries was first formalized in a proclamation issued on 2 January 1551. Physicians who did not comply could be punished at the presiding judge's discretion. The importance of punishing doctors who failed to notify the authorities in a timely manner is emphasized in major criminal treatises of the period—most notably in *Praxis et Theoricae Criminalis* (1594–1614) by the jurist Prospero Farinacci.[168]

This requirement to alert the authorities was an attempt to align medical knowledge with political power, aiming for a broader and more effective invasive control of public order. It clearly reveals the ambivalence inherent in a doctor's vigilance toward visible and invisible symptoms. Whenever issues of social order were at stake, the same attentiveness required for treatment could, paradoxically, work against the patient's best interests.

This ambivalence regarding medical vigilance is particularly evident in relation to sodomy. Throughout the early modern period, sodomy was harshly condemned—legally, as a crime, and morally, as a sin. Political and ecclesiastical authorities' extreme aversion to sodomy stemmed from the belief that it was considered an "act against nature," its purpose not being procreation but rather the quest for sheer sexual pleasure. To use Priori's definition:

> Sodomy is a nefarious vice, which occurs when intercourse is performed against nature, and it is committed in three ways [...]. The first is through touching [...]. The second is when engaging in carnal acts with a male, or also with a female in a way that is against nature [...] The third is when a man engages sexually with a beast, a dead body, a Jew, or an infidel.[169]

Along with blasphemy, sodomy was thus strictly condemned as an offense against God. Ultimately, fear of divine retribution being unleashed on a particular city—

---

167 ASVe, Maggior Consiglio, Liber Comunis Secundus, f. 103r: "et teneantur isti medici dicere et manifestare quinque de pace percussem quem habuerit in cura infra duos dies et si esi videbitur quod predictus percussum staret pro illa percussion in periculo mortis, teneatur manifestare dominis de nocte quam cicius poterunt bona fide." Cited in Ruggiero, The Physicians and the State, p. 159.
168 Savelli, *Pratica Universale*, p. 203.
169 Priori, *Prattica criminale*, p. 165f: "Sodomia è un vizio nefando, ch'è quando il coito si fa contro natura il quale si commette in tre modi [...]. Il primo è quando si usa col toccamento [...]. Il secondo è quando s'usa carnalmente col maschio, ed anco con la donna contro natura. [...] La terza è quella quando l'uomo usa con un animal brutto, con un corpo morto, con un ebreo o un infedele."

such as the outbreak of an epidemic—led Italian political authorities to enact a series of extremely punitive laws against these practices. A central aspect of these laws was the requirement for doctors to report patients believed to have sustained anal injuries from sodomy. The desire to effectively control the moral order of citizens made it necessary to issue a series of directives against doctors who failed to comply. For instance, on 12 October 1578, a decree—later reissued in 1623—imposed a fine of 100 *ducats* and banishment from Venice on surgeons who failed to report violent injuries.[170] This obligation to notify the authorities became particularly crucial during plague epidemics, as the preventive function of medicine took precedence over the curative. The gradual segregation of those infected was largely based on reports submitted by doctors themselves. Given the severity and danger that pestilence posed to public health, physicians who violated this duty were subject to corporal punishment, as mandated by local health authorities.[171]

5 Medicine and jurisprudence: a complicated relationship

While the physician's role in court constituted an integral part of early modern criminal procedure, the relationship between physicians and jurists was not without conflict. Even within academia, a long-standing rivalry persisted between these two professions, each asserting its epistemological superiority. De Castro explores this contentious debate in Chapter XII of Book I of the *Medicus Politicus*, where he compares law and medicine [*Iurisprudentiae et Medicinae Comparatio*].[172] Advocates for the primacy of jurisprudence, according to de Castro, argued that its superiority stemmed, above all, from its function in administering justice. Unlike medicine, which focused on the health of the individual, jurisprudence concerned itself with public affairs. Medicine was criticized as a purely technical and manual art, concerned only with material matters [*de terra*] while jurisprudence's raison d'être was regarded as more noble—even divine [*de caelo*].[173] Faced with such accusations, physicians countered that their discipline was a rational science, while jurisprudence, they claimed, was in a state of servitude. Whereas medicine adhered to the universal rules of reason, jurisprudence could only respond to civil laws—historical constructs devised by the simple will of men. Medicine, they argued, addressed divine matters, since nothing was more sacred than keeping the

---

170 Pastore, *Il medico in tribunale*, p. 151.
171 Cipolla, *Contro un nemico invisibile*, pp. 21–24.
172 De Castro, *Medicus Politicus*, pp. 42–53.
173 Ibid., p. 42f.

body healthy and free of disease. Furthermore, it should be regarded as a science: to acquire insight and cure any ailment, one had to investigate invisible and deeper causes, for which the study of logic and philosophy was indispensable. By contrast, lawyers were said to rely on "grammar" to practice their profession; their knowledge was aimed at regulating practical events and they did not engage in speculative inquiry.

Medicine alone could be recognized as a science capable of providing certainty. In response to claims that they were nothing more than "mechanics," physicians conceded that although surgery had a mechanical aspect to it, it was both practically and epistemologically distinct from the broader field of medical knowledge and therefore held a lower rank.[174] The relationship between the macrocosm (the universe) and the microcosm (the human body) endowed medicine with almost religious significance. The nobility of the human body reflected the honor bestowed on those charged with monitoring and diagnosing its health or illness, thereby elevating the physician's art within the intellectual hierarchy of the time.[175]

Similar arguments were presented in Paolo Zacchia's *Quaestiones medico-legales*, particularly in the third title of the sixth book [*De Praecedentia inter Medicum, & Iurisperitum*].[176] Zacchia acknowledged that critics rightly targeted false doctors who were nevertheless officially recognized as members of the medical profession. The Roman protophysician echoed common stereotypes about physicians, who were often accused of avarice, arrogance, and ignorance. Medicine was criticized as an impious practice, as it disregarded religious dictates—such as allowing patients exemptions from fasting during Lent. Above all, it was denounced as a mechanical art in which experience often outweighed reason.[177]

Critics of jurisprudence similarly contended that any law created by humans was not natural, but historically constructed. They pointed to the multiplicity of laws and constitutions across Western Europe as evidence that laws were born not of Divine Will, but from unsuccessful attempts to curb human transgressions. Given that jurisprudence did not teach anything about the natural world or require manual practice, they contended that it was neither an art nor a science.[178] While acknowledging the importance of jurisprudence in regulating human society—since humans are, by nature, social animals—advocates for the healing arts made it clear that medicine should be ranked higher than jurisprudence

---

[174] Ibid., pp. 4345.
[175] Ibid., p. 34.
[176] Zacchia, *Quaestiones*, pp. 431–461.
[177] Ibid., p. 436.
[178] Ibid., p. 440.

[*medicina iurisperitiae anteponenda*]. They argued that the health of the human body—and by extension, the soul—was derived from medicine. For this reason, the practice of healing was considered noble in itself, requiring no external justification—such as mediating conflicts between litigating parties.[179]

Thus, the relationship between doctors and lawyers fluctuated between mutual complementarity and professional rivalry. Despite their clear differences, some theorists proposed a correlation between the two fields. For example, the 14th-century jurist Baldo degli Ubaldi (1327–1400) identified a series of procedures common to both fields, even asserting: "a judge is just like a doctor." Both, he claimed, had to proceed by means of "conjectures" and gradual approximations to achieve any degree of certainty in their investigations.[180]

As Alessandro Pastore notes, early modern academic medicine occasionally attempted to appropriate jurisprudence's probabilistic framework, endeavoring to combine "possible arguments and accredited doctrines."[181] Yet the idea persisted that medicine, unlike law, could never reach "truth," only "verisimilitude."[182] Both disciplines remained subject to social criticisms, particularly accusations of avarice and arrogance.[183]

Rather than resolving their bitter rivalry, physicians and lawyers were often compelled to collaborate in court. While the role of the medical expert was generally recognized and accepted—particularly in cases of serious injuries leading to permanent disability—this collaboration often led to friction between the two professions. Lawyers struggled to acknowledge that doctors had their own independent sphere of expertise in areas traditionally dominated by the legal profession. As a result, when physicians sought a more prominent role in court, aiming for greater social recognition and professional status, their efforts were met with hostility. Lawyers view such ambitions as an encroachment on judicial authority.

These objections reflected a broader desire to maintain a clear separation between the two professions—one rooted in the era's strict social hierarchies. To maintain this distinction, it was deemed necessary to establish a well-defined division of competencies to prevent overlap between their roles. In contrast, some Italian doctors argued that the medical expert should no longer be seen just as an adviser to be consulted, but rather as a professional with real prerogatives and technical knowledge, even within the legal sphere. This gave rise to intense debates over whether expert opinion should be binding on judicial decisions.

---

179 Ibid., p. 457.
180 Pastore, *Il medico in tribunale*, p. 16: "iudex est sicut medicus."
181 Pastore, *Le regole dei corpi*, p. 98.
182 De Renzi, Witnesses of the Body, p. 223.
183 Siraisi, *Medieval Renaissance Medicine*, pp. 20–25.

5.1 Medical expertise as *consilium sapientis*

In Western Europe, the custom of seeking expert opinion has been common since the Middle Ages, particularly when judges faced uncertainty in cases requiring technical expertise. This procedure, known as *consilium sapientis*, involved presenting expert testimony in court at the formal request of the judge, or of one or both parties. As Mario Ascheri notes, the earliest theoretical foundation for this tendency to consider expert depositions as highly reliable can be traced to a decretal by Pope Gregory VIII (1187), later inserted in *De probationibus* under the title *Liber Extra* (X 219 c.4). In this text, it was argued that the testimony of seven midwives [*matronae*], who agreed with the woman under investigation, held greater evidentiary weight in canon law than the husband's oath to the contrary in proving a woman's virginity in a matrimonial case.[184]

Although expert testimony was generally accepted as reliable, a fierce debate among jurists raged from the 14th century onward about whether a judgment based on medical expertise could be overturned in light of fresh expert testimony. While no consensus prevailed, the majority of jurists still believed it could be. The non-binding nature of doctors' opinions was justified by the belief that, due to the partiality of their knowledge, they testified not *de veritate* [based on truth] but only *de credulitate* [based on belief or assumption].

As a result—contrary to the arguments put forth by physicians from the 16th century onward—their opinion could not, legally speaking, be considered sufficient to establish *proba plena*, the standard required to determine truth. Instead, their depositions were seen instead as merely probable.[185] The debate over the relationship between expertise and testimony also remained unresolved. The prevailing view tended to be that the two roles should not be conflated, arguing, for example, that if a doctor certified that a patient had died of apoplexy, they should then not be called upon to testify on the definition of apoplexy.[186]

5.2 Antonio Maria Cospi's *Il Giudice criminalista*

In his treatise *Il Giudice criminalista* (1643), the Florentine jurist Antonio Maria Cospi explicitly denounced the growing role of the medical expert in criminal courts. Written during Cospi's tenure as a judge at Bologna's *Torrone* criminal

---

[184] Ascheri, Conslium sapientis, p. 534.
[185] Ibid., p. 536 f.
[186] Pastore, *Le regole dei corpi*, p. 94.

court and later as secretary to Otto di Balia in Florence, the work was published posthumously by his nephew Ottaviano Carlo Cospi, a knight of the Knights of St Stephen.[187] Much like the medical treatises on the "good doctor," Cospi's treatise is essentially a programmatic work outlining the attributes of a "good" criminal judge. Its primary objective was to equip judges not only with a model of behavior but, more importantly, with enough foundational knowledge to enable them to critically evaluate expert opinions—particularly those of physicians—so that they would not accept such testimony uncritically. The underlying conviction was that judges should command an encyclopedic knowledge, allowing them to function as partial experts in any given subject over which they might be required to arbitrate in court. This, in turn, would shield them from being overwhelmed by technically-derived expert opinions.[188] According to Cospi, the remit of such experts was—as belonging to a specific category of witnesses—confined to explaining certain complex issues and performing technical tasks such as collecting and presenting data and information for the judge. Ultimately, it was the judge's responsibility to draw conclusions based on their own judgment. In addition to advocating for a clear distinction between the roles, functions, and prerogatives of judges and experts, Cospi directly accused experts of dishonesty, noting that some, motivated by the prospect of profit, had often made false statements in court.[189]

Cospi's treatise, written in the vernacular, is divided into three parts. The first is devoted to the ethical and deontological qualities of a good judge. The second examines a wide range of crimes, with close attention to those falling under both civil and ecclesiastical jurisdiction such as heresy and witchcraft. The third part—alongside discussing subjects like forgery and *simulatio*, particularly in relation to alchemy and necromancy—delves into key aspects of criminal procedure. Notably, the final section contains no fewer than 36 chapters devoted to subjects that could be classified as "forensic medicine."[190] This level of attention reflects Cospi's desire to equip criminal judges with a breadth of expertise beyond the legal sphere, so that they would not be at the mercy of medical experts. It also sought to enable judges to assess the competence of these experts and the reliability of their reports and evidence. For instance, Cospi shows marked skepticism toward medical opinion, particularly in the chapter on poisons. He emphasized that the judge needed to understand the nature, characteristics, and effects of the var-

---

[187] Ibid., p. 85.
[188] De Renzi, Witnesses of the Body, p. 225.
[189] Cospi, *Il giudice Criminalista*, p. 5.
[190] Ibid., pp. 359–469.

ious poisons available given that poisoning was often difficult to prove and considered an "occult crime." Physicians, he argued, were prone to attribute death to "natural causes" rather than poisoning. To prevent the crime from going unpunished, the judge, Cospi insisted, had to be capable of determining whether poisoning was the true cause of death—"leaving Doctors, Physicians, and Surgeons to the curative part."[191]

Medicine's merely secondary role vis-à-vis jurisprudence is perhaps most evident in the sections devoted to corpse dissection. Only surgeons were authorized to perform autopsies, supervised by physicians. Paradoxically, this official recognition of medical expertise only served to reinforce medicine's disrepute by reaffirming its status as a manual practice at the service of the law. Even when a medical expert performed the autopsy and identified potential signs of death, the responsibility for formulating the final judgment on the findings rested entirely with the presiding judge who, armed with the requisite knowledge, had to interpret independently the relevant evidence. Cospi further argued that inquiries into the victim's lifestyle and the questioning of relatives—tasks typically within the doctor's purview in the therapeutic field—should fall under the authority of the judge and his notary at the trial, particularly during the identification of the of corpse.[192] Cospi portrays the judge's careful engagement with expert knowledge as a hallmark of honor and nobility. Expertise, when used judiciously, helped prevent deception by specialists and facilitated a quicker confession from suspects—thus sparing the need for torture.[193]

6 A new self-awareness: the publication of the *Methodus Testificandi*

In contrast to the hostility of jurists and the overall neglect to the crucial role played by medical practitioners in court, some Italian physicians in the 16th century started publishing medical treatises aimed at establishing diagnostic criteria to be applied in legal investigations of the accused's presumed pathological conditions. While medical expertise had become an indispensable element in the judicial process, no scientific concepts or methodologies had yet been developed for the medico-legal practice. As a result, expert opinions were often rushed and superficial. To address this shortcoming, they produced works that systematically ex-

---

191 Ibid., p. 375.
192 Ibid., pp. 406–409.
193 Pastore, *Le regole dei corpi*, p. 89.

amined matters of "forensic medicine," grouped under the general heading *Methodus testificandi* [Methods of Testifying].

The authors' overriding objective was to underscore the paramount importance of the medical expert, and to propose a model of exemplary professional conduct. The treatises served as a series of evolving perspectives and solutions to various and often challenging medico-legal problems. Their purpose was not only to assist doctors in delivering expert opinions but also to guide lawyers and judges, who required at least a general understanding of medicine to perform their duties effectively. These works did not seek to replace the jurist's authority but to support it by offering technical insights into matters beyond legal expertise. For instance, the 1602 treatise published by the Sicilian physician Fortunato Fedeli served as a practical manual for doctors summoned to testify in court, including suggestions and templates for drafting expert reports to be submitted to judicial officers. While Fedeli outlined the many domains in which medical competence was essential, he also acknowledged the inherent objective limits of the field— for example, the difficulty of offering conclusive opinions in cases of suspected poisoning or in interpreting certain autopsy findings.[194]

Regarding the figure of the *optimus medicus*, Fedeli emphasized that the technical expertise of the *optimus peritus* should be grounded in the triad of prudence, circumspection, and attention, and must be accompanied by strong moral integrity. Far from competing with magistrates, he insisted that physicians should seek to collaborate with them, using clear, evidence-based reasoning and language. For this reason, medical reports were to be written in Latin—or in the vernacular, if requested by the judge for the sake of greater clarity—while avoiding overly technical jargon, and steering clear of digressions and irrelevant issues.[195]

In addition to expressing a desire for reconciliation with jurists, these treatises were clearly crafted in response to a growing ambition to redefine the role of medicine in early modern European society. The call for a more active and central presence of the physician in court was rooted in the assumption that it was up to the physician himself to investigate and determine the facts in any suspicious case —thereby allowing him to "grasp the truth."

The earliest treatise on forensic medicine appears to be *Methodus dandi relationes pro mutilatis, torquendis aut a tortura excusandis,* written by the Sicilian protophysician Giovanni Filippo Ingrassia. The work remained in manuscript form and was likely unknown until it was published posthumously in 1914,

---

[194] Fidelis, *De Relationibus*, p. 333 f.
[195] Ibid., pp. 349–352.

after a handwritten copy dated 1632 was discovered in Palermo's municipal library by Francesco Garsia, an heir of the author.[196] Although long overlooked, this treatise is notable as an early indication of a wider social and cultural shift toward recognizing forensic medicine as an autonomous field.[197]

Ingrassia's work is considerably shorter than the other *Methodus Testificandi* published in the 17th century, both in length and scope. What it shares, however, with the other treatises is that it outlines the methodology to be followed when writing reports for the criminal magistrates and addresses the medico-legal aspects surrounding mutilations resulting from injuries and judicial torture. Central to Ingrassia's approach is his belief that doctors—particularly when testifying in court—must exercise great caution to avoid error. To do so, a solid understanding of medical theory, especially in human anatomy, was essential. Echoing the treatises on the *optimus medicus*, Ingrassia devotes a few pages to portraying the ideal medical expert as a professional rigorously trained in his art, always vigilant against acting recklessly or without reflection.[198]

The second significant work is *Methodus Testificandi* by the physician Giovanni Battista Codronchi, likely written in 1595 and published in a single edition in Frankfurt am Main in 1597.[199] Codronchi's text is often regarded as the world's first treatise on forensic medicine—as its author himself qualifies, "as far as I know."[200] It was intended to offer explanations and solutions to new and challenging medico-legal questions of interest not only to doctors engaged in drafting expert reports—whether young physicians still in training or older ones often caught unprepared—but also to lawyers and judges required to adjudicate such matters. Codronchi's objective was to establish medical expertise in legal proceedings as a discipline anchored in written doctrine and characterized by scientific coherence.

The preface opens with a robust defense of the role and status of medicine, noting that not even emperors and pontiffs would shy away from submitting certain cases to the physician's judgment, aware that judicial decisions in forensic contexts often hinged on medical opinion.[201] Codronchi denounced the careless and hasty way in which medical expertise was generally carried out, often entrust-

---

[196] Perrando, *Prefazione*, p. VI.
[197] di Renzo Villata, Paolo Zacchia, pp. 42–44.
[198] Ingrassia, *Methodus dandi relationes*, p. 27 f.
[199] Puccini, *Il Methodus Testificandi*, p. 87
[200] Ibid., pp. V–VII.
[201] Ibid., p. 5 f.

ed to barbers and midwives, or other practitioners who lacked even minimal formal training.[202]

The treatise itself follows a schematic and logical approach: topics are identified and introduced in relatively autonomous chapters, each constituting a separate discussion. In total, the work comprises seventeen chapters, addressing a wide array of topics such as disease simulation, general methodology of expertise, traumatology, toxicology, and advice on how to treat certain illnesses.[203] Rather than delving into a detailed analysis of each chapter, this discussion will focus on Chapter III, which outlines the general methodology for composing medical reports in court. Here, the caution required when delivering a prognosis before a judge is directly compared to the care a conscientious doctor must take during a bedside examination.

Beyond methodological prudence, Codronchi emphasizes the moral profile of the ideal medical expert. The expert must express his opinion with scrupulous care, without being overwhelmed by his passions. He is to tell the judge the truth—nothing more and nothing less. He must avoid both arrogance and ignorance, always bearing in mind that medicine is a conjectural art, and, as such, should express his opinions with the utmost caution. Above all, Codronchi insists that loyalty to the public good should always take precedence over any potential gain, including bribes or favors promised by the accused.[204]

In Palermo in 1602, Fortunato Fedeli (1551–1630) published *De relationibus medicorum*. Apparently unaware of Codronchi's work, which had been published just five years earlier, Fedeli believed himself to be the first to address the subject.[205] This treatise is divided into four books, each addressing a distinct set of medico-legal concerns: the sanitary conditions of various locations—an essential concept in humoral theory; congenital and acquired functional limitations with sections on disease simulation and the considerations to be made before subjecting a witness to torture; sexual and obstetric problems; and finally, injuries and violent deaths. A chapter on the attributes a doctor should display in court is also included, appearing in the eighth section of Book II titled *De erroribus eorum qui Medicinam faciunt* [On the Errors Committed by Those practicing Medicine]. Once again, contemporary physicians were denounced for their overreliance on book learning, which—lacking practical application—inevitably led

---

[202] Ibid.
[203] Ibid., pp. 89–91.
[204] Ibid., pp. 11–13.
[205] Ibid., p. 88.

to ignorance and recklessness, resulting in rushed and often erroneous conclusions.[206]

The uniformity of the subjects covered "for the first time" by the three authors suggests that these were the key issues on which doctors were frequently called upon to provide evidence in court. This is why I have chosen not to dwell on the content of the various chapters. While none of the authors appears to have been aware of each other's work—or at least claimed not to—their occasional references to one another indicate a shared effort to systemize and codify traditional practices and knowledge. It is also plausible to hypothesize that their common regional backgrounds—Palermo in the Spanish Empire and Imola in the Papal States, both familiar with the inquisitorial system—help explain the early development of such reflections on the subject, in contrast to other parts of the Italian peninsula, which would have to wait for Zacchia's work in the first half of the 17th century.[207] In any case, rather than focusing on the content of the treatises, it is the very existence of the treatises that merits attention, as they signal the emergence of a new and more mature medical self-awareness.

The work of the Portuguese physician Rodrigo de Castro represents an ideal convergence between the literary tradition focused on the characteristics of the *optimus medicus* and the *methodus testificandi*. The need to offer contemporary physicians an ideal model of conduct to which they could aspire is grounded in the belief that medicine—as the title of de Castro's work clearly indicates—serves a political function, and, as such, must contribute optimally to the public good. The civic role is predicated on the assumption that medicine, in de Castro's words, is "an art based on reason and experience to obtain and preserve health" with deeply ancient origins, taking the form of a gift from God to mankind.[208] The passage the Lusitanian doctor referred to is taken from the Old Testament, specifically from *Ecclesiasticus* (38:1–15): "Give doctors the honor they deserve, for the Lord gave them their work to do. Their skill came from the Most High, and kings reward them for it. Their knowledge gives them a position of importance, and powerful people hold them in high regard. The Lord created medicines from the Earth."[209] Written around 180 BC, *Ecclesiasticus* is one of the Bible's Wisdom books. Clearly influenced by Greek philosophy, it was frequently cited in Christian medical circles to justify medical practice. However, as the text emphasizes, a doctor was simply God's instrument: it is God who creates the medicines

---

206 Fidelis, *De Relationibus Medicorum*, pp. 311–336.
207 De Renzi, Witnesses of the Body; Pastore, *Il medico in tribunale*.
208 De Castro, *Medicus Politicus*, p. 4 cited in Arrizabalaga, Medical Ideals, p. 116: "ars cum ratione, et experientia faciendae conservandaeque sanitatis."
209 De Castro, *Medicus Politicus*, pp. 29–34.

and performs the cures, albeit through the doctor's hands.[210] De Castro invoked this biblical passage to support his thesis on the elevated status of medicine. In particular, he links the medical profession to tasks associated with governing a State, asserting medicine's centrality as an instrument of governance and discipline at the service of the political authorities. Viewed from this perspective, the figure of the court physician becomes pivotal.

Book IV addresses illegal practices—such as magic spells and love potions—but, more importantly, centers on medico-legal expertise. Chapters IX to XI are particularly relevant to the present study, with Chapter XII focusing on diagnosing a woman's virginity and a man's putative sterility. Chapter IX is devoted to the simulation of diseases [*Qua ratione morbum simulantes deprehendi queant*]. This chapter's content is particularly original, as de Castro introduces personal experience, presenting concrete cases in which he exposes deception, such as the case of a prostitute who faked an abortion, or Portuguese sailors who feigned illness to avoid fighting against the English navy.[211] Chapter X deals with the methods for identifying poisoning [*Testificandi methodus circa eos, quibus venenum fuit exhibitum*], while Chapter XI offers guidance on how to judge the severity of injuries and deaths by drowning [*Testificandi ratione in vulneribus capitis: & in iis qui aqua fuerunt suffocati*].[212] Although de Castro was largely reiterating the teachings of his colleagues, the very fact that topics of a forensic nature found their way into a treatise whose ultimate aim was to assert medicine's predominant political function in pre-modern European society is significant.

It is also worth noting that in de Castro's work, the defense of the public utility of learned medicine—and of its practitioners—assumes an even greater significance. Born in Lisbon to a family of Jewish converts, de Castro moved to Hamburg in the late 1580s due to the growing pressure of the Portuguese Inquisition against Jews and Christian converts. His name appears in the city's list of Jews, even though he had his children baptized.[213] His work, therefore, not only vindicates the political role of medicine, but also the political function and utility of doctors of Jewish origin in an increasingly anti-Semitic context, such as that of 1614 Hamburg.[214]

Although he would later go on to publish more extensively, Paolo Zacchia (1584–1659) is still regarded as the father of modern forensic medicine.[215] Author

---

210 Minois, *Il prete e il medico*, p. 22.
211 De Castro, *Medicus Politicus*, p. 251.
212 Ibid., pp. 254–259.
213 Arrizabalaga, Medical Ideals, pp. 111–114.
214 Ibid., p. 122.
215 De Renzi, Paolo Zacchia, pp. 53–56.

of numerous medical treatises, he is best known for his *Quaestiones medico-legales*, published in nine volumes between 1621 and 1635, and reprinted in several editions by Italian and European printers until the late 18th century.[216] Presented as a veritable summary of contemporary medical and legal knowledge, this work specifically targeted an audience of physicians and jurists, seeking to assert the legitimacy of the physician's presence in the legal world as an expert worthy of the highest consideration. Zacchia's methodology for the medical *peritiae* is fundamentally based on treating individual *quaestiones*—addressing the content of the individual cases in their own right by choosing the specific concepts and methods necessary for solving the concrete cases at hand. More significant than the content of the individual *quaestiones*—which largely reiterate what had been asserted by the previously mentioned authors, all of whom Zacchia cites—with the exception of Ingrassia's unpublished work—is the revolutionary tenor of the treatise, given the boldness and the outright radicality of the positions expressed therein.

Zacchia's *Quaestiones medico-legales* was intended to position the medical practitioner in court—not merely as a simple expert to be occasionally consulted on matters of criminal justice, but as a meticulous professional figure who assumed, in his words, "the prerogative of prescribing if, how, and when any act must be performed in order to be valid in its legal consequences."[217] Far from being an art [*ars medica*] subservient to jurisprudence, forensic medicine asserted its full legitimacy by establishing itself as a distinct science, employing interdisciplinary methods. It was Zacchia who formulated the earliest definition of what would later come to be known as forensic medicine: the examination of medical issues through the lens of the law [*rebus medicis sub specie iuris*].[218] This was not merely a matter of providing clinical opinions on legal questions outside the bounds of medicine, but rather a specialized field in which medical and legal knowledge were required to interact and collaborate. Superseding the works of his predecessors, Zacchia's *Quaestiones medico-legales* was to become the European benchmark for medico-legal practice until the 19th century.

---

216 di Renzo Villata, Paolo Zacchia, p. 12f.
217 Zacchia, *Quaestiones*, p. 342.
218 Pastore, *Il medico in tribunale*, p. 8.

7 A question of loyalty: the physician's dilemma

Among the multiple topics addressed in the various *Methodus Testificandi*, one in particular held a dominant ethical and deontological significance: the problem of the *secretus* [secrecy or privacy]. This issue is central to any further analysis of the ambivalent nature of the doctor-patient relationship. As previously noted, in theory, both doctors and patients were regarded as having a moral obligation to always tell the truth, especially to ensure the patient's good recovery. Consequently, the patient was never permitted to lie, while the doctor was only exempt from this obligation in a limited number of circumstances, such as when a patient's condition was so severe that informing them of their imminent death would likely hasten their demise.

Furthermore, the Hippocratic Oath required physicians to remain silent about what they had discovered during medical examinations, in accordance with what might be termed professional secrecy.[219] The Oath states: "What I may see or hear in the course of the treatment or even outside of the treatment in regard to the life of men, which on no account one must spread abroad, I will keep to myself, holding such things shameful to be spoken about."[220] Yet this principle lost much of its force whenever physicians acted in an institutional capacity, such as when called as a court-appointed medical expert [*peritus*] to provide opinions in judicial proceedings, or when examining patients under suspicion.

The importance of professional secrecy is underscored by the fact that Zacchia concerns himself with the topic on three occasions, in Books II, III, and V.[221] As noted earlier, the professional secrecy to which a doctor was compelled to adhere was first theorized in the Hippocratic Oath. Western Christianity adopted the Hippocratic Oath from its pagan origins, and, inspired by Christian mores, the vow to confidentiality assumed an ever deeper meaning during early modern times. The physician's practice, concerned with the care of the body, had frequently been compared to that of a priest, who cares for souls. While these two spheres should not be conflated, it is worth noting that before delivering a prognosis of death, a doctor was obliged to consult a priest to ensure that the patient had confessed and received Extreme Unction. The two roles were clearly complementary. The diagnostic examination thus took the form of confession, with the patient revealing intimate details to the doctor, who—like the priest—was not supposed to disclose what he had learned to any third party.

---

[219] Nutton, Medicine in Greek World, p. 29.
[220] MacKinney, Medical Ethics, p. 31.
[221] Laugier, *Paul Zacchias*, p. 21.

Although numerous authors equated the violation of medical secrecy with a mortal sin, Zacchia openly argued that there were certain situations in which this rule could not—and indeed should not—be upheld. For him, the true mortal sin for a doctor was to give false testimony about a patient's or a defendant's symptoms when acting as an expert or a witness in court.[222]

This ambivalence inherent in the doctor-patient relationship is most clearly reflected in the duty to report, to which medical personnel were subject in the early modern period. If the medical examination revealed signs of involvement in activities considered illegal or immoral—often two sides of the same coin—the physician was legally obliged to breach professional secrecy and notify the political authorities. While this obligation was externally imposed, it only became truly effective when doctors themselves recognized its significance for maintaining order and social stability. Undoubtedly, the threat of physical or financial punishment for failure to comply with the obligation served as an incentive for doctors to cooperate with the authorities. Yet coercion alone was not enough to achieve total obedience; a degree of personal assent was also required.

Thus, while the primary goal of medical practice has always been to ensure the patient's good health, a physician's loyalty often aligned more closely with those exercising political power than with the patient—especially when that patient was deemed "suspect" due to their actions, morals, or religion.

In this regard, the galleys offer a particularly revealing context for examining the complex nature of medical vigilance and, by extension, the doctor-patient relationship. Physicians assigned for rowers were responsible not only for maintaining their health—vital to the vessel's operations—but also for supervising men who were often slaves or convicts. This dual role blurred the boundaries between healing, discipline, and vigilance. These tensions between care and control lie at the heart of the complex medico-legal dynamics explored in the following chapter.

---

[222] Ibid., p. 21 f.

Chapter 2
Early Modern Galleys as Spaces of Vigilance: Livorno and Civitavecchia from the 16th to the 18th Century

1 Vigilance aboard galleys: punishing, controlling, and correcting deviance

While doctors were a marginal group aboard the galleys, at least in numerical terms—typically numbering one or two per galley serving a crew of around 300 men—their function was undeniably essential: they were not only responsible for providing health care but also for exerting control over the crew's bodies and behavior. Their duties extended beyond the crew's welfare for the practical purpose of ensuring successful navigation; they also played a key role for maintaining order and punishing criminal behavior, such as fighting or sodomy.

In this context, the Battle of Lepanto can be seen as marking a turning point for Italian galleys and military forces more broadly. The intensity of the battles prompted a series of reforms in crew management and discipline for the armies fighting the Turks in the Cyprus War (1570–1573). At the same time, the Battle of Lepanto made the authorities aware of the shockingly poor levels of medical care aboard sea-faring vessels. This realization created a sense of urgency for better sanitary conditions, creating an ideal environment for developing innovative hygienic and medical organizational measures. Thus, medicine's dual role—treating illness and controlling the body—became an essential tool for successful navigation and combat.

This chapter will focus on the galley as a space of vigilance, examining the pervasive and systematic oversight exercised over galley slaves and convicts in the Tuscan galleys of Livorno and the Papal galleys of Civitavecchia, from the 16th to the 18th century. In these fleets, convicts and slaves made up more than half of the galley crews.[1]

Sources testify to the overriding concern of those officials managing the *galeotti*, as they were often aware that they were dealing with morally ambiguous figures. Most galley sentences were given to repeat offenders, considered incorri-

---

[1] Lo Basso, *Uomini da remo*, pp. 345f, 391. In 1570, forced oarsmen made up 53% of the crew of the Tuscan galleys; the situation later changing to a fleet primarily consisting of slaves. In the Papal State, forced rowers made up 79.8% of the crew.

gible; there was a constant suspicion that convicts would continue their criminal behavior aboard. Enslaved rowers were similarly viewed as immoral due to their heretical, blasphemous beliefs, particularly the Muslim faith. As a result, a series of rules and instructions were established to maintain constant surveillance over forced rowers who, as written in the *incipit* of various manuscript orders for the galleys issued in the late 16th century, were "morally inclined to licentious living, or accustomed to live badly."[2]

Crew members, including the *galeotti* [galley rowers], were tasked with ensuring that the entire crew behaved in the most appropriate manner, not only from a technical perspective, to guarantee successful seafaring, but also, and more importantly, from a moral and behavioral standpoint. Sending convicts to serve on the penal ships was not merely a means of providing free labor for the shipping industry, but was also a way to punish them and, in a sense, "re-educate" them in the name of Christian doctrine and military discipline by forcing them to row. As for the slaves, military discipline took on the character of true religious instruction, with the hope of converting them and preventing them from influencing the convicts to renounce Christianity and embrace Muslim. The ultimate goal was to cultivate a crew of obedient and disciplined rowers, with the twofold objective of ensuring successful navigation and avoiding divine wrath—two aims regarded as inseparably linked. Once the practice of deploying slaves and convicts to complete galley crews was officially established, galley service evolved into a fully-fledged penal and servile system in which the correction and re-education of rowers took center stage. It seems fair to say, then, that aboard early modern galleys, punishment, subordination, and military discipline were interwoven facets of the same phenomenon.

For this reason, there was constant and meticulous monitoring of the convicts' and slaves' behavior aboard. To make this system of control fully effective, it had to rely on an efficient denunciation apparatus and, even more crucially, on a system of mutual surveillance. These multiple control strategies clearly illustrate the ambivalence surrounding vigilance, which could disrupt any sense of equilibrium. For example, in certain cases, to facilitate effective "horizontal control," some rowers were assigned specific tasks that required them to monitor their fellow inmates. This approach, in turn, "verticalized" relationships that were otherwise horizontal—between two individuals of equal status and, theoretically, equal standing and duties. In particular, oarsmen were tasked with keeping a watchful eye not only on their galley-bench mates, but also on those rowers in front of and

---

[2] ASF, MP, 2131, dossier 6, *Ordini da osservarsi nelle Galere del Ser.mo Gran Duca di Toscana*, cc. n.n.: "i quali sono moralmente inclinati ò dalla Licentia, o dall'habito confermati à mal vivere."

behind them. For instance, in the regulations governing the detection and punishment of blasphemous offenders aboard the Papal galleys, officers were instructed to question rowers from neighboring benches to identify those committing blasphemy.[3] Similarly, convicts and slaves were encouraged to monitor from "below" whether their officers were behaving properly and, if not, to report instances of abuse or harassment. It is noteworthy that the most common punishment for officers who failed to carry out their duties was being sent to the galley themselves, thus becoming convicts and joining that very social category they had once been tasked to supervise. Additionally, a form of vigilance could be observed among the officer class—both toward their colleagues and themselves. The galley thus became a space in which everyone was to become subject to strict behavioral codes and under permanent scrutiny from other crew members. Obviously, it was no easy feat to keep an eye on 300 men all at once, even in such a confined space as the galley, typically only 40 meters long, five meters wide, and less than two meters high.[4] This may explain why a whole series of strategies were devised to ensure that each crew member effectively remained the object of observation by others within the confined space they shared.

The deep concern for what might be termed the "moral correction" of rowers was actually part of a broader stratagem to ensure military discipline—a responsibility entrusted to religious orders, particularly from the latter half of the 16th century. The need for disciplined and God-fearing soldiers in conflicts against heretics rendered an extensive catechization program of military forces imperative.[5] The ideal of the Christian soldier [*miles christianus*] could only be realized through a rigorous system of controlling soldiers in both land and sea forces. Furthermore, the need for diligent and virtuous conduct aboard the galleys was reinforced by the hope of achieving victory through divine intervention.[6] In the fleets, the deployment of convicts and slaves to man the crews necessitated tighter control over behaviors deemed criminal, immoral, or deviant. The demand for the

---

3 AAV, Misc. Arm. IV/V, b.54, f. 105: "Che non sia persona alcuna di qualsivoglia stato, ordine e condizione, che ardisci bestemmiare il Santissimo Nome di Dio, Gloriosissima Vergine Maria, e de' Santi, sotto pena della Galera per cinque anni, & essendo Forzato in vita, di cento bastonate, e della morte ad arbitrio, & acciò un sì grave, e pernicioso delitto più facilmente si scopra, ordiniamo, e comandiamo ad ognuno delle trè Bancate, cioè à quella, dove sarà stata detta la bestemmia, à quella di sopra, & à quella di sotto di rivelare subito il Delinquente all'Officiale della Galera."
4 Aymard, Chiourmes et galères, p. 73.
5 On this topic see Lavenia, *Dio in uniforme*.
6 Filioli Uranio, I cappellani delle galere, p. 219 f.

unwavering and rigid observance of military discipline thus justified the implementation of an extensive program aimed at disciplining and re-educating the forced rowers in the name of Christian morality, enforced through strict surveillance and regulation of their conduct. The ultimate goal was to impose an efficient, rigid military discipline on board.[7]

Even from the perspective of moral correction, the Battle of Lepanto is often regarded as a turning point, heralding the beginning of a stable and substantial presence of chaplains aboard galleys, whose religious zeal was deemed pivotal in igniting the crusading spirit in soldiers and thus, in winning battles.[8] As early as 1570, Pope Pius V had planned to have 27 Capuchin Friars embark aboard the vessels to provide pastoral care for sick soldiers, and in 1571, 29 of them did embark; with the papal brief *Cum dilectus filius*, issued in March of the same year, they were granted full faculties to celebrate the sacraments aboard.[9] The captain of the Papal fleet, Pantero Pantera, devoted several pages of his treatise *L'Armata Navale* (1614) to detailing the crucial role played by chaplains, whose main task was to persuade sinning rowers not to succumb to temptation, thereby averting God's wrath.[10] Despite this initial presence of the Capuchins, the spiritual care of the galleys after the Battle of Lepanto was entrusted to secular priests, as well as to the Dominicans. It was not until 15 May 1684 that the Capuchins were officially designated as responsible for the pastoral care of the galley crews, as well as that of the galley hospitals.[11]

The need to enforce strict moral and behavioral discipline aboard arguably aligns seamlessly with the broader stratagem of "disciplining" and "confessionalization" that characterized the Italian states following the Council of Trent (1545–1563).[12] Within the more confined spaces of the galley, we can observe an attempt to put into practice the ideals upon which the foundation of the Christian *societas perfecta* was to be built. Evidence of the imposition of the Counter-Reformation framework can be found even aboard the penal ships, in a system aimed at trans-

---

[7] This reflection has been primarily inspired by the positions formulated in Civale, *Guerrieri di Cristo*; Lavenia, *Dio in uniforme*; Filioli Uranio, I cappellani delle galere.
[8] Filioli Uranio, I cappellani delle galere, pp. 225–227.
[9] Criscuolo, *I Cappuccini*, p. 121. The presence of the Capuchins aboard the papal galleys was later confirmed by Innocent IX in 1684 as we read in De Polis, *Storie di galee e soldati*, p. 13.
[10] Pantera, *L'armata Navale*, p. 116.
[11] Angioini, Leone, *La Guerra di Morea*, in De Polis, *Storie di galee e soldati*, p. 19.
[12] However controversial the terms "disciplining" and "confessionalization" may be, they are effective for analyzing the specific phenomena of Counter-Reformation Catholicism, as attested by the works of Paolo Prodi. Therefore, as much as I reserve the right to use them, I am aware that it is always necessary to do so wisely and carefully.

forming soldiers and *galeotti* into members of the ideal Catholic army.[13] Hence, the significant presence of Capuchins, and subsequently Jesuits, aboard Italian galleys. Their responsibilities included overseeing living conditions, providing spiritual care to the crew, managing field hospitals, and tending to the sick.[14] The various control strategies employed against the early modern *galeotti* reflect broader power strategies that defined the Italian states during the Counter-Reformation. In this context, the galleys could be defined as a kind of "Other Indies" [*Otras Indias*]: barbaric, uncivilized, and superstitious spaces requiring the religious orders to undertake a mission of evangelization and an initiation into the values of Christian morality. The Capuchins, and later the Jesuits, assumed responsibility for this evangelical mission.[15] The Council of Trent marked a decisive turning point, ushering in an unprecedented phase in the concept of "mission," one imbued with an increasingly functional significance for a project to thoroughly educate and discipline the populace. The expression *Indias de por acá* [the Indies over here] thus served to legitimize efforts to spiritually conquer Europe, employing strategies not so dissimilar to those implemented in the New World.[16] Aboard the galleys, we can therefore find the presence of missionaries in the true sense of the term—religious brethren specialized in conversion and evangelistic activities "among the infidels or natives of those distant regions where Christianity had never been preached or where all reference to the primitive preaching had now been definitively lost."[17] After all, the *galeotti* were nothing more than enslaved, condemned rowers, and therefore were seen as blasphemous, criminal, and immoral people. Thus, missions conducted within galley life were configured as missions *ad intra*,[18] akin to those missions conducted across continental

---

13 Filioli Uranio, I cappellani delle galere, p. 229.
14 Here, too, the Church's involvement in care for the body is still profoundly strong. For the long-standing tradition of hospital care by the Capuchin Fathers, see De Castelsangiovanni, *L'assistenza ospedaliera*.
15 See Prosperi, *Tribunali*, pp. 551–559; *La vocazione*; Broggio, *Evangelizzare il Mondo*; Pavone, *I gesuiti*. As will be shown, the Jesuits organized a mission aboard the papal galleys in the first half of the 18th century. See Ginzburg, Folklore; Faralli, *Le missioni*, pp. 97–116; Prosperi, "Otras Indias," pp. 205–234; *Missionari*.
16 Broggio, *Evangelizzare il mondo*, p. 17; Camaioni, *Pulpiti*, p. VII. On the idea of slave conversion projects in the Mediterranean as an evangelizing mission, see also Lavenia, Schiavi.
17 Cantù, preface to Broggio, *Evangelizzare il Mondo*, p. 13.
18 Whereas the *missiones ad extra* were directed toward the so-called savages of the New World, the *missiones ad intra* took place within the "Indies of over here"—remote and quasi-dechristianized territories within Europe, whose inhabitants were often likened to the indigenous peoples of the Americas due to their perceived ignorance of Catholic doctrine.

Europe. This same inward-facing evangelical impulse extended to rural communities as well:

> [W]here heresy or the sad condition of peasant life in rural settings, degraded by ignorance and poverty, had introduced a dangerous departure from doctrinal orthodoxy, from righteous sacramental practice, from Christian morality, and from the customs and traditions rooted in the tradition of the faith.[19]

This hypothesis is supported by archival sources, and particularly the reports and instructions compiled by the Jesuits—who took over the management of the convict fleet from the Capuchins at the end of the 18th century—concerning the religious missions conducted aboard the galleys of Civitavecchia, typically held before Easter Week. The explicit purpose of these missions was to provide religious instruction to the rowers, hear their confessions, and comfort the sick in the hospital.[20] These missions typically lasted about a week, with the days structured around prayer and catechism. In addition to preaching aboard the ships, the missionaries visited the port hospital and organized long processions in which the inhabitants of Civitavecchia took an active part, thus imbuing the activities with both solemnity and public significance. These missions, with the core objective of educating the rowers in Christian values and persuading them to abandon sin and crime, were deemed a genuine success whenever they culminated in the conversion of slaves who—to be enslaved—had to be either heretics or Muslims. A notable example was the 1716 mission, which resulted in the conversion of one Zwinglian and three Lutherans to Catholicism.[21] Particularly significant was the mission of 9 November 1783, organized at the behest of Cardinal Marcantonio Colonna. Over the course of 22 days, the ten religious chosen for the task were re-

---

[19] Francesca Cantù, preface to Broggio, *Evangelizzare il Mondo*, p. 13. The idea that the spaces inhabited by slaves were "Other Indies" has been by now officially accepted by historiography. See, for example, the studies by Colombo and Sanseverino. However, Italian historiography has not considered convicts in these kinds of analysis.

[20] See ARSI, Rom.138, *Breve relazione delle missioni fatte in Civitavecchia il 1716 per ordine di nostro signor papa clemente XI. Missione alle galere pontificie*, ff. 61–63, 65–73; Rom. 132.I, *Breve relatione della Missione fatta alle Galere Pontificie in Civitavecchia da quattro padri della compagnia di Gesù l'anno 1649*, f. 248 f; Fondo gesuitico, manoscritti, 10, *Breve instruttione Per quelli che vanno Alla Missione delle Galere*; Instit. 50, *Breve informat. O instruttione per quei, che vanno a Civitavecchia alle Galere*, ff. 157–168. It is important to point out that the Capuchins also carried out processions with the crews of the galleys, which can be considered as real missions, and which usually ended with some conversions and baptisms. Unfortunately, there is less evidence of this, but some useful information can be found in Angioni, Leone, *La Guerra di Morea*, pp. 21–27.

[21] ARSI, Rom.138, *Breve relazione delle missioni fatte in Civitavecchia il 1716 per ordine di nostro signor papa clemente XI. Missione alle galere pontificie*, ff. 61–63.

sponsible for communicating with, providing religious instruction to, and hearing the confessions of a total of 3,000 rowers. Sources particularly highlight Father Bartholomew of St. John's zeal; not only did he die from a malignant fever and pneumonia upon completing the mission,[22] but he was also credited with numerous conversions, including those of seven "Turkish" slaves.[23]

It is noteworthy that the conversion of slaves was a point of contention between the religious and naval authorities. The latter were opposed to converting slaves, as it would necessarily entail an—albeit slight—improvement in their conditions of captivity, without fundamentally altering their status as slaves.[24] Such conversions could potentially lower the likelihood of their ransom, and, consequently, the associated economic gain. Conversely, for the chaplains, the conversion of heretical slaves was considered one of the primary objectives of their mission aboard the convict ships.

Although there were fewer chaplains on galley vessels of the Tuscan fleet, they were significantly more numerous in the *Bagno de' forzati*, or *Bagno degli schiavi*—the prison compound in which forced rowers were confined when not sailing. The fleet was under the authority of the Order of Saint Stephen, which had been founded specifically to attack the commercial and military interests of the Berbers and the Ottoman Empire by means of corsair raids.[25] While sources from Livorno mention fewer missions than those from Civitavecchia,[26] the life of the *galeotti* in the *Bagno* can be viewed as a genuine religious mission, carried out daily after the Capuchins were placed in charge in 1677. Their first superior, Father Ginepro (born Cristofano Cestoni, 1630–1709), a missionary who had previously worked in the outlying rural areas of Bologna, was appointed "chief missionary of the *Bagno* and galleys of Livorno."[27] Evangelical preaching—a hallmark of the Capuchin Order since its earliest directives—was characterized by a pronounced theatrical rhetoric intended to instill fear and enforce moral discipline among the faithful.[28] The Capuchins' arrival at the *Bagno* was marked by a strik-

---

22 During the mission, the missionaries shared the same living conditions as the rowers, in extremely poor hygienic conditions, at the mercy of the elements.
23 Vitalini Sacconi, *Civitavecchia*, pp. 200–202.
24 Benedetti, Servi introvabili, p. 63. Differences in treatment were minimal, consisting mainly in receiving some extra food and clothing. For Livorno, see Lo Basso, *Uomini da remo*, p. 353.
25 For a somewhat dated but accurate history of the birth of the Tuscan navy and the Order of St. Stephen see Guarnieri, *I cavalieri di Santo Stegano*; for a more recent volume see Angiolini, *L'Ordine di Santo Stefano*.
26 The only references I have found to a mission established for galley crews refer to the mission of January 1694, and are contained in ASF, MP, 2102, cc.n.n.
27 Cavallo, Ginepro da Barga, p. 32.
28 Camaioni, *Pulpiti*, pp. XII–XIII.

ing procession in which the friars walked barefoot, wearing a crown of thorns on their heads, a hempen rope around their waists, and a crucifix around their necks, chanting the psalm *Miserere mei, Deus*.[29] A vivid spectacle of repentance and penance, this procession was intended to ignite a sense of Christian zeal among the *Bagno*'s inhabitants. The time spent in the *Bagno* became marked by prayers and religious services designed to instruct as many *galeotti* as possible in Christian doctrine, with the goal of guiding Christian convicts back to the righteous path and converting the slaves.[30]

Thus, galleys were depicted as isolated, self-contained universes in which crew members were, in theory, separated from the rest of society.[31] Peter Burke has noted how the maritime world represented a distinct subculture, separate from the popular culture of early modern European societies. Endowed with their own set of beliefs, practices, and language, seafarers were often viewed with suspicion by the general population, frequently likened to other itinerant social categories such as vagrants, and were thus completely marginalized.[32] I propose examining the *galeotti* through the lens of the "foreigner"—an individual relegated to the margins of society due to their professional, economic, religious, or social status.[33] The galley crew, therefore, may be regarded as the very embodiment of deviance, confined to a restricted space that, in many respects, was detached from the structures of official social life. Not only did *galeotti* share a culture exclusive to the maritime world, but they were also aboard the galleys as convicts—criminals—or as slaves of another faith, most likely Muslim. With the practice of assigning convicts and slaves to crew duties, the sailors' subculture was further enriched with new layers. The sources thus depict the galleys as a true microcosm of violence and illegality—albeit one governed by its own specific rules and codes of conduct.[34]

This project's aim to control the *galeotti's* morals and conduct—though it is important to note it was intended for the entire crew—was not limited to the salvation of their souls and did not envisage their social reintegration. The primary objective was to make the crew reliable, disciplined, and God-fearing individuals, for navigational and military purposes. Doctors also played a key role in this

---

29 Bernardi, *Relazione*, p. 26
30 On the element of spectacle during religious missions, see Alfieri, Forme memoria, pp. 344–346.
31 We will see how galley rowers continued to maintain close links with the rest of society and how many outsiders, sometimes even women, could come aboard the vessels.
32 Burke, *Cultura popolare*, pp. 44–49.
33 Cerutti, *Étrangers*, pp. 21–24; Pomara Saverino, *Rifugiati*, pp. 19–21.
34 Santus, "*Il turco,*" p. 48.

broader effort to detect—and thus punish—any deviation in conduct on board. Medical vigilance was seen as complementary to the task of disciplining the crew. Indeed, it is notable how care and control of the body were seen as inextricably linked to, and conducive to, salvation of the soul. Medicine's attempt at secularization was still in its nascent phase.[35] Doctors were not only tasked with making sure that the sick confessed their sins and received the last rites on their deathbeds, but also with using their expertise to ensure that the patients they treated and examined showed no signs of engaging in activities deemed illicit. What was implemented aboard early modern galleys could be described as an attempt "to conquer the soul through a cure extended to the body."[36] Preventing illicit behavior served not only a militaristic function but also a more specific religious one, aimed both at ensuring the rowers' spiritual salvation and averting the risk of incurring divine punishment. As is often the case, however, the hoped-for outcome diverged from reality, and the imposition of strict discipline and control frequently led the *galeotti* to devise innovative responses in their bid to maintain even the smallest margin of freedom.

2 Early modern galleys and rowers from the 16th to the 18th century

In the early modern Mediterranean, the galley was the quintessential seafaring vessel for maritime warfare. Known since the Roman Empire, its enduring success can be attributed to three key factors: simplicity of construction, low cost, and, above all, its adaptability to the Mediterranean's unique topographical and meteorological conditions. As Lo Basso explains, the persistence of galleys in the Mediterranean should not be seen as a sign of technological backwardness, but rather as an adaptation to the region's geographical and climatic features.[37] The Mediterranean is characterized by a dense geography of narrow passages, islands, inlets, and similar obstacles, making rowing the most effective way to maneuver through such confined spaces without risking damage to the vessel. In addition, the region's weather patterns, alternate between strong winds and calm. In periods of calm, sailboats struggle to navigate, sometimes for days at a time. In this context, rowing provided a highly practical solution to this problem.[38]

---

[35] On the strict relationship between medicine and religion, see Donato, Berlivet, Cabibbo, Michetti, Nicoud (études réunites par), *Médecine et religion*.
[36] Lavenia, *Dio in uniforme*, p. 139.
[37] Lo Basso, *Uomini da remo*, p. 11.
[38] Ibid.

The decision to deploy forced rowers to man galley crews appears to have been primarily driven by technical and military considerations. Between 1530 and 1540, there was a shift from the *alla sensile* rowing system known in Italian as *galloccia* or *scaloccio*. The *alla sensile* system required each oarsman to maneuver his own oar, with typically three rowers per bench—though there could be as many as five. A typical galley using the *sensile* system would generally have between 150 and 164 rowers. In contrast, rowing *a scaloccio* involved the use of a large single oar operated by multiple oarsmen seated on the same bench, usually three in a typical galley, and between five and seven in larger ones.[39] The number of oarsmen required increased considerably from 192–265 to around 350 oarsmen aboard the Tuscan galleys, arranged in 51 benches: 26 on the starboard side and 25 on the port side. According to Lo Basso's calculations, this change in rowing technique increased the number of rowers aboard each galley by between 16% and 43%. This new navigational system—which required more men but did not demand specialized skills—thus became both the precondition and the inevitable consequence of the use of convicted and enslaved rowers to fill the crews. *Scalocio* rowing enabled the rapid manning of the galleys by mixing experienced and inexperienced rowers, thus eliminating the need for the latter to undergo long training periods, as had previously been the case.[40] This innovative rowing system thus allowed convicts and slaves to be deployed in the galley crews, circumventing the increasingly difficult task of recruiting oarsmen from among freemen. Given that rowing was an exhausting, poorly paid, and socially degrading job, only those in desperate circumstances were likely to accept it.[41]

Early modern galley crews consisted of three types of *galeotti:* free rowers, convicts, and slaves. Together, they constituted the generic mass of *huomini da remo* [oarsmen]. Free rowers were known in the Ponentine navies as *buonavoglia* [men of good will], and were often considered a marginal and suspect category—on par with convicts and slaves—despite, or perhaps because of, their willingness to serve in the galleys. As many scholars have argued, the challenge of recruiting crews was closely linked to pauperism. The economic crises of the 14th century, the "price revolution" of the 16th century, and, more broadly, the collapse of the medieval feudal economy and the resulting transformation of traditional rural society were among the key causes of the widespread proletarianization of the rural population.[42]

---

39 Candiani, *Dalla galea*, pp. 192, 279. There were about 23–25 benches in the regular galleys and about 27–28 benches in the larger ones.
40 Lo Basso, *Uomini da remo*, pp. 13–20.
41 Angiolini, La pena della galera, p. 85.
42 Geremek, *Poverty*, pp. 76–109.

Lacking the resources to secure a basic existence—let alone a stable livelihood —the rural poor migrated to the cities in search of work, or at the very least some form of relief. The dramatic increase in the number of impoverished individuals, combined with demographic shifts and greater mobility, posed a serious challenge to cities, which were still closely tied to the traditional corporate economy and struggled to respond adequately to such an influx. From the 1520s onward, urban poverty and—even more so, vagrancy—became a mass phenomenon reaching unprecedented levels and compelling society to view it as a genuine threat to the prevailing public order.[43] This amorphous and haphazard mass of impoverished individuals was to become—especially in the maritime cities—a valuable resource for naval recruiters. By offering substantial enlistment bonuses, or recruiting indebted men through gambling, it became possible to draw large numbers of oarsmen from the ranks of the city's destitute. As Bronisław Geremek wrote: "In times when the harshest misery reigns [...] it is not difficult to enlist volunteer *galeotti*. It is enough to choose among the city's wretched, deprived of every means of subsistence and any possibility of work. [...] It is evident that when many unemployed are eager to find work, their needs are modest and the wages low."[44]

Alberto Guglielmotti, in his pivotal 19th-century treatise on the papal navy, also described the *buonavoglia* as robust, healthy young men "obliged by necessity, or debt, or gambling, or any other—even honest—reason, such as helping their parents, or providing a dowry for a sister; in short, anyone who needed money at that time." He particularly highlighted the widespread prevalence of gambling during this period and how easy it was to find individuals in debt and unable to repay what they owed.[45]

The practice of filling crews with the most destitute and desperate people in society—those willing to sell their freedom—was already noted by Pantero Pantera in his naval treatise.[46] There, the captain describes the *buonavoglia* pri-

---

43 Geremek, *Inutiles au monde*, pp. 144–147.
44 Geremek, *Il pauperismo*, pp. 695–696. "Nei periodi in cui regna la miseria più nera [...] non è difficile arruolare galeotti volontari, basta scegliere tra i miserabili della città, privi di ogni mezzo di sussistenza e di ogni possibilità di lavoro. [...] è evidente che quando i disoccupati desiderosi di trovar lavoro sono molti, le loro esigenze sono modeste e i salari bassi."
45 Guglielmotti, *La Guerra dei Pirati, Volume Primo*, pp. 298–304: "Un giovane robusto e sano, stretto dal bisogno, o dai debiti, o dal giuoco, o da qualunque (anche onesta) ragione, pognamo di soccorrere i genitori o di dotare una sorella; in somma chiunque voleva denaro per quei tempi, purché fosse robusto e giovane."
46 Pantera, *L'armata Navale*, p. 136: "impresa molto difficile persuadere gl'huomini liberi à maneggiare un remo, & porsi alla servitù d'una catena, & alle battiture, & à gl'innumerabili incommodi della galea, i quali se la necessità, ò la sciocchezza di molti vagabondi, & per altro

marily as either convicts who had completed their penal sentences but were still in debt to the state and were required to row until the debt was repaid, or vagabonds who had sold their freedom to survive and had agreed to serve onboard. Despite their circumstances, *buonavoglia* were always highly valued in galley life and considered essential to its effective operation. In particular, they were often sent ashore when needed, for example, to collect provisions or ammunition. Ex-convicts were given special consideration, especially those who had served long sentences, as this allowed them time to train and develop into skilled rowers. This led to a tendency to find any means possible to retain them aboard once their sentences had elapsed.

The *buonavoglia's* marginal social character is also emphasized by Capuchin Father Filippo Bernardi of Florence who, in his account of the Capuchin management of Livorno's *Bagno de' forzati*, describes the galley crews as divided into "Turks" [*turchi*, i.e., Muslim slaves], convicts [*forzati*], and lastly *buonavoglia*, for whom contempt is both explicit and patently deliberate. Noted for their trademark mustaches and beards,[47] the *buonavoglia* were known as such because they "for the sake of a few coins voluntarily subject themselves to such a laborious and disreputable, standard of living." They were considered worse than brutes, for not even beasts, though irrational, would have sold their freedom, "more valuable than any rich treasure," for a few coins.[48]

Being sent to the galleys was one of the primary forms of punishment and repression in pre-modern Europe. According to Franco Angiolini, the earliest evidence of convicts and slaves as rowers on galleys dates back to the late 14th and early 15th centuries. In 1331–1332, we find sailors who, after being imprisoned for owing money to Venetian shipowners under a law of 1329, were placed on galleys without pay to settle their debts. Similarly, over the course of the 15th century, the Aragonese Crown resorted, when necessary, to convicts known as *reos galeotes* to fully man the crews. These were men convicted of minor crimes who, as an alternative to capital punishment, served as rowers for a set period of time. Although there were numerous antecedents, these remained sporadic

---

mancipij vilissimi, & abiettissimi de i vitij, non gli conducesse à termini di vender se medesimi, si può credere che non si troverebbe mai huomo alcuno, che spontaneamente volesse sottomettersi ad una vita così infelice, & piena di tanti miserabili & horribili incidenti."

47 Ibid., p. 131 f. On the contrary, convicts and slaves had to shave their faces completely.
48 Bernardi, *Relazione*, p. 13 f: "[…] ed I terzi pùr Cristiani, sono le buona Voglie, così detti, perchè volontariamente per interesse di pochi soldi si soggettano ad un tenor di vita così laboriosa, ed infame. Mostransi costoro d'essere di peggior condizione de' Bruti, i quali benché irrazionali, fuggono nulladiméno il soggettarsi; e l'huomo, tralasciando i migliori discorsi dell'intelletto, per prezzo di pochi denari vénde la propria libertà più stimabile di qualunque ricco tesoro."

measures. The first instance of the massive and regular use of convicts on board galleys appears to date to 1443. In that year, the French shipowner Jacques Cœur obtained permission from the King of France to use convicts to arm his galleys—despite opposition from the Parisian courts and the Church. Later, in 1490, Charles VIII (1438–1498) issued an edict obliging royal judges to send to the galleys all those sentenced to death or other corporal punishment, as well as those deemed "unfit for social life." In this way, many individuals convicted of crimes—especially vagrants, Romani, and others considered "unsocial"—were removed from society and sent to the galleys.[49]

Italian navies soon followed the French example. In Genoa, the presence of *galiotti per forza* is recorded as early as 1473. The presence of men chained to oars is also mentioned in the account of a shipwreck near Oneglia in 1492.[50] Given the shortage of Muslim slaves to man the penal fleet of the Papal States in 1502, shipowners turned to using nearly all of Rome's prisoners, as well as a large number of vagrants begging in the city streets.[51] In his treatise on the Papal Navy, Guglielmotti explicitly mentions "a very ancient usage" of sentencing criminals to row. According to him, it was customary for those sentenced to serve for a year at the oars in place of execution. This arrangement was thought to benefit all parties involved: those condemned would believe that they had succeeded in extracting one more year of life from the authorities, while society would, at the same time, be compensated for any damages caused by their crimes.[52] However, these were also episodic measures, and it was not until the latter half of the 16th century that galley punishment was officially codified in Italian legislation.[53] Initially, galley service targeted those individuals who would otherwise have faced corporal punishment. The first to be condemned were those whose conduct was considered deviant in multiple respects, rendering them candidates for exclusion from society.

Over time, a galley sentence assumed an increasingly exemplary function, designed to deter the population at large from committing similar offenses through

---

49 Angiolini, La pena della galera, p. 79 f.
50 Lo Basso, *Uomini da remo*, p. 232.
51 Guglielmotti, *La Guerra dei Pirati, Volume Primo*, p. 21.
52 Ibid., p. 116 f: "L'uso antichissimo del condannare I malfattori alla pena del remo viene espresso nel capitolo decimottavo, con una giunta straordinaria. Si tratta di mettere a remigare per un anno anche i condannati a morte; ai quali senza ingiuria pensavano di poter concedere un anno di vita, perché alla società oltraggiata dai loro misfatti venisse compenso con qualche servigio di pubblica utilità."
53 The Papal State was an exception, given that galley punishment was formally introduced in the first half of the 16th century.

the threat of forced labor and deprivation of liberty. For this reason—alongside the practical need to find rowers—the range of crimes punishable by such service expanded during the 16th and 17th centuries.[54] In the aforementioned naval treatise by Pantera, convicted *galeotti* are described as rowers sentenced by the courts to serve aboard for a limited period—or even for life. He argued that the use of convicted criminals was the ideal solution to the problem posed by the difficulty of persuading free men to row, to "submit to the bondage of chains," and, more generally, "to the torture and misery of life on the galleys."

He contended that it was praiseworthy for criminal judges to commute not only corporal punishment, but also fines, turning them into galley sentences of reasonable duration, proportionate to the crime committed. In this sense, those sentenced to death would serve in the galleys for life. Similarly, all vagrants found begging in the city streets were dispatched to the galleys and put to public use.[55] As a distinguishing mark, they were required to shave off both their hair and beards and to wear a red beret. A convict's most obvious sign of servitude, however, was the iron clasped to his leg, used to chain him to the oar bench, ensuring he could not escape.[56] Although the tasks, way of life, and conditions of forced *galeotti* were largely consistent across the Mediterranean, the main point of divergence lay in the length of their sentences. For example, in the Papal States, a life sentence in the galleys was considered acceptable and widely practiced. In contrast, in the Serenissima, an edict of the *Maggior Consiglio*, proclaimed on 15 January 1588, stipulated that the sentence could only range between a minimum of eighteen months and a maximum of twelve years: a life sentence was not permitted. The rationale behind this decision rested on the belief that the lack of hope of future freedom would lead the *galeotti* to despair, thereby reducing their efficiency.[57] Nevertheless, convicts were still required to repay the Republic all the expenses they incurred during their years of service, thus accumulating a debt that was unlikely ever to be repaid. As a result, they ended up serving onboard until the end of their days.[58]

---

[54] Angiolini, La pena della galera; Filioli Uranio, *La squadra navale Pontificia*.
[55] Pantera, *L'Armata Navale*, p. 137f.
[56] Ibid., pp. 130–132. "Gli sforzati son quelli, che da i tribunali della giustitia sono stati condannati per i loro delitti à vogar in galea à tempo limitato, ò à vita, & non si lasciano uscir mai di galea, ne di catena per qualsivoglia occasione, fin che non hanno finito il tempo della loro condannagione […] Sono distinti da gl'altri con questo segno, che portano il capo, et la barba tutta rasa."
[57] Lo Basso, Condannati alla galera, p. 123.
[58] Lo Basso, *Uomini da remo*, p. 141f. On indebtedness as a strategy used by masters to keep servants and laborers in their service even after the end of their contracts, see Campbell/Stanziani, *Debt and Slavery*; Stanziani, *Lavoro coatto*.

The last category of rowers consisted of enslaved individuals, most of whom were Muslims, and were generically referred to as slaves, or "Turks."[59] These individuals were primarily prisoners of war, captured mainly during naval battles. In theory, they were to be freed and repatriated at the end of the wars—either after a ransom had been paid or in exchange for a Christian prisoner. In reality, however, efforts were made to keep them on the galleys for as long as possible, often without ever releasing them. The term "slave" was widely used in the Ponentine navy. In Venice, however, slaves were referred to only as "Turks," in order to emphasize their status as prisoners of war rather than as slaves—likely with the aim of maintaining a less antagonistic relationship with the Ottoman Empire.[60] In the Papal Navy, in contrast, the term "Turks" referred to a category of slaves which also included "Moors" and "Blacks." As Pantera wrote:

> Slaves are those Turks who are captured or purchased, and they fall into three categories: Moors, Turks, and Blacks. The Moors are the best, and among them, the most excellent are those taken aboard galleys, brigantines, galiots, or other corsair vessels, for having become accustomed to hardships and toil of the sea and of rowing, they are better than the others and are perfect oarsmen. But by nature, they are so proud, brutal, treacherous, and seditious that their behavior must be closely monitored, as they have, on occasion, gone so far as to kill their masters. The Turks are not as good, nor as suited for rowing and hard labor as the Moors, yet they are far gentler and more docile. The Blacks are the worst of all, and most of them die purely from melancholy. They can be distinguished from the others by the tuft of hair they wear on the crown of their heads, being otherwise completely shaven. They receive the same food and clothing as the convicts.[61]

---

59 See Di Nepi, *Confini salvezza*, p. 163; Ricci, *I turchi*, p. 66.
60 Lo Basso, *Uomini da remo*, p. 29.
61 Pantera, *L'armata Navale*, p. 131: "Gli schiavi sono quei Turchi, che si pigliano, ò s comprano, & sono di tre sorti, cioè Mori, Turchi, & Negri. I Mori sono i migliori, et tra loro ottimi quelli che si pigliano sopra le fuste, ò sopra i brigantini, ò galeotte, ò galee, ò sopra altri vascelli da corso, i quali per haver fatto l'habito ne i patimenti, et nelle fatiche del mare, et del remo, sono migliori de gl'altri, et sono perfetti vogatori: ma sono per natura talmente superbi, bestiali, traditori & seditiosi, che bisogna osservare bene i loro andamenti, che alcune volte si è condotta fino ad ammazzare i patroni. I Turchi non sono buoni, ne atti al remo, ne alle fatiche, come i Mori, ma sono ben più mansueti, & più docili [...] I Negri sono peggiori di tutti, et muoiono la maggior parte di pura malinconia [...] si conoscono da gli'altri per la ciocca di capelli che portano nella sommità della testa, essendo nel resto tutti rasi. Hanno il vitto, & il vestito come gli sforzati." It is noteworthy that Bartolomo Crescentio, writing about fifteen years before Pantera, describes slaves in a similar fashion, but does not speak of "blacks"; instead he refers to *morlacchi*. This likely refers to the Morlachs, a rural Christian population from the Balkans, subjected to the Ottoman Empire and enslaved and sold prior to sub-Saharan Africans. Indeed, it should not be forgotten that medieval slavery primarily involved populations of the Balkan and the Black Sea regions, see Barker, *Most Precious Merchandise*; Schiel, Slaves' Religious Choice, pp. 23–45. Crescentio, *Nautica mediterranea*, p. 95: "Schiavi parte sono Mori, parte Turchi, parte Morlacchi. I Turchi, & i Mori

The distinction between these three categories of enslaved rowers remains somewhat unclear, with differentiation likely based on skin color and geographic origin. The term "Turks" was often used interchangeably with "Muslims," though in this context it might also refer to North African men. It seems reasonable to conclude that the term "Blacks" described slaves from sub-Saharan Africa, bought by slave traders from Spain, Genoa, and Portugal, and later northern Europe.[62] More uncertain, however, is the meaning of the term "Moors." In later sources, "Moors" and "Blacks" are frequently used interchangeably. It's possible that the primary distinction between these two categories of enslaved individuals was the darker skin color of the latter, while "Moors" likely referred to those of Maghrebi origin.[63] Adherence to the Muslim faith—or any other religion differing from Catholicism—was a defining characteristic of Mediterranean slavery. It was, in fact, a prerequisite for becoming a slave, given that the practice of Mediterranean *captivitas* originated in the Papal edict *Romanus pontifex* of 8 January 1454, by which Pope Nicholas V (r. 1447–1455) granted the King of Portugal the power to enslave "Saracens, pagans, infidels and enemies of Christ."[64] Enslaved individuals were dressed like convicts and subjected to similar tasks. They differed from the *forzati* primarily in their original faith and in the distinctive knot of hair arranged atop their heads, though they were otherwise completely shaven.[65] Interestingly, convicts were often referred to as "slaves for punishment" [*servi/schiavi di pena*].[66] The galley sentence—despite some notable differences—aligns with the category of servitude by punishment, as theorized in Roman law.[67] As the *Scrivano Generale* [Chief Scribe] of the Tuscan galleys wrote in 1699: "it scarcely matters whether one says Slave or forced rower, as even the forced rowers are said to be a slave of punishment for the time they are onboard."[68]

---

pigliati su le loro Fuste sono megliori, che quei che in terra se pigliano. Morlacchi la maggior parte muoiono di melanconia & ostinatione." See also ASF, MP, 2077.

[62] Hershenzon, *The Captive Sea*, p. 19.

[63] On the problem on distinguishing between Moors [*mori*] and Blacks [*negri*] in the Italian context, see Boccadamo, *"Mori Negri"*; De Lucia, *History of a Black Saint*.

[64] Benedetti, Servi introvabili, p. 54.

[65] On the visual signs of identification of Muslim slaves in the early modern Mediterranean, and in particular on the topknot as the key sign of identification of Muslims, see Martin, Weiss, *The Sun King at Sea*.

[66] *Servi* and *schiavi* are used in the sources as synonyms and used interchangeably.

[67] Fiume, *Schiavitù Mediterranee*, p. 17.

[68] ASF, MP, 2108, cc. n.n.: "poco per altro importando, che si dica Schiavo, ò forzato, mentre anche il forzato si dice schiavo della pena, per quel tempo che deve servire alle Galere." See Chizzolini, Navigating Ambiguities.

On early modern Italian galleys, the status of "slave" hardly differed from that of "convict." In either case, rowers had their liberty curtailed—either permanently or temporarily. They were subject to similar duties, meals, and accommodations, and even their clothing differed only in the smallest details. Above all, both categories of rowers were considered deviant and marginal on account of their behavior, their religion, and, to some extent, their personal inclinations. Belonging to a different geographical or religious context only served to amplify their already marginal status, simply by virtue of being forced aboard such vessels.

3 Juridical discourse on the galley sentence

Although there are precedents in antiquity, galley punishment represented a novel development in the practice and theory of early modern European criminal jurisprudence. Recognizing this punishment's innovative nature, some 16th- and 17th-century jurists sought to develop an increasingly complex and articulated corpus of law that would position galley punishment within the broader spectrum of crimes and their respective punishments. To fully legitimize this new form of punishment, it had to be integrated into the legal cultural heritage of the time, as well as into the existing legal-historical categories.[69]

As Franco Angiolini noted, the first jurist to attempt to establish a historical and doctrinal antecedent for galley punishment was Tiberius Deciani (1509–1582), who argued that the Athenians had employed this practice. Despite Deciani's influential status, later scholars did not widely accept this hypothesis. The criminal judge Giulio Claro (1525–1575), while acknowledging its novelty, likened service aboard the galley to forced labor, specifically the *damnatio ad metallum* in Roman law. In fact, Claro proposed that galley service be categorized among capital punishments involving the loss of liberty and citizenship [*civitas*], thereby reinforcing its character as a form of penal servitude. For this reason, the galley sentence was ranked in severity immediately below capital punishment and above penalties that only involved the loss of civic status. While Claro's thesis—linking the punishment to *damnatio ad metallum*—gained widespread acceptance, many jurists argued that it more closely resembled punishments such as deportation, banishment, and forced labor, collectively associated with *in insulam deportatio* in Roman Law. Doctrinally, galley punishment was thus equated with both the second degree [*damnatio ad metallum*] and third degree [*in insulam de-*

---

[69] Angiolini, La pena della galera, p. 87.

portatio] of capital punishment. Though formally less severe than death—and thus more compatible with Christian teachings—it remained a harsh penalty, intended both to punish the condemned and serve as a deterrent.[70] According to Marcantonio Savelli, a judge of the Florentine *Rota Criminale*, galley sentencing was the civil counterpart of canonical excommunication. Both were considered substitutes for the now-obsolete *interdictio ignis & aquae*—a Roman legal sanction that stripped offenders of their citizenship and, by extension, their civic rights, rendering them unworthy of membership in the Roman community.[71]

As the jurist Luigi Cremani (1748–1838) observed in the late 18th century, no doctrinal framework for galley punishment was ever developed.[72] It had largely escaped jurisprudential reflection, which accounts for its multifaceted character: at once a form of confinement, a means of public labor, and an arbitrary punishment [*custodia loco depositi, opus publicum,* and *poena extraordinaria*].[73] The notable exceptions are two legal treaties, both shaped by the Venetian context and thus specific to it: Lorenzo Priori's *Prattica criminale* (1663), and Antonio Barbaro's *Pratica Criminale* (1739). According to Priori, a galley sentence was viewed as a corporal and arbitrary punishment, imposed solely at the judge's discretion.[74] A century later, Barbaro categorized punishments into corporal and non-corporal, with the former further divided into capital and non-capital forms. As noted, Priori had already classified the galley as a form of corporal punishment, but Barbaro refined this classification, placing the galley sentence in both categories on the basis of its duration. Capital punishments included death, banishment, perpetual ignominy, life imprisonment and ten years' service in the galley. Conversely, non-capital punishments encompassed amputation, whipping, temporary banishment, as well as perpetual ignominy, imprisonment, and a galley sentence of fewer than ten years.[75] Despite the lack of sustained, in-depth theoretical reflection, galley punishment would ultimately become an integral part of criminal legislation across European jurisdictions. As Franco Angiolini suggested, it may have been precisely this very absence of theoretical underpinnings that enabled the galley to thrive as an effective and flexible instrument of repression—

---

[70] Ibid., pp. 88–91.
[71] Savelli, *Pratica*, p. 167.
[72] Angiolini, La pena della galera, p. 91.
[73] Alessi, Pene e remieri, p. 250.
[74] Priori, *Prattica criminale*, p. 123. "Ordinary," on the other hand, referred to the punishment established by laws or statutes.
[75] Barbaro, *Pratica criminale*, capo XXXVI—*Delle pene*, cited in Lo Basso, *Uomini da remo*, p. 141.

one fundamentally rooted in the deprivation of liberty. It thus emerged as a key institution for both punishing and correcting criminals.[76]

As much as the specificity of galley punishment lay in its provision of forced labor—especially through the grueling labor of rowing—it is essential not to overlook the central role played by the curtailment of liberty. The need to maintain a high degree of discipline onboard—vital to effective navigation—formed the basis for a correctional program and, in a sense, a process of re-education for convicts grounded in the denial of their freedom. Stripping individuals of their liberty enabled the exercise of total control over them. In galley life, this control was directed at the profound transformation of behaviors and tendencies deemed deviant and criminal, with the aim of turning convicts into disciplined individuals. Contemporary sources—both implicitly and explicitly—consistently emphasize this objective of correction. From the late 16th century onward, galley punishment came to be seen as one of the best tools to manage social deviance. It served to eliminate the perceived threats posed by individuals deemed harmful to the social order by removing them from society and attempting to reform them—albeit without any provision for their eventual reintegration into society. Suffice it to say that the earliest galley condemnations in early modern Italy were specifically directed at social categories deemed disruptive to moral and civic norms. In Tuscany, for example, the galley penalty was first applied in 1542 to blasphemers and sodomites, while in Venice, it was first used in 1545 against vagabonds and Romani people.[77]

One could argue that the galley sentence was thus conceived as complementary to—and, in some ways, continuous with—imprisonment. In many cities, the earliest galley sentences were applied as commutations for death penalties or prison terms. Conversely, incarceration was often offered as an alternative to the galley if a medical examination at the time of sentencing deemed the condemned unfit to row. The most-well documented case is undoubtedly that of Venice, where a fixed conversion scale was established to determine the equivalent prison term: eighteen months at sea corresponded to three years in prison, three years in the galley to five years in prison, and so on.

The stark disparity between the duration of galley service and imprisonment underscores the perceived inherent difference between these two institutions. The galley was widely regarded as a harsher punishment—almost akin to the death penalty—due to the inhumane conditions and the grueling labor. Yet, I would argue that the fact that these two punishments were often treated as interchange-

---

76 Angiolini, La pena della galera, p. 91.
77 Ibid., pp. 79–82.

able suggests a fundamental structural similarity. When the galley was docked, convicts were confined to specially designed holding facilities intended primarily to prevent escapes. The most notable of these—complementary to the galley both in nature and function—was the *Bagno de' forzati* in Livorno.[78]

Ultimately, life as a convict aboard the galley scarcely differed from that of a convict on land, with the main distinction being the emphasis of forced labor for public works at sea. The 1542 proclamation that introduced the first appearance of the galley sentence in Tuscan legislation, explicitly stated that the punishment could entail confinement either aboard a galley or in Florence's Stinche prison.[79] By the 19th century, prisons had already formally adopted forced labor for inmates, mirroring the fate of those confined to the galleys. A prime example is the Stinche prison itself, which, under Austrian rule, was converted into a penal colony where inmates sentenced to public works were held. Like their counterparts at sea, these prisoners were easily identifiable by their red caps and the chains around their feet—clear symbols of their convict status.[80]

Against this hypothesis, many scholars have argued that, prior to the Enlightenment, imprisonment—or more precisely the legal deprivation of liberty—was viewed primarily as a custodial measure rather than a punitive one. This interpretation largely draws upon legal sources of the time, particularly the jurist Ulpianus (170–228), who asserted that "prison is meant to guard, not to punish."[81] However, recent research has challenged this view, showing that local prisons during the Middle Ages, far from being restricted to canon law,[82] sometimes served punitive and even socially rehabilitative functions—albeit with a fully developed objective of reintegrating convicts into society.[83] In light of these new perspectives, I argue that the galleys and the institutions associated with them—such as the *Bagno* in Livorno—were deliberately designed to enforce discipline. These institutions

---

[78] See Frattarelli Fischer, Il bagno delle galere; Santus, *"Il turco."*
[79] Cantini, *Legislazione toscana*, p. 212.
[80] ASF, Gabinetto, 111, dossier 2, cc.n.n. Regarding the continuities and discontinuities between galley sentence, prison, *ergastolo*, and forced labor, please refer to the Ph.D. project by Andrea Giuliani, *Punishing the Convicts of Serious Crime: the Evolution of the Italian Ergastolo between Forced Labor and Detention 1769–1890*, defended in 2024 (Università di Roma Tor Vergata), which aims to analyze the process through which *ergastolo* was established as a penal institution based on forced labor, in order to understand the trajectories through which penal practices have evolved, in Italy and elsewhere, since the age of reform.
[81] Geltner, *La prigione medievale*, (translation) p. 83: "carcere ad custodiendum, non ad puniendum."
[82] On this topic see particularly Paglia, *La Pietà dei carcerati*; Mibillion, *Riflessioni sulle prigioni*; Benedetti, Dalla galera all'ergastolo, p. 18 f.
[83] In this sense, the study by Geltner, *La prigione medievale* has been absolutely inspiring.

aimed not only to reform convicts but also to exert a broader societal influence through the constant threat of the deprivation of liberty.

Throughout the early modern period, imprisonment increasingly became a punishment for specific social groups: the nobility, the clergy, as well as women and children. This shift is evident in the use of imprisonment as an alternative to the galley sentence on account of the convict's elevated status. This appears to contradict the views of jurists who upheld the idea that galley punishment, in theory, should be indiscriminate—that even nobles and clergy could be subjected to such a fate. Yet, as is often the case, theory diverged from practice. When members of the nobility or the clergy were incarcerated, steps were typically taken to prevent them from mingling with lower-class criminals. This disparity is particularly evident in the Papal States, notably following the establishment of the Corneto "Ergastolo"—a facility specifically designed to house members of the clergy sentenced to galley service.[84] With the notable exception of the nobility and the clergy, the rest of the male population remained effectively subject to galley punishment.

In Rome, the death penalty and galley sentences for minors were often commuted to terms of confinement in Saint Michael's penitentiary, founded in 1693 by Pope Innocent XII. Notably, in this case, imprisonment was not regarded as an alternative to the galleys but rather as its precursor. Once the condemned youths were of age, they were taken to Civitavecchia and assigned as rowers on the Papal galleys. This practice supports the view that prison was often treated as little more than a holding space prior to the execution of a sentence.[85] My intention is not to refute this widely accepted interpretation, but to suggest that it represents only one of the many functions that imprisonment could fulfill. In this case, once again, we see that imprisonment functioned in a complementary relationship with galley service. Importantly, detention at St Michael's was also explicitly aimed at re-education and frequently responded to parents' requests to confine particularly unruly or disobedient children.[86]

In conclusion, both galleys and prisons constituted forms of punishment rooted in confinement and the deprivation of liberty. It is telling that the petitions of the *galeotti* sought not only release from rowing service but also, and perhaps more significantly, the restoration of their freedom and their rights as citizens. Even before the Enlightenment's penal reforms, the notion of freedom as a primary

---

[84] On this topic, see in particular Benedetti, Dalla galera all'ergastolo, pp. 15–69.
[85] Baldassarri, *Bande giovanili*, p. 28 f. On the history of the institute, see Lucrezio Monticelli, S. Michele a Ripa, pp. 397–420; Coccoli, La Casa di correzione.
[86] Baldassarri, *Bande giovanili*, p. 31.

human prerogative—and thus a supreme value—was already deeply embedded in early modern thought, both in preventive and repressive contexts. As noted earlier, Father Filippo Bernardi's disdain for the *buonavoglia* stemmed precisely from the fact that they—by consenting to be recruited in exchange for paltry pay—had sold what he saw as a human being's most precious attribute: freedom.[87]

That the early modern penal treaties make little or no mention of the deprivation of liberty should not lead us to conclude that discussing its rehabilitative or moral dimension is anachronistic. On the contrary, this silence highlights the frequent disjunction between juridical theory and penal practice.

3.1 "There is nothing so dear in this world to man as freedom"

Nearly every scholar is now familiar with Cesare Beccaria's stance in his treatise *Dei delitti e delle pene* (1764) against capital punishment and torture, which he deemed inhumane practices—contrary to Christian values—and utterly useless from a practical standpoint. Beccaria's criticism was directed primarily at the death penalty, which, rather than deterring crime, often provoked, indignation and contempt for the executioner. Initially, this cruel spectacle of justice had the merit of attracting curious onlookers and imparting a lesson to the public. However, prolonged exposure to such a spectacle inevitably led the audience to empathize more with the condemned, fostering pity for the criminal. A criminal justice system that relied on excessive severity and bloodshed could not maintain control over the populace: over time, the population became desensitized to violent, rendering punishment less effective as a deterrent. As an alternative, Beccharia proposes life imprisonment as the most severe—and the most effective—penalty a State could impose. His work was widely read throughout Europe, and became a symbol of Enlightenment intellectual renewal. It also played a pivotal role in shaping the "prison revolution" of the 19th century during which incarceration was thoroughly reimagined not only as a punitive tool for but also as a rehabilitative measure for social integration.[88]

However, as has been demonstrated over the past two decades, referring to a "19th-century revolution" is misleading.[89] Confinement and deprivation of liberty had already been used as punitive and rehabilitative measures since the Middle

---

87 Bernardi, *Relazione*, p. 13f.
88 Ironically, the first printed version of Beccaria's work appeared in Livorno in 1764, in the building that once housed the *Bagno de' Forzati*, which was then converted into a printing press.
89 Lucrezio Monticelli, *Prigioni e rappresentazioni*, p. 2f.

Ages. Early medieval Christianity introduced a "medicinal" conception of punishment—aimed primarily at correcting the offender and rectifying their offenses through imprisonment.[90] Canon law, from the 4th to the 13th centuries, provided for solitary confinement in monasteries as a form of penance. This practice evolved into a broader system of monastic isolation for religious offenders. Solitary confinement thus took on a dual purpose: as a temporal punishment which involved depriving the religious of their liberty and as a spiritual punishment aimed at re-educating the prisoner.[91]

In civil law, while contemporary jurists did not formally legitimize punitive imprisonment, such practices are widely documented in medieval statutes and judicial records.[92] Guy Geltner's thesis is noteworthy: he argues that the use of prison as a primary punishment represents a shift from "deportation"—understood as isolation from the rest of society—to "containment" of offenders. The purpose of incarceration was not to eradicate social deviance but rather to contain and monitor it, maintaining a visible reminder of the consequences of criminal behavior. Furthermore, as early as the mid-16th century, a specifically correctional strategy for imprisonment—distinct from mere detention and punishment, began to emerge.[93] Throughout the early modern period, the concepts of "correction" and "deprivation of liberty" were not seen as oppositional but were closely linked. As several historians of penal and penitentiary regimes have argued, the Judicial Enlightenment—of which Beccaria is the most prominent exponent—contributed to a decisive shift in penal thinking, culminating in the emergence of an unprecedented "prison discourse." This period marked the creation of a new approach to punishment shaped by a renewed interest in the penal question and public debate. While the penal system itself was by no means an exclusive creation of the Enlightenment, its real innovation lay in its systematic use of such punishment as a tool for both punishment and reform—and in making it a central topic in public discourse.[94]

A similar reasoning can be applied to galley punishment, which, at its core, was rooted in the same principles as imprisonment: deprivation of liberty in service of public labor. In this context, I will focus on the previously cited report by Father Filippo of Florence concerning the Capuchins' work at the *Bagno* in Livorno. In the introduction to his account, Father Filippo reflects on the prince's need to have recourse to a strict and unyielding justice—punishing the guilty

---

[90] Benedetti, Dell'Ergastolo, p. 10.
[91] Paglia, *La pietà*, pp. 7–11.
[92] Geltner, *La prigione medievale*, pp. 84–86.
[93] Coccoli, La Casa di correzione, p. 12.
[94] Lucrezio Monticelli, Prigioni e rappresentazioni, p. 2.

while rewarding the virtuous. The prince's mercy, he argues, must be balanced by the "terror of punishments." Criminal justice must not only penalize wrongdoers but also set an example for the broader population—discouraging them from committing crimes and encouraging them to become better citizens. The idea of strict and inflexible justice is underscored by the proverbial expression: "one is punished, a hundred are warned."[95] Particularly fitting is Father Filippo's image of the gardener, who must uproot weeds so that healthy plants may flourish.

Without going so far as to theorize the futility of the death penalty, Father Filippo acknowledges that corporal punishment often proved excessively severe and ineffective for certain crimes. Since antiquity, he reminds us, galley service has served as a substitute for corporal punishment, with the added advantage of being more effective: like death, it entailed the loss of honor and freedom. This deprivation of liberty, then, represented the gravest punishment—second only to the death penalty—as it amounted to a form of social death. As Father Filippo suggests, this innate love of liberty is not unique to humankind; even animals, though lacking reason, have often demonstrated a preference for death over bondage.[96]

It is possible to argue that the widespread use of galley punishment—driven in part by the need for rowers—demonstrates how confinement functioned as a form of penal discipline. Yet, galley service went beyond mere confinement: it entailed strenuous physical labor. From the 15th century onward, the use of forced labor as a corrective measure became increasingly common, supported by the belief that physical work could transform the idle individual—"curing" him, so to speak—and re-educating him in the core values of society.[97] As Bronisław Geremek has argued, by the 15th and 16th centuries, a genuine work ethic had emerged: manual labor was not just idealized as virtuous but treated as an obligation for the lower classes, and a tool for moral correction and social rehabilitation. The aim was to prepare the poor for society once they had been "cured" of laziness and criminality.[98] In this context, galley service was considered particularly well-suited for vagrants, whose idleness and perceived "uselessness" made them prime candidates for reformation.[99]

---

[95] Bernardi, *Relazione*, p. 11: "Suol dirsi fin per Proverbio, Che uno se ne castiga, e cento se n'ammonisce."
[96] Ibid., pp. 11–13: "Non ha l'huomo cosa più cara al mondo della libertà; anzi gli animali stessi, che pur non conoscendo le loro prerogative, incontrano più tosto la morte che cadere in servitù."
[97] This was to become the norm from the mid-16th century, thanks in particular to the English example of the Bridewell House of Correction, founded by Henry VI in 1552.
[98] Geremek, *Inutiles au monde*, p. 60.
[99] Viaro, *La pena della galera*, p. 400 f.

The ruling classes, both on mainland Europe and in England, recognized the galley's preventive and rehabilitative nature. Admiral Sir William Monson (1569–1643), praised the "terror of galleys" in his posthumous work *Naval Tracts*. He contended that not only were they the best deterrent to crime and the most efficient instrument with which to control and discipline, but they had also "saved so much blood that is unfortunately shed by the execution of thieves and criminals."[100] Notable examples that illustrate the corrective and disciplinary function of deprivation of liberty linked to coerced labor include the Bridewell House of Correction, founded in London in 1555, and the Amsterdam *Rasphuis* established in 1596, both of which served as models for the Parisian *Hôpital Général* in the following century.[101] The link between imprisonment and imposed labor is also evident in Tuscany after the abolition of galley labor in 1750 when convicts still serving their sentences were transferred to fortresses, where they continued to perform their sentences.[102]

The significance attributed to the curtailment of liberty aligns with the theories of the social contract, such as those outlined by John Locke in the *Second Treatise on Government* (1689). Thus, for example, an edict published in 1784 on the management of the papal galleys—focused on the large number of forced rowers attempting to escape—explicitly stated that freedom was a natural human right. This right, however, could be legitimately revoked from any individual whose actions violated and threatened public order. Furthermore, the edict further emphasized that individuals had, in effect, renounced their natural right to freedom once they decided to form a society. As a result, attempting to escape the galley was not only a rejection of the punishment legally imposed but also a breach of the social contract.[103]

In the same vein, while the concept of the deprivation of liberty as a punitive and re-educational tool was already present in early modern Italy, it is important to stress that re-education aboard the galleys was primarily focused on ensuring successful seafaring and military performance. Although in theory, convicts were expected to return to normal life once they completed their sentences, many ended their days in the galleys or remained excluded from society, labelled as

---

100 Cited in Angiolini, La pena della galera, p. 84.
101 Alessi, *Il processo penale*, pp. 141–143.
102 Angiolini, La pena della galera, p. 110.
103 ASR, Camerale III—Comuni, Civitavecchia, b. 846, f. n.n.: "Conciosiachè si oppongono a sì fatto titolo le primordiali convenzioni del contratto sociale, su cui poggia ogni civile società, e per cui il contravventore della legge legittimamente condannato, e caduto nella Forza del Prencipe suddito diviene, e Servo della pena avendo rinunciato alli diritti della naturale libertà per li vantaggi della vita sociale."

"disgraced" individuals. The absence of reintegration efforts is particularly evident in the *Proclamation on the Government of the Papal Galleys*, issued on 19 December 1705. Article 70 of this document explicitly forbade any freed rower from entering Civitavecchia, arguing that their presence posed a threat to public order and tranquility.[104] Indeed, fully restoring these individuals to society was virtually impossible, as the very notion of reintegration presupposed the existence of structured resettlement policies which simply did not exist at the time. Nevertheless, this absence did not diminish the significance of depriving individuals of their liberty as a corrective tool in the penal practices of the early modern period.

4 Cultural discourse concerning the galleys

Galley life was widely perceived as a miserable existence. In popular imagination, *galeotti* were often depicted as wretched individuals, condemned to row in chains under inhumane and exhausting conditions. Contemporary sources, particularly the institutional reports, confirm the harsh reality of daily life for these convicts and slaves, with a high degree of violence and exploitation. These documents reveal a pressing need to maintain order on board, not only among the rowers as well as the officials, who were frequently accused of brutalizing and harassing the crew. Beyond this extreme violence, additional hardships included rough seas, diseases, and the perils of naval battles.[105] The sea's unpredictability was a constant cause for fear and anxiety. Impossible to control or predict, bad weather could strike suddenly, often catching crews off guard. Even Saint Paul, as noted by Pantera, in his First Letter to Titus, emphasized the necessity for mariners to keep themselves in God's good grace because of the danger of being at sea, constantly in a state "between life and death."[106] A prime example of this peril is the

---

[104] AAV, Misc. Arm. IV/V, b. 54, f. 106: "70. Considerando quanto sia perniciosa alla pubblica, e privata quiete la dimora in Civita Vecchia di quelli forzati, che dopo aver terminato il tempo della condanna sono stati sciolti dal Remo, volendo perciò opportunatamente provedere d'ordine espresso della Santità sua ordiniamo, e comandiamo à qualunque Forzato, che posto in libertà debba subito sfrattare da Civitavecchia, e suo Territorio, sotto pena della galera per cinque anni da incorrersi irremisbilmente in caso di contraventione, considerate però le circostanze, e qualità delle persone, e cause di essersi ivi trattenuto."

[105] AAV, Fondo Pio, b.112, *Discorso in materia della salute di qual si voglia armata di mare e di terra*, f.333: "tre potentissimi nemici sono che ognuno di loro è bastante a ruinare una armata, per fare potente che ella sia il primo de quali è il turbato mare quando à fortuna, il secondo è l'infermità quando regnano, et il terzo è il nemico col quale si guerreggia."

[106] Pantera, *L'armata Navale*, p. 101.

storm that struck the Marseilles coast between 19 and 22 April 1569, resulting in the loss of five of the ten Florentine galleys anchored there.[107]

Furthermore, the cramped conditions and long periods at sea often led to extremely poor hygienic conditions. Diseases were common, stemming both from poor sanitation and the practical impossibility of effectively separating the healthy from the sick, thus promoting contagion and the spread of epidemics. The situation worsened during times of war. During the Battle of Lepanto, for example, the sick rowers were made to lie directly on the deck, leaving them exposed to being trampled by the rest of the crew.[108] Finally, it should not be overlooked that the galleys were primarily military vessels, equipped to engage in naval warfare.

4.1 An infernal place: depiction of life onboard the galleys

The reconstruction of galley life—or at least its depiction—is also possible through the study of non-institutional sources. While rare, there are cultural sources such as memoirs and poems written in the first person by rowers themselves, evoking daily life within the galleys: a miserable, painful, and even infernal existence. Unquestionably, the most widely known of these is the autobiography by the Frenchman Jean Marteilhe, who was condemned to the galleys in the 18th century on account of his Protestant beliefs. The picture he paints is one of a painful existence, comparable to hell itself.[109] Although less well-known, there are also comparable accounts from Italy. The earliest dates back to 1577: an autobiographical manuscript report by Aurelio Scetti, addressed to Francesco I de' Medici. Scetti was a Florentine musician who, in 1565, was sentenced to life at the oar for the murder of his wife. His manuscript recounts the successful voyages of the Tuscan fleet during those years of his forced service. The author, through this narrative, explicitly sets out to incite pity and win the favor of the Grand Duke in hopes of securing his release.

Apart from Scetti's fascinating details of his personal trajectory, what is particularly interesting is how he describes his condition aboard the galleys. Life there is portrayed as desperate and painful, yet not without purpose. In line with his intention of pleasing the political authorities, his suffering as a galley rower becomes the starting point for the author's account of an experience of

---

107 Monga (translated and edited by), *Aurelio Scetti*, p. 39.
108 AAV, Misc. Arm. II, b.110, f. 386.
109 Marteilhe, *Mémoires d'un galérien*.

growth, spiritual renewal, and catharsis. Just as the galley's misery had brought the narrator back to the spiritual light by freeing him from the darkness of his sins, Scetti hoped to be saved and reintegrated into the light—into society. To this end, he adopts the use of the third person, transforming an autobiographical narrative into a more universal exemplary tale of purification and conversion.[110]

A century later, in 1654, another convict aboard Lieutenant Lomellino's papal vessel described his galley-bound journey to the Levant—likely with the same intent of seeking grace and redemption in exchange for his suffering, thus also testifying to the effectiveness of life at the oar. While the author remains anonymous, the use of imagery and the Christian lexicon of mission and salvation prominently feature alongside descriptions of seafaring and battles against the Turks. It is reasonable to assume that this text was also written with the purpose of demonstrating to the authorities the writer's military and religious zeal, in the hope that it might help him in his quest for freedom. Indeed, the text appears to have been composed with the intent of being read, if not published, as evidenced by the author's direct appeal to the "benevolent readers."[111] To my knowledge, there has been no study or critical edition of this manuscript memoir.

Reflections on the harshness and misery of forced service in the galleys—often likened to death—can be found in a variety of literary works, occasionally in the most unexpected places, such as in literary collections primarily intended for entertainment. Particularly noteworthy in this regard are two rhymed poems preserved in the Biblioteca Apostolica Vaticana. The first, published in the 19th century under the title *Sforzadi alle galere* [Forced Convicts in the Galleys], is actually part of a larger composition on the Mediterranean arsenals by a Venetian named Battista Baldigara, who likely penned the work around 1562, during the early stages of galley punishment. Even at that time, galley service was considered the worst possible torture—some even argued that death would be a preferable alternative. The oarsman's life is depicted as sheer hell on earth, both because of the harshness of the conditions in which he had to live and his deprivation of freedom. He was reduced to an animalistic state, yoked to an oar and deprived of all agency—chained to his bench ever at the mercy of his masters' blows. The poem is made even more pitiful by the use of the first-person singular. Unfortunately, we have no biographical information about the author, though it seems safe to assume that his portrayal was more a product of poetic license than a genuine account from a former convict at the oar.[112]

---

110 Monga, *Aurelio Scetti*, pp. 1—9.
111 BCR, 34D11, *Scritture diverse e notizie appartenenti alle galere pontificie*, p. 262, ff. 167–251.
112 BAV, Stamp.Cappon.V.681(int.3), Baldigara, *Ragionamento di maraviglie*.

The second rhymed poem is the anonymous *Vita miserabile che fanno li poveri forzati delle galere in ottava rima. Ad esempio delli sfrenati giovani* [The Miserable Life of the Poor Convicts of the Galleys in Ottava Rima. As an Example for the Unrestrained Youth]. The poem's stated purpose—"that every unbridled youth may escape so much pain"[113]—was to highlight the harshness of the convict's life at the oar, aiming to warn society, and young people in particular, against committing crimes that could result in galley service. The poem is structured as a discourse addressed directly to an ideal reader.

Once again, galley life is depicted as a space of utter degradation,[114] where physical torment and moral despair converge, mortal sin reigns supreme and social distinctions are obliterated in the face of justice.[115] What makes this poem especially intriguing is that the author clearly knew the realities of everyday life in the galleys. He describes the clothing, the food—and in what quantities—the daily tasks, as well as the navigational difficulties they had to confront. Typically, the oarsmen were not sailors and were unaccustomed to life at sea. Their pleas were ignored and they were assigned the most grueling tasks—ones for which they were entirely unprepared and often lacked the necessary stamina.

Strict discipline was enforced, with officials even resorting to cutting off the noses and ears of thieves.[116] Though not explicitly stated, we can infer that the author had been a convict—not only on account of the granular detail, but also because of the use of the first-person plural, especially when describing the excitement and desperation that gripped the crew as they set sail—emotions so intense that they made the rowers appear "frenzied."[117]

If this were indeed the case, the text would have served an overriding didactic purpose—intended to convey repentance. Like Scetti's manuscript, it was likely written to win the favor of the political authorities by demonstrating that the author had learned his lesson and repented. Furthermore, it seems likely that the text was part of a broader political agenda to make citizens truly aware of what galley service entailed, using exemplary and brutal punishments as a deterrent. It is probable that the text was later edited and published, indirectly indicating how first-hand accounts of galley life had become integral to early modern popular culture, forming literary motifs. Indeed, the mutual references in these

---

113 BAV, Stamp.Cappon.V.681(int.16bis), s.n., *Vita miserabile:* "Acciò ogni sfrenato giovinetto, Possa fuggir da lui tanto dolore."
114 Ibid., p. 78: "Se ritrova un'inferno in questo mondo, che un altro inferno non se può trovare, salvo che quello che giù nel profondo, Donde fa capo il peccator mortale."
115 Ibid., p. 78 r–v.
116 Ibid., p. 79r–v. On the practice of *denasatio* see Groebner, *Defaced*, pp. 68–86.
117 Ibid., p. 78r.

texts confirm the persistence of cultural patterns of representation. This hypothesis is supported by comparing the texts preserved in the Biblioteca Apostolica Vaticana with a Spanish anonymous poetic composition in stanzas of eight lines written in the early 17th century and published by Benedetto Croce under the title *La vita infernale delle galere* [The Infernal Life of the Galleys].[118]

Suffering became even harsher when life at the oar was experienced in captivity. It is noteworthy, for instance, that as early as the 15th century, Teseo Pini's *Speculum Cerretanorum* (1484–1486)—a work focused on depicting various categories of false beggars—highlighted the deep compassion evoked by the conditions endured by galley slaves. Among the many categories described in his work is that of the *accattosi*—beggars who feigned enslavement aboard Turkish galleys, claiming to have escaped and returned home, or who asserted that they had relatives captured by the Turks and sought alms to secure their release. Once again, Pini's description of daily life at the oar is vivid and detailed, underscoring the widespread awareness of how such a miserable fate could potentially strike anyone:

> They begin to cry out, "*Allah, Allah, Allah, heber, elhemdu, lillahi, la illah, illelhac,*" and other words in such a strange tongue, while displaying long chains and irons, claiming to have been bound and to have escaped from the galleys. They seek to convince the common people that they endured severe beatings daily at the hands of the Turks, enemies of the Christian faith, showing certain marks on their flesh that they have skillfully fabricated. [...] They claim to have subsisted on dry bread, black biscuits as dark as the earth, and to have drunk water infested with vermin.[119]

As for captivity in the Italian context, we can consult the memoirs of William Davies—an English Protestant captured by the Tuscans in 1598 while sailing in the Aegean Sea as a ship's doctor on an English vessel carrying Turkish goods, and thus exchanged for an Ottoman ship and eventually sent to the galleys. His memoirs, published in London in 1614, recount nearly nine years of captivity— three years in the shipyards of Livorno, and later as a galley surgeon for almost six years, sailing as far as the Amazon in search of the mythical Eldorado.[120] He

---

118 Croce, La vita infernale delle galere, pp. 83–92.
119 Camporesi, Lo *"Speculum cerretanorum,"* p. 115: "Cominiano a gridare 'Allah, Allah, Allah, heber, elhemdu, lillahi, la illah, illelhac', et altre parole con sì strana lingua, e mostrare longhe catene e ferri con cui dicono essere stati legati e dalla galera fuggiti, danno ad intendere al volgo d'aver ricevuta ogni dì grandissima quantità di bastonate da' Turchi, inimici della fede di Cristo, mostrando certi segni che artificiosamente hanno fatto nelle carni. [...] Dicono d'aver mangiato pane secco, biscotto nero come la terra e aver bevuto acqua verminosa."
120 Davies, *Captivitie of William Davies*, pp. 13 f.

likened his time at sea—marked by physical and mental suffering—to hell on earth. "The misery of the galleys is beyond all judgment or imagination, no one would think that such torture or torment exists in the world, only those who suffer it."[121] Chained to their benches, the oarsmen were constantly exposed to the elements and to the violence of the officers day and night. During battles, they were compelled to endure gunfire by holding a cork in their mouths to stifle their screams of pain.

In this regard, a remark by the Tuscan commander Giovanni Paolo del Monte is noteworthy. In his report of the voyage to Negroponte in 1623, he recounts the battle with two Turkish galleys: "Where they fought for two hours, the sound of artillery and musketry was an incredible horror to our ears, the moans of the wounded aroused great pity, the darkness of the night increased their misery."[122] The suffering, the violence, and the hardships of the sea voyages were so unbearable that many slaves attempted suicide or plotted to murder their officers. Few followed through, however, for fear of the severe punishment that failure would bring.[123] Furthermore, as Prosperi noted, the horrors of Tuscan captivity are exacerbated by the fact that Davies—a Protestant surgeon—viewed Italians as a superstitious and idolatrous people, of a deceitful nature, and devoted to murder and lust—comparable, in his eyes, to the savages of the Amazon.[124]

5 Galleys and vigilance practices in early modern Tuscany: the case of Livorno

Throughout history, the Tuscan port of Livorno has come to symbolize the already high level of mobility and interconnectedness that marked early modern Europe. The image of Livorno as an open and cosmopolitan port has, over time, become something of a myth.[125] As one of the central hubs of the early modern Mediterranean economy, Livorno quickly acquired the status of a "free port." With the

---

121 Partner, *Corsari e crociati*, p. 105 f.
122 The report is conserved in BAV, Stamp.Ferr.IV.8532 (int.4), *Fedel relazione mandata dall'Illustrissimo Signor Balio di Cremona C. Bernardo Vecchietti generale delle Galere della Sacra Religione Gerosolimitana Del viaggio, e presa delle tre Galeotte, Fuste, e vascelli d'Infedeli. Fatta dalle medesime Galere in Levante*, in Roma, appresso Ludovico Grignani, 1641: "ove si combattè per due ore con horrendi urli, al suon d'Artigliarie, e moschetti faceva nelle nostre orechie un terrore incredibile, li lamenti delli feriti eccitavano compassione grande, il buio della notte ne accresceva l'affanno."
123 Davies, *Captivitie of William Davies*, p. 20.
124 Prosperi, *Presentazione*, in Davies, *Captivitie of William Davies*, p. 11 f.
125 See Prosperi, *Livorno 1606–1806*; Santus, *"Il turco"*; Trivellato, *Familiarity of Strangers*.

promulgation of the *Livornine Laws* (1591–1593), the Grand Dukes of Tuscany committed themselves to welcoming all people of non-Catholic faith—"merchants of any nation, Levantines, Ponentines, Spaniards, Portuguese, Greeks, Germans and Italians, Hebrews, Turks, Moors, Armenians, Persians and others"[126]—treating them with justice and protecting them from religious intolerance.[127] The city thus became home to a notably large Jewish,[128] Protestant, and Muslim presence—the latter found in the port both as merchants and, above all, as slaves captured and employed as rowers in the galleys.

In the early 17th century, Livorno also became a fortified settlement, a center for military raids against the infidels carried out by the Knights of St. Stephen. Cesare Santus' book *Il "turco" a Livorno* (2019) successfully highlights the city's paradoxical character—at once a fortress of Christianity and a free port. While the "Turk" continued to be an object of prejudice, stereotyping and discrimination, the opportunities for encounters and cultural exchanges between Muslim slaves and Catholic citizens formed an integral part of everyday life in the city, illustrating that their integration into Tuscan society was indeed possible.[129]

The *Bagno de' Forzati* was a privileged site for such exchanges. In theory, its purpose was to isolate the galley slaves and convicts from society during periods when the vessels were not at sea. However, interaction with society was a daily occurrence, and slaves, in particular, enjoyed considerable freedom of movement—in fact, some were permitted to run shops and taverns in the city harbor. Livorno presents a compelling case for examining surveillance practices directed at the *galeotti*, precisely because of this relationship with the slaves and convicts, as well as the presence of the *Bagno de' forzati*—a facility that, by its very nature and function, complemented the galley system. It thus represents an ideal context in which to explore the liminal status of *galeotti*, set apart from society and yet constantly in contact with it.

---

[126] The text is quoted in Frattarelli Fischer, La Livornina, pp. 44–49: "mercanti di qualsivoglia nazione, Levantini, Ponentini, Spagnoli, Portoghesi, Greci, Todeschi et Italiani, Hebrei, Turchi, Mori, Armeni, Persiani et altri." On the Livornine laws see Felici, *La Livornina*.
[127] Ibid. However, the supposedly cosmopolitan and tolerant character of Livorno has been debated in recent years. As Frattarelli Fischer and Villani highlighted in *"People of Every Mixture,"* religious co-existence seems more appropriate than tolerance to describe the situation.
[128] See Trivellato, *Familiarity of Strangers*.
[129] Santus, *"Il turco,"* p. 10 f.

5.1 The rise of Livorno and the birth of the Medici fleet

Throughout the 17th century, Livorno emerged as the Grand Duchy of Tuscany's most important commercial and military port. This rise, however, stemmed from a strategic initiative launched a century earlier by Cosimo I (r. 1537–1569). Before this, Livorno had been little more than a fortified village of five hundred inhabitants—a garrison supporting the now-silted port of Pisa. After its brief occupation by Genoa (1406–1421) the site was sold to Doge Tommaso di Campofregoso for the enormous sum of 100,000 florins. Despite the swampy terrain, the House of Medici soon recognized its strategic and economic potential, and set about revitalizing and improving the city. Construction work began under Cosimo the Elder (r. 1434–1464), but it was only under Cosimo I that a comprehensive Medici project for Livorno took shape—driven in part by the need to compensate for the loss of access to the Pisan Gulf. In an effort to boost the local population, the Medici began granting economic incentives and legal immunity to those who relocated to the territories of Pisa and Livorno, beginning in the 1540s. This initiative formed a central part of their strategy to transform Livorno into a key hub for Mediterranean trade.[130]

After a brief interlude in the 15th century, Cosimo I, recognizing the strategic importance of galleys in the Mediterranean's geopolitical landscape, decided to rearm the fleet. On 13 April 1547, the first galley of the Medici fleet docked at the Pisa arsenal. Initially, the Medici relied on the Ligurian navy for their technical and organizational skills. The Tuscan galleys were built by Genoese shipwrights until the end of the century, and their crews were composed almost entirely of Ligurian officers and oarsmen. The fleet's initially non-Tuscan character was further emphasized by the fact that Cosimo I, in order to equip the galleys, resorted to the *appalto*—also known as *assento* or *asiento* on the Italian peninsula—a system of private management whereby, by signing a contract, one party undertook to manage, arm, and finance the galleys, although the vessels themselves remained the property of the Grand Duchy.[131]

During its early years, the Tuscan navy, still composed of three-banked galleys —known as triremes—rowing *alla sensile*, followed a recruitment model that relied heavily on a majority of forced rowers, given that corsair activity remained relatively low. In 1555, the five galleys of the Tuscan fleet had a total crew of 768 oarsmen, of whom 535 were convicts and 234 were slaves. The first Tuscan

---

130 Santus, *"Il turco,"* p. 16f. On the history of the rise of the port of Livorno, see Frattarelli Fischer/Papi, *Studi di storia*; Vaccari, *Il porto*, pp. 302–323.
131 Lo Basso, *Uomini da remo*, p. 337f.

prisoners were condemned to the galleys in October 1547, while many more came from the territories of the Papal States. A law issued on 8 July 1542 ensured that blasphemers and sodomites were punished by being sent to the galleys.[132]

Throughout the early modern period, blasphemy and sodomy were harshly condemned both legally, as crimes, and morally, as sins. Both were viewed as acts against God, and as such, needed to be punished to prevent divine retribution from being unleashed on the Grand Duchy's cities.[133] However, galley service was primarily reserved for repeat offenders and was typically not imposed for a first offense.[134] Repeat blasphemers were sentenced to two years at the oar, while recidivist "active" sodomites faced life sentences.[135] The sources make no mention of Tuscan convicts before 1547.[136] In 1547, galley service was extended to Romani people and vagrants, and gradually came to cover a broader range of serious offenses.[137] The most commonly punished crime was theft (20%), while it is interesting to note that acts of sodomy and vagrancy were relatively rare (1.1% and 0.6%, respectively).[138] The expansion of galley sentences was further facilitated by lowering the age of majority in criminal cases; it was reduced from 25 to 18, following a law passed in 1561.[139] Furthermore, many of those sentenced were tried by the Inquisition or came from the territories of the Papal States, although these were predominantly vagrants.[140]

While crews aboard the Medici fleet later came to be largely composed of slaves, galley service continued to play a major role in Tuscan legislation. Over time, it was applied to an ever-widening range of offenses, and remained in force until 6 February 1750, following the final disarmament of the fleet's last two galleys in 1748. The law that abolished galley service also outlined the penalties that would replace it. A life sentence at the oar was replaced by branding the offender on both shoulders, public flogging, and perpetual exile. And yet, those

---

[132] *Legislazione toscana*, vol. I, pp. 210–221, *Bando sopra la Bestemmia e la Sodomia del dì 8 Luglio 1542 ab incarnazione*.
[133] Ibid., p. 210: "che li vitii al tutto si spenghino, quelli massime che sogliono provocare a ire el sommo & onnipotente Dio: & conoscendo che la bestemmia è peccato che più offende sua Maiestà che li altri, dal quale procedono nel mondo turbolentie, & inopinati flagelli"; p. 211: "nefando vitio della Sodomia […] Et volendo al tutto estinguerlo per la grande offese che se ne fa al sommo, & onnipotente Iddio, & al dishonore che ne resulta nell'universale."
[134] Ibid., p. 211 f.
[135] Grassi, *Sodoma*, p. 16.
[136] Lo Basso, *Uomini da remo*, table p. 341.
[137] Angiolini, La pena della galera, p. 95.
[138] Lo Basso, *Uomini da remo*, p. 341.
[139] Angiolini, La pena della galera, p. 99.
[140] ASF, MP, 2099, c. 21r, 19 giugno 1680.

originally condemned to the galleys for a limited term had their sentences commuted to an equal duration of public service, with the remaining time to be served in prison.[141]

In 1558, the Medici government, unable to conclude a new and profitable *asiento* treaty, decided to create an order of knights to manage their fleet. Negotiations began in 1560, and the agreement was signed by Pope Pius IV (r. 1559–1565) on 1 October 1561. The Pope approved the statutes at the end of January 1562, and on 15 March in Pisa, Cosimo I was officially proclaimed Grand Master of the Order of Santo Stefano.[142] Created with the aim of targeting the commercial and military interests of the Berbers and Turks, the Order quickly began to finance itself through piracy. By capturing prisoners of war and enslaving them, corsair activities also provided a solution to the persistent and enormously difficult problem of finding enough manpower to enlist.[143] In 1570, slaves made up 23.4% of the rowers and convicts accounted for 53%. By 1680, however, only the *Capitana* galley had a crew composed of 56% slaves and 30.8% convicts.[144] According to Franco Angiolini, up until the first half of the 17th century, the Tuscan fleet was a major player in Mediterranean corsairing: between 1563 and 1693, some 16,000 slaves were captured by the Medici galleys.[145] In 1601, Turkish slaves accounted for about 20% of Livorno's population, a figure that dropped to 8% by the early 1640s, steadily declining as conflicts with the Muslim world diminished, along with piracy.[146]

5.2 *Bagno de' Forzati*

The *Bagno de' Forzati* in Livorno was constructed between 1598 and 1604 by order of Grand Duke Ferdinando I de' Medici; its design was based on the model of the *Bagni*[147] of Constantinople and Algiers. While no similar facility is documented in other continental Catholic ports, the concept of a *Bagno degli schiavi* was typical in large Mediterranean ports and was advocated by Giorgio Vasari in his treatise

---

141 Angiolini, La pena della galera, p. 95.
142 Lo Basso, *Uomini da remo*, p. 342 f.
143 Frattarelli Fischer, Il bagno, p. 69.
144 Lo Basso, *Uomini da remo*, p. 345 f.
145 Angiolini, Il Granducato di Toscana, pp. 39–61; Slaves and Slavery, pp. 67–82.
146 Santus, "Il turco," p. 42.
147 Probably, the English term that corresponds most closely to the original meaning is "penal colony."

on the ideal city of 1596. Vasari offered an erudite justification for such a facility—one that did not exist elsewhere in Christendom outside of Malta.[148]

The *Bagno de' Forzati* functioned primarily as a prison compound where *galeotti* were confined at night and during periods when they were port-bound. Between 1598 and 1604 alone, roughly 3,000 forced rowers were imprisoned there.[149] Both the nature and function of this prison were complementary to those of the galley. As Father Filippo noted, it was constructed with the dual purpose of improving sanitary conditions for convicts—who would otherwise be "cramped like dogs in the galleys day and night"—and ensuring that they could not escape at night.[150]

The entire compound covered an area of 6,000 square meters and consisted of two main buildings: the *Bagno* itself and the *Biscotteria*. The *Bagno* was a trapezoidal quadrangle built around a central courtyard, which housed a freshwater well and a cistern for washing clothes. On the ground floor of each side were dormitories for the crew of each of the four galleys. At the rear of these dormitories—except for St. Anthony's dormitory—was a mosque where Muslim slaves could gather. A Catholic chapel was located within St. Anthony's dormitory.

Above the dormitories were quarters for those officers in charge of the *Bagno:* the scribes, the captain, and also the doctors and apothecaries. The *Biscotteria*, in contrast, was the factory where the food for the convicts—a type of hardtack called *biscotto*—was produced by milling wheat stored in underground pits. In addition to food for the crew, white bread and other refined products were manufactured there and sold to the inhabitants of Livorno.[151] Eventually, Livorno's *Bagno* closed on 13 February 1750, following the peace treaties concluded with the Ottoman Empire and the Berbers. The slaves were freed, while the convicts were branded and put to forced labor in the city's *Fortezza Nuova* and the Pisan Arsenal.[152]

---

148 Frattarelli Fischer, Il bagno, p. 79.
149 Ibid., p. 70.
150 Bernardi, *Relazione*, pp. 13–15: "che que' poverelli stessero tutto l'anno di giorno, e di notte abbrancati come cani, e ristretti dentro una Galera [...] a fine dunque d'accomodare la ciurma delle Galere, e [perché] non potessero di notte tempo tentar la fuga, come facilmente può avvenire stando in Galera, [si decise] la costruzione d'una gran fabrica isolata per ogni parte e circondata da alte muraglie a guisa di fortezza."
151 Santus, "Il turco," pp. 35–39.
152 Ibid., p. 39.

5.3 Life aboard a Tuscan galley: Control over convicts and slaves

We can reconstruct, to a certain extent, daily life aboard Tuscan galleys from the available sources, primarily letters, reports, travel diaries, and instructions written by and addressed to the galley officers. Most of these sources are housed in Florence's State Archives—in the *Mediceo del Principato* fond—under the heading "Livorno." While numerous, these sources provide only a partial picture of reality, with few first-person accounts from the *galeotti*, so the records primarily reflect the perspectives of those in charge of managing the convicts. Due to the variety of documents in this archive, scholars such as Luca Lo Basso, Lucia Frattarelli Fischer, and Cesare Santus have partially analyzed this collection to study the construction techniques, navigation, and organization of the vessels and Livorno's *Bagno de' Forzati*.

The governors' overriding concern was that the naval crew and officers should behave appropriately. Strict discipline was enforced, guaranteeing that each individual could perform his duties and fulfill the position entrusted to him. The guidelines for operating the galleys during navigation emphasized that disorder on board should be strictly avoided: no fighting, gambling, blackmail, swearing, or any other misconduct, and above all, no escapes.[153] This focus on preventing fraud or violence equally applied to the officer class. Numerous orders were issued requiring them to treat the crews well, as they were considered the galleys' driving force and primary source of manpower. In particular, soldiers were explicitly ordered—under threat of being condemned to the galley themselves—not to beat convicts and captives without just cause.[154]

Among the many forms of mistreatment convicts endured, extortion was likely the most widespread. Numerous *galeotti*, in exchange for a fee, were able to secure their release and depart with relative ease.[155] As Santus observed, similar "mechanisms of oppression and corruption typical of prison institutions and the servile condition" were prevalent within the galleys and at the *Bagno*.[156] Recognizing the gravity of the issue, ten of the thirty articles in *Costitutioni, et ordinationi dell'offitio del Capitano del Bagno di Livorno* specifically addressed the prohibition of exploitation and corruption.[157]

---

[153] ASF, MP, 2131, dossier 3, cc.n.n.
[154] Ibid.
[155] BCR, Ms, 34D18, *Memorie e scritture diverse appartenenti alle galere Pontificie e condannati nelle medesime Principalmente in tempo del tesorierato di mons. Lorenzo Corsini poi cardinale, e papa col nome di Clemente XII*, f. 35.
[156] Santus, "Il turco," p.119.
[157] ASF, MP, 2132, dossier 7, cc.n.n.

In those cases examined by Santus, the most common motive for illegally extorting money from inmates was the desire for their release from chains, enabling them to leave the *Bagno* at their own discretion. Even more scandalous, however, were the demands for money from convicts and slaves who wished to engage in sexual relations with younger inmates.[158] In an effort to prevent disorder, a deeply paternalistic system of control emerged—one that balanced punishment, discipline, and the welfare of the crew. To ensure its effectiveness, a stringent system of vigilance was put in place to monitor the rowers, alongside the enforcement of a harsh justice, intended to make sure that "for fear of punishment, they would restrain themselves from committing any harm."[159] Mutual surveillance and denunciation of wrongdoings—with severe and exemplary penalties for those who violated them—formed the twin pillars of the disciplinary framework governing rowers aboard Tuscan galleys.

Given that the various regulations concerning the control of convicts aboard Tuscan galleys largely repeat one another over the years, my observations will focus primarily on the manuscript volume contained in fascicle 2131 of the aforementioned fonds *Mediceo del Principato*, titled *Ordini da osservarsi nelle Galere del Ser.mo G. Duca di Toscana* (c. 1626). The issuing of these directives stemmed from the desire to prevent rowers and officers from behaving licentiously.

Aboard the vessels, the *galeotti* were under the authority of the *aguzzino* [overseer], whose primary responsibility was to supervise them. He was often assisted by a *mozzo* [deckhand], typically selected from among the slaves and convicts. To maintain order and discipline aboard, the authorities sought to discourage crew members from committing crimes. To this end, they aimed to establish a permanent system of mutual control and denunciation, and to introduce an extremely harsh regime of corporal punishment. For example, mariners or soldiers who dared to blaspheme were punished with a minimum of two months and a maximum of one year at the oars, while convicts and slaves faced flogging and, in cases of recidivism, tongue-piercing. No exceptions were allowed: punishment had to be rigidly enforced. To identify both offenders and their crimes, all crew members were instructed to report any offenses they witnessed. Rowers chained to the benches were tasked with watching over their fellow oarsmen, as well as those seated on the benches directly in front, behind, and beside them. Any violation of this directive resulted in two months' imprisonment. It is noteworthy

---

158 Santus, *"Il turco,"* p.130.
159 ASF, MP, 2131, dossier 3, *Ordini da osservarsi nelle Galere del Ser.mo Gran Duca di Toscana*, c.n.n.: "nell'avvenire per timor delle pene s'habbino a ritenere di far male, e in questo modo provedere alla sicurezza de buoni e reprimere l'insolenza de cattivi."

that, while blasphemy and sodomy were among the earliest crimes in Tuscany to be punished by time in the galleys, sodomy receives little or no mention in the directives that have been analyzed. Conversely, recommendations against blasphemy were consistently present.[160]

Furthermore, it was strictly forbidden to provoke conflicts—whether verbal or physical. Fights in particular were strongly discouraged, and if they erupted aboard, both the galley captain and the admiral had to be notified. To prevent rowers from escaping, continuous shifts of watchmen were required both during the day and at night. The sailors on watch were forbidden to allow any unauthorized person to board or leave the vessel, and anyone who disobeyed orders or, even worse, was negligent, such as falling asleep during their watch, faced punishment—including loss of pay and other corporal penalties such as a galley sentence.

In a similar vein, regular patrols aboard were scheduled to ensure that no convict had gone missing. Officers were instructed to report any absence due to illness, death, or escape.[161] To ensure greater horizontal control, spies—known as trombi or buonomini—were recruited from among the forced rowers, with the specific task of communicating to the captain everything that transpired amongst the crew.[162] What was being implemented was a decentralized system of information focused on the actions and behavior of rowers, grounded in a system of horizontal and vertical denunciation.

Any rower who saw, heard, or knew of a fellow convict who had violated any orders was compelled to notify his superiors; failure to do so would result in punishment akin to that meted out to other offenders. On the vertical level, this system was based on the information gathered by figures such as the *aguzzini*, as well as by soldiers, officials, doctors, or surgeons—all of whom were tasked with communicating any relevant information to the captain. This obligation to report, together with the active participation of all crew members, was fully recognized as essential to establishing an effective system of control.

When not at sea, slaves and convicts were confined to the *Bagno*, though they were not required to remain locked up at all times; in fact, they enjoyed numerous opportunities to go outside. This was especially true for slaves, who, being unchained, could partake in daytime economic activities such as running taverns and workshops in the dockyard. Convicts, however, were primarily used for forced labor, especially in the city's arsenal.

---

160 Ibid.
161 Ibid.
162 Lo Basso, *Uomini da remo*, p. 353.

This arrangement underscores the transitional nature of the *Bagno* and its inhabitants, as well as their physical and figurative isolation from society, even as they continued to maintain close relations with it. At the end of each working day, however, everyone had to return to sleep in the compound, where the gates were locked and monitored by the warden—the *Custode del Bagno*.[163]

Although originally conceived as a male-only setting, women were often forced to stay there as well. These women were slaves—captured or bought on behalf of the Grand Duke—who were temporarily housed in a wing of the *Bagno* while awaiting sale to private individuals. Ideally, the women were to be kept apart from the men. However, at least on two occasions—likely due to space constraints—women and men were locked in together, leading to sexual promiscuity and violent assaults on the enslaved women.[164]

As Lucia Frattarelli Fischer noted, the administration of the *Bagno*, like that of the galleys, was rooted in militaristic discipline and the systematic use of institutionalized violence and corporal punishment.[165] According to the Bagno's rules, guards were responsible for maintaining law and order, following directives from the *aguzzini*, higher officials, and the *Bagno*'s captain. The *Bagno*, as outlined in these instructions, was to operate as a fortress, with its guards expected to be exceptionally vigilant, for, in this case, the enemy was within. The prisoners were considered the worst of the worst and were constantly seeking to escape.[166]

Despite strict supervision and corporal punishment, disorder and violence were rampant. An examination of the edicts governing daily life within the institution reveals that stolen and smuggled goods such as opium, salt, and tobacco circulated freely within the compound.[167] The crimes committed there were similar

---

[163] Bernardi, *Relazione*, p. 16.
[164] Salvadorini, Traffici e schiavi; Herzig, Slavery and Sexual Violence; Cultural Commemoration; Chizzolini, Enslavement.
[165] Frattarelli Fischer, Il bagno, p. 82.
[166] ASF, MP, 2132, dossier 7, *Constituzione, et ordinazione dell'Offitio del Capitano del Bagno*, cc.n.n.: "L'offizio del Capitano del Bagno è di grandissima importanza, poiche alla diligenza e fede sua, viene commesso notabilissimo numero d'infedeli, et anco cristiani mal fattori, ritenuti in servitù, della quale procurano sempre con ogni industria di liberarsi, e perche il Bagno, è simile a una fortezza, dove si ritiene la sudetta Gente destinata al servizio pubblico, e delle Galere, il Capitano di questo luogo può aguagliarsi più presto a castellano di fortezza che a custode di carcere, poiche questo tiene li rei in custodia a fine solamente che i colpevoli possano punirsi, e gl'innocenti liberarsi, e quelli che già giudicati cristiani ò infedeli Schiavi tutti applicati all'uso delle Galere [...]e dev'essere ancora più vigilante e diligente, visto che il castellano per guardarsi dai nemici di fuori si avvale di soldati e gente d'honore, e giurata, et il capitano del Bagno vi tiene dentro l'inimici infedeli, e mal fattori, con pochi, e deboli strumenti di chiavi, e catene."
[167] Frattarelli Fischer, Il bagno, p. 83.

to those aboard the galleys: violence, gambling, blasphemy, and so forth. Murders and brawls between slaves and convicts were frequent, especially those triggered by excessive drinking. Wine was included in the *galeotti's* rations, and even the "Turks," forbidden by their religion from drinking, partook.[168]

In April 1651, the slave Solfara di Moratto from Scio, without provocation, brutally stabbed the slave Mostafà di Arsano di Cecimeche twice while the latter was shaving another convict, leaving him critically wounded. When asked for a motive, Solfara simply answered: "Wine." The perpetrator was punished severely at the commissary's discretion.[169] In another incident, in 1692, a drunken slave in the Turkish dormitory stabbed a comrade in the kidneys after being told to stop drinking. As the wound was not life-threatening, the punishment was limited to a hundred lashes and a foot ring, for there was no greater humiliation for a slave than being unable to leave the *Bagno*.[170] Tension within the institution was so intense that violence often erupted for the most trivial of reasons, such as saying a single word too many. Reports indicate that sodomy was common within the *Bagno* due to the pervasive promiscuity and the relative freedom among the prisoners.

In addition to behavioral control, there was the pressing need for strict spiritual oversight, as the rowers were depicted as highly immoral and blasphemous individuals. Religious guidance was particularly urgent because of the large number of Muslim slaves within the *Bagno*, as the authorities feared that they might influence—or rather, "infect"—Christian convicts with Islamic heresy. The stereotype of the diabolic and infidel "Turk," devoted to lust and black magic, was widespread.

Consequently, Muslim slaves had to be rigidly controlled and, when necessary, punished. To judge moral and religious offenses, the Pisan Inquisition's tribunal was established circa 1560, remaining in place until its abolition by a *motu proprio* initiated by Grand Duke Peter Leopold on 5 July 1782. The Pisan Sant'Uffizio was based in the Convent of St. Francis and presided over by judges of the Order of the Friars Minor. During the 1620s, a General Vicariate was created with extensive powers and autonomy, and was located in a chapel within the *Bagno*.[171] The effort to enforce Christian moral teachings reached a critical turning point with the arrival of the Capuchins, led by Father Ginepro of Barga, on

---

168 Santus, *"Il turco,"* p. 123.
169 ASF, MP, 2132, part II, dossier 8, *Liburniensis Iurisdictionis*, n. 14, *sommario* 6.
170 ASF, MP, 2102, c.n.n.
171 Santus, *"Il turco,"* pp. 55. On the origin of the Roman Inquisition, see Prodi, *storia giustizia*, pp. 92–97.

1 December 1677. Their core objective was to "induce those wretched people to live as good Christians and in fear of God."[172] The Capuchins' approach in the *Bagno* combined popular evangelization and missionary activity.[173]

5.4 *Vigilanti di Maria*

The need for greater and more consistent control of the *galeotti*—to regulate their conduct in the *Bagno*—led the Capuchins to create, on 8 December 1678, a special corps of convicts dedicated to denouncing their fellow inmates, known as the *vigilanti di Maria* [Vigilantes of Mary]. Comprising only four prisoners, they were described as being "among the most God-fearing," who, under the leadership of a Chief Vigilante, were entrusted with monitoring their comrades day and night. Their duty was to report any illicit actions, speech, or even thoughts they became aware of, ensuring that those who committed criminal or immoral acts were duly punished.

The *vigilanti* were specifically tasked with preventing "nefarious sins" such as blasphemy, lechery, sodomy, and heresy.[174] In serious cases, they had to notify the Vicar of the Inquisition, who would then initiate a trial. Minor blasphemies were punished with the pillory, while sodomy was penalized with shackles and the cane. They were also held responsible for the infirmary in the *Bagno* compound, ensuring that sick rowers were treated with compassion and were carefully guarded at night to prevent escape.[175] This vigilance corps extended its duties to the galleys during the sailing season, with each vessel equipped with four vigilantes responsible for reporting any "dangerous" behavior or language they were aware of. The *vigilanti* themselves, however, were not beyond scrutiny. They were subject to strict monitoring by a guardian and a governor from one of the four companies of clerics operating in the compound, who ensured their diligence and moral conduct.

---

172 Bernardi, *Relazione*, p. 32: "ridurre quella misera gente a vivere da buoni Cristiani e col timore di Dio."
173 See also Santus, "*Il turco,*" pp. 57.
174 Bernardi, *Relazione*, p. 33: "Hebbe il buon P(ad)re notizia, che tra quella Ciurmaglia del Bagno vi si commettevano molti peccati nefandi, per ovviare a' quali prese spediente con consenso di S.A.R. di costituire quattro Forzati per Bagno, de' più timorati di Dio, e più esemplari, et accorti, i quali giorno, e notte invigilassero."
175 Ibid., p. 42.

In a report by Father Philip, the *vigilanti* are described as trustworthy, faithful, and sincere. Yet given that they were condemned criminals, it is reasonable to question how much these commendatory terms reflected reality, or to what extent they served propaganda purposes? Father Philip's account concludes with a celebration of the Capuchins' religious zeal, extolling their dedication to both the physical and spiritual well-being of the inmates. According to his report, they were so effective that "it almost no longer looked like a prison—a receptacle for infamous and dangerous people—but a Cloister of modest and devout Religious, both in their speech and in their deeds."[176] Here, the report's idealistic and eulogistic tones become apparent, particularly when set against the countless cases of violence that continued to surface in the archives, even after the Capuchins' arrival.

5.5 Punishing crimes in a liminal context: a clash of jurisdictions

The image of Livorno that emerges from the archival sources—particularly from the criminal records—depicts a typical early modern seaport: a hub not only for merchants but also a haven for soldiers, galley slaves, convicts, and prostitutes, where crime was commonplace. One of the primary causes of unrest on Livorno's streets was the copious quantities of wine consumed in the port's numerous taverns,[177] and violent conflict among galley rowers was, therefore, not confined to the galleys or the *Bagno*—it frequently spilled into the city streets.

Convicts and slaves had numerous opportunities to venture out and actively participate in Livorno's daily life. Though integrated into society, the *galeotti* remained a marginalized group, and this ambiguous status often led to jurisdictional conflicts between the overlapping authorities charged with managing them. Formally, the *galeotti*, as public slaves, were under the authority of the Grand Duke's navy officers. When not at sea, they were placed under the control of the *Auditore delle Galere* and, while in the *Bagno*, under the authority of the *Capitano del Bagno*. However, the existence of an autonomous authority independent from that of the city's governor was resented by the *Capitano del Bagno*, as Livorno's governors frequently claimed the right to judge and prosecute those *galeotti* who committed crimes on urban soil.

---

[176] Ibid., p. 113 f: "che quasi non più rassembra un Bagno di Galeotti, che vale a diré un Ridotto di gente infame, e per lo più facinorosa; ma ben sì un Claustro di modesti, e divoti Religiosi, tanto nel parlare, che nell'operare."
[177] Santus, *"Il turco,"* pp. 120–122

To clarify roles and avoid ongoing confusion between the civil and military authorities, Cosimo II issued a decree on 3 July 1609, specifying that the galleys' commissioner had full authority over crimes committed aboard the vessels and within the *Bagno*, while the governor would preside over crimes committed in Livorno itself. Even so, the governor's jurisdiction was limited: in cases of minor infractions—those not involving death or mutilation—the parties involved could resort to an out-of-court settlement. Only if that process failed did it become necessary to appeal to the governor.[178]

The 1609 Edict did not completely resolve the issue, and even in 1613,[179] and again in 1635, Ferdinand II issued a *motu proprio* reaffirming the limitations to be placed upon the governor's authority. Even in those cases falling within his jurisdiction, he still had to operate in conjunction with the *Auditore delle galere*.[180] Despite the Grand Dukes' interventions, the matter was never definitively resolved, and the two authorities continued to clash even into the 18th century. A case from 1731 illustrated the enduring conflict. A storehouse slave was fatally wounded and was arrested by Spanish soldiers, who detained him in their guardhouse. Whereupon the *Bagno*'s officers requested the slave's return so he could be punished as an example for the entire crew. The Spaniards, however, refused. Instead, they handed him over to the Governor who imprisoned him and prepared him for trial. The galley authorities then appealed to the Grand Duke, arguing that, because the incident only involved slaves and no citizens, it should it not fall under the Governor's authority. Most importantly, the authorities wanted to avoid the publicity associated with the governor's trials, expressing a desire to punish the misdeeds of slaves and convicts, whatever their nature, "without publicity and solemnity," away from the indiscreet eyes of the population.[181]

To complicate matters, there was a third authority to which galley rowers were subject: the ecclesiastical. The Catholic Church maintained a presence within the *Bagno*, with the Vicar of Livorno reporting to the Inquisition of Pisa. Since Cosimo I's reign, relations between secular and ecclesiastical authorities had been fraught. Despite an apparent rapprochement between the two, the Tuscan Church sought to maintain a degree of autonomy from the Papal Curia by creating an ecclesiastical structure aligned with the Medici. Bishops were officially appointed by the Pope but chosen from a shortlist of three candidates submitted by the dukes, who thereby had the opportunity to present individuals more closely

---

178 Ibid.
179 ASF, MP, 2130, cc.n.n.
180 Santus, *"Il turco,"* pp. 120–122.
181 ASF, MP, 2116, cc.n.n.: "risolto senza solennità."

aligned with their interests and thus secure valuable allies.[182] Nonetheless, ecclesiastical courts consistently sought to preserve their authority, and it was not uncommon for them to interfere in the affairs of both the Governor and the *Auditore delle Galere*. Because jurisdictional boundaries among civil, military, and ecclesiastical authorities were never clearly defined, criminal punishment in Livorno remained a liminal, contested practice.[183]

6 The Papal States and the galleys of Civitavecchia

From the late 15th century onward, after a series of alarming events such as the Ottoman conquest of Otranto in 1480, the Papal States found themselves in conflict with the Ottomans and the Berbers, and increasingly felt the need to arm a small squadron of galleys to defend their coastline. One of the first measures taken was the promulgation of a decree in 1486 by Pope Innocent VIII (r. 1484–1492) to arm a galley to defend the merchant navy against pirate raids in the Tyrrhenian Sea.[184] The threat of corsair aggression, along with the possibility of organizing a crusade against the Ottomans—who by then had advanced into the Balkans—led the Pope in 1492 to order the arming of six more galleys in Civitavecchia, which soon became the Papal States' leading military port.[185] This laid the foundation for what would later be called the "permanent squadron."

Once the fleet was established, the usual problems associated with recruiting crews arose. To make up for the small number of volunteers, the papal crews were mostly composed of forced oarsmen. The stated aim of defending Christianity against the threat of the "infidel Turks," as well as the high percentage of convicts aboard, led to the decision to provide each galley with chaplains responsible for caring for the sick, enforcing discipline, and ensuring the crew's religious education.[186] Thus, the Papal galleys became the setting in which, for obvious reasons, the program of moral reform and behavioral discipline for the rowers—aligned with Catholic doctrine—was most clearly enacted. The *galeotti* were placed directly under the authority of the Roman Church, and, in particular, under the spiritual authority of the Cardinal Vicar,[187] who entrusted them to the daily pastoral care of

---

182 Cavarzere, *Cosimo I*, p. 82.
183 Ibid., pp. 52–55. See Chizzolini, Navigating Ambiguities.
184 Filioli Uranio, *La squadra navale*, p. 44.
185 Ibid., p. 48 f.
186 Guglielmotti, *La Squadra Permanente*, p. 23.
187 On the jurisdiction of the Court of the Cardinal Vicar of Rome, see Bonacchi, *Il governo delle anime*; Rocciolo, *Della giurisdittione del Vicario*; Il tribunale del cardinal Vicario, pp. 175–184.

the Capuchins. Aboard the galleys, the temporal and spiritual dimensions were perfectly aligned. Care and control of the soul, and care and control of the body, were two sides of the same coin. This was perfectly in keeping with the profound importance—both real and symbolic—attached in early modern Rome to the concrete realization of earthly discipline and justice. Given the dual nature of papal authority, earthly justice in the Papal State became an anticipation of divine justice; and disciplining the body was both a consequence of, and a prerequisite for, disciplining the soul.[188]

6.1 The port of Civitavecchia and the *squadra permanente*

Civitavecchia was the leading port of Papal States, not only commercially but also politically and militarily, serving as the training ground for the Papal Militia.[189] Known in ancient times as *Centum Cellae*, Civitavecchia has been a prominent town since Etruscan times due to its strategic location.[190] During the mid-15th century, it was enlarged and given new defenses by Popes Callistus III (r. 1455–1458), Pius II (r. 1458–1464), and Paul II (r. 1461–1471). After the discovery of alum shale in the Tolfa Mountains in 1462, the papacy began systematically modernizing the port and making it more accessible. Between 1508 and 1513, during the pontificate of Julius II (r. 1503–1513), the architect Donato Bramante was commissioned to work on the fortress and the harbor. While all the popes were committed to improving the port's facilities, it was Pope Clement VII (r. 1523–1534) who initiated a systematic program of improvements, describing the port as the "portal to the Papal States."[191]

In 1621, to ensure Civitavecchia's further development and future prosperity through trade, Pope Urban VIII granted the city "free port" status. This stipulated that no ships or goods landing in the port could be subject to customs duties or *gabelle* [taxes], and that all merchants could store their goods in warehouses without having to pay any additional charges, except for a minimal payment to the *custodi dei magazzini* [warehouse keepers]. Moreover, no one would be prosecuted

---

188 Prodi, *Storia giustizia*, pp. 62–64. On the dual nature—temporal and spiritual—of the Papal monarchy, see also Prodi, *Il sovrano pontefice*.
189 Calisse, *Civitavecchia*, p. 235.
190 Moroni, *Dizionario*, Vol. 13, p. 298 f.
191 Filioli Uranio, *La squadra navale*, pp. 126–128.

for debts incurred or sentences received outside the territories of the Papal States.[192]

As early as the 16th century, the popes adopted the *assento* system to maintain the small squadron of galleys defending the Roman coastline. The first contract, signed in September 1511, did not specify whether the vessels were state-owned or captain-owned. Regardless, the Apostolic Chamber granted the captain the right to levy a 2% tax on all goods passing through Roman ports and seaboards, in exchange for the obligation to patrol and defend the coast against piracy.[193]

The foundation of what would later be known as the *marina pontificia* [Papal navy] dates back to the pontificate of Pope Sixtus V (1585–1590). The Battle of Lepanto revealed the need for Christendom to rely on a permanent and sufficiently large fleet, with an adequate bureaucratic and fiscal apparatus. On 22 January 1587, the Pope issued a decree establishing a congregation tasked with creating and maintaining a permanent fleet. Having ensured the "defense of public tranquility on land," it was now time to guarantee "common security at sea" against the threat of piracy.[194] This was commemorated on a medal minted in 1588 to honor Pope Sixtus V's zeal for arming the navy, with the inscription: "*Sicurezza per terra e per mare* [Security on land and sea]."[195]

Thus, ten galleys were armed, and the pope introduced the "galley tax" that same year, distributed across the provinces and capitals.[196] From 1606 to 1644, the Papal navy came under the direct control of the *Reverenda Camera Apostolica* [the Apostolic Chamber], with the management of the fleet entrusted to private individuals via a system of contracts. However, this management system was short-lived, and by 1670, the pontiffs resumed delegating galley operations.[197] For the Pope, the establishment of a squadron of galleys had twofold significance: on one hand, to defend the temporal territories against external aggression; on the other, to control the seas, the primary vector for trade.[198] Although the fiscal management of the galleys evolved over time, the moral and military oversight of the

---

[192] AAV, Misc. Arm. IV/V, b.54, f. 32: *Privilegii, & Essentioni concesse à quelli che condurranno mercantile nel porto di Civita Vecchia.*
[193] Lo Basso, *Uomini da remo*, p. 383 f.
[194] Guglielmotti, *La squadra permanente*, p. 22: "e restituita oramai dentro terra la pubblica tranquillità, sì che ciascuno senza tema può riposare all'ombra del frascato nel suo campo, fa mestieri adesso provvedere alla sicurezza comune dalla parte del mare, perché le nostre riviere superiori ed inferiori, per quanto è possibile, sieno garantite dalla prepotenza dei pirati e ladroni."
[195] Ibid., p. 38.
[196] Filioli Uranio, *La squadra navale*, p. 94.
[197] Lo Basso, *Uomini da remo*, p. 388 f.
[198] Ibid., p. 95.

rowers was strictly regulated from the outset, ensuring that "the conduct of the Roman fleet might be praised everywhere."[199]

6.2 Condemned to the Papal galleys

The legal provision for a galley sentence in the Papal States was established relatively early compared with other Italian states. Though not yet systematic, references to the use of convicts to man the galleys can be found as early as February 1502. In the absence of Muslim slaves, Cesare Borgia (r. 1475–1507), the famous *condottiero* and son of Pope Alexander VI (r. 1492–1503), ordered that every prisoner in Rome be conscripted as a rower. Additionally, he decreed that the large number of vagrants arrested for begging in the city streets should also be made to serve.[200]

Throughout the early modern age, the galley sentence became one of the primary punishments for vagrants, who were often regarded as a serious social "plague" to be eradicated.[201] The official provision for using convicts to man the crews of the Papal galleys dates back to 1511. In the edict issued by the Captain of the Galleys, published that same year, the various courts of the Papal States were explicitly instructed to commute the death penalty to a sentence requiring the condemned to man the oars for at least one year.[202] Over time, the galley sentence evolved into a formal punishment, a sentence that could be imposed for a limited period, or even for life—as evidenced by the *Libro dei condannati* [Convicts' Book], in which galley captains were required to record each convict's full name, along with the duration and reason for their sentence.[203]

Thus, from 1523 onward, the forced *galeotti* became the primary component of the crews in the Papal galleys. In 1655, for example, out of 1,512 rowers, 79.8% were convicts, while 12.6% were slaves, and 7.6% were volunteer rowers. Similarly, in 1726, convicts made up 70.8% of the crew, slaves 24.5%, and the *buonavoglia*, only 4.7%.[204]

---

[199] Guglielmotti, *La squadra permanente*, p. 23: "in ogni luogo tutti abbiano a lodarsi del contegno della squadra romana."
[200] Guglielmotti, *La guerra dei pirati*, Vol. I, p. 21.
[201] On the medical metaphor of social deviance as a plague to be eradicated, see Pastore, *Le regole dei corpi*, pp. 19–21.
[202] Guglielmotti, *La guerra dei pirati*. Vol. I, pp. 96–109.
[203] Ibid., p. 256.
[204] Lo Basso, *Uomini da remo*, p. 391. As we read in BCR, Ms, 34B13, *Raccolta di notizie e scritture diverse sopra le galere pontificie, armamento di vascelli, fatto dal papa Alessandro VII per soccorso*

The convicts had been sent from all the courts in the Papal States. Accompanied by magistrates, they were brought to the collection point set up in Civita Castellana and, after no more than three days, continued their journey to Rome to await their final sentence. From there, they were sent to Civitavecchia.[205] The transfer to Civitavecchia took place during daylight hours, through the streets and alleys of the city, providing an exemplary demonstration of justice to the rest of the population. Often, however, the effect was counterproductive: frequent attacks by the populace aimed to free the condemned, with whom they sympathized.

By the late 18th century, it was decided to abandon this public spectacle for greater security, and the condemned were instead brought to Civitavecchia at night.[206] Following the establishment of the *Carceri Nuove* in 1656, criminals sentenced in Rome were temporarily locked up in one of the prison's cells, specifically called the *Galeotta*, while awaiting transfer.[207]

6.3 Life aboard the Papal galleys

From the fleet's earliest years, sources attest to the extreme disorder and lack of discipline aboard the galleys. Desertions were frequent, as were violent quarrels that often led to bloody confrontations. As in Tuscany, much of the violence on the Papal galleys was attributed to excessive drinking.[208] Despite the strict control to which they were supposed to be subject, sources indicate that rowers experienced an unusual degree of freedom. Apparently, the ship's guards were easily corrupted, and often, for a fee, convicts were free to leave the ship at any time, and even to sleep with whomever they pleased—a liberty that frequently gave rise to sinful acts. Convicts and slaves often spent their days gambling by playing cards or dice games. Gambling was considered an abominable practice that not only disrupted order aboard but also acted as a precursor to unrest, as rowers, caught up in the heat of the games, were prone to lose all self-control, both in words (blasphemy)

---

di veneziani contro il turco, fortezza e porto di Civitavecchia, Instruttione del proveditore delle galere, f. 942, in 1656 on the five Papal galleys the total number of convicts was 1063, slaves was 153, and *buonavoglia* was 113.
205 Benedetti, Tribunali e giustizia, p. 508.
206 Fosi, *La giustizia del papa*, p. 156 f.
207 Calzolari/Di Sivo/Grantaliano, *Giustizia e criminalità*, p. 31 f.
208 Calisse, *Civitavecchia*, p. 327.

and deeds (brawls). Fighting—whether armed or unarmed—was strictly forbidden and punished severely.[209]

Indiscipline was not confined to rowers alone; officers were also often accused of taking advantage of the oarsmens's' misery, extorting money from them, as well as beating and humiliating them without just cause. Consequently, the Superintendent of the Galleys was tasked with not only making sure that the galleys were well-maintained and adequately supplied, but also with making sure the galley officers exercised their duties honestly and diligently. As one directive emphasized:

> It is essential that the Superintendent, with prudence, vigilance, and diligence, continually oversee and provide for whatever is necessary or may arise from the various and ongoing contingencies that might occur at any moment, and he must always remain vigilant [...] to ensure the galleys are kept in good order and fully supplied.[210]

The guards, and in particular the *aguzzini*, became a significant source of concern. Given their direct responsibility for the custody of the rowers, they were expected to be vigilant day and night to ensure that those in their charge lacked nothing and, above all, that they behaved properly and did not attempt to escape. They were cautioned, however, not to abuse the power entrusted in them, and only to use force when absolutely necessary.[211] Even Guglielmotti, in his treatise, notes how it was often necessary for higher authorities to intervene to address officers' abuses, frequently in response to denunciations and petitions from the crew themselves.[212]

The General of the Galleys of Civitavecchia, in his annual orders and regulations, explicitly warned galley masters, captains, and officers of every rank not to extort money from the *galeotti* with promises of a reduction in punishment or in exchange for more favorable treatment. Similarly, galley officials were strictly forbidden to accept any offers from rowers, as this was considered bribery, punishable by dismissal from service, and could subject them to an arbitrary amount of

---

209 AAV, Misc. Arm. IV/V, b. 54, f. 85.
210 BCR, Ms, 34B13, *Raccolta di notizie e scritture diverse sopra le galere pontificie, armamento di vascelli, fatto dal papa Alessandro VII per soccorso di veneziani contro il turco, fortezza e porto di Civitavecchia, Instruttione del provveditore delle galere*, ff. 1–56 (11 giugno 1652): "Però è necessario che esso provveditore con la prudenza, vigilanza e diligenza supplischi invigilando continuamento quello bisogna e può occorrere per gli diversi e continui accidenti che ogni momento possono nascere e deve anco tenere sempre avvisato [...] che siano sempre le galere bene in ordine e proviste quanto è necessario."
211 ASF, MP, 2131, dossier 6: *Ordini da osservarsi nelle Galere del ser.mo G.duca di Toscana*.
212 Guglielmotti, *La squadra permanente*, p. 48.

corporal punishment—including sentencing them to the galley itself for ten years. Furthermore, the prohibition of improper behavior such as blasphemy, gambling, fighting, and so forth extended to all crew members. Ironically, but not surprisingly, the primary form of punishment was a sentence to the galleys.[213]

Among the Papal galley officers' methods of harassment, we find mention of the practice of forcing galley oarsmen to undress and pose—probably for a price—as freaks for the amusement of visitors, mostly women. Unfortunately, the sources do not specify the nature or status of these women.[214] In an attempt to remedy this desperate situation, numerous decrees were issued to ensure stricter discipline aboard, both from a moral and behavioral standpoint—two aspects which, though conceptually distinct, were inextricably intertwined. In fact, as we read in the preface to the annual proclamations on the management of the Papal galleys: "Good governance of the pontifical galleys require an exact and punctual observance of both the Christian religion and good morals, as well as of the discipline and duties proper to each one."[215]

To achieve this goal, it was necessary to discipline the crew and punish any transgressions with maximum severity and exemplary force. The orders issued aboard the Papal galleys differed little from those on the Tuscan galleys. Here, too, the captains had to ensure that disturbances such as theft, escape, fights, gambling, or sodomy did not erupt. Likewise, the guards were required to be upright and refrain from extorting any money from the convicts in exchange for favors. The mechanisms used to enforce discipline were similar: the creation of an efficient apparatus of vertical and horizontal control and denunciation, and the threat of financial or corporal punishment to discourage crime.

For example, if an officer—whether a captain, a *comito*, or an *aguzzino*—failed to notify the Governor of Civitavecchia of a brawl or a riot, he could, at

---

[213] AAV, Misc. Arm. IV/V, b.54, ff. 85–87.
[214] BCR, Ms, 34D18, *Memorie e scritture diverse appartenenti alle galere Pontificie e condannati nelle medesime Principalmente in tempo del tesorierato di mons. Lorenzo Corsini poi cardinale, e papa col nome di Clemente XII*, f. 35f: "Quando si conducono donne o signore a vedere le galere fanno spogliar nudi li forzati e gli fanno far diversi giochi, ò come dicono essi li buffoni, e burattini per dare spasso alle signore."
[215] In the AAV, all the texts of the proclamations published at the beginning of each new governor's mandate for the period 1668–1715 are preserved in Misc. Arm. IV/V, b. 54, ff. 85–123. The incipit of these edicts is always the same and, in case, changes only in minor details: "Il buon governo e cura delle Galere pontificie richiede più esatta, e speciale osservanza così della religione Christiana, e buoni costumi, come anche dell'obbedienza al servitio, e fattioni, alle quali ciascuno sarà eletto, e destinato, onde con la piena & assoluta autorità del nostro offitio, e per ordin espresso havutone à bocca della Santità di nostro Signore papa Clemente Nono ordiniamo, e commandiamo sotto le infrascritte pene."

the governor's discretion, be fined 50 *scudi* and sentenced to five years in the galleys. A similar punishment was imposed on galley surgeons and their assistants who, having treated prisoners with wounds, infirmities, or mutilations clearly caused by fighting, failed to report them.[216] In cases of bribery and extortion committed by officers, the penalty—beyond dismissal from duty—was a ten-year sentence to man the oars. To ensure compliance with this prohibition, governors often encouraged convicts to denounce offenders, in return for which they were promised impunity, mercy, or a financial reward if they came forward voluntarily.[217]

In response to the increasing number of escapes, time-limited sentences for those caught were doubled, while life-sentenced slaves or rowers faced the death penalty.[218] The fear of escape and of convicts gaining freedom unlawfully became a near-obsession. Ultimately, this led, in 1770, to the formation of a new corps of guardians, tasked with monitoring the crew both in port and at sea, day and night.[219]

Despite the numerous and meticulous regulations, order and discipline aboard the galleys were apparently lacking, as can be inferred from the series of proclamations issued by the Commissioner of Galleys between 1668 and 1715. The earliest of these was published on 22 February 1668, contained 68 articles;

---

216 Ibid.: "8. 8. Li Comiti, Sottocomiti, Agozzini & altri respettivamente debano, succedendo alcun delitto ò scandalo, denunciarlo alla nostra Corte subito sotto pena di scudi 50. E di anni cinque di galera, & altro ad arbitrio &c nelle quali pene incorreranno li Barbieri delle Galere, loro Barbierotti & altri che cureranno feriti, stroppiati, e altri offesi non li denunciando nel modo sudetto."

217 Ibid.: "17. E perché si è trovato che alcuno benché ne Offitiale, ne Minsitro è sì temerario, che per estorcere denari, promesse, & obblighi, ò altre cose à forzati, & altri di Galera promette farli ottenere gratie, libertà, ò in tutto, ò in parte, ò altri allegerimenti di pena, ò pure officij, ò cariche col fingere, e dare intendere haver emzzi, & amicizie atte à tali impetrationi; Si prohibisce à qualsisia di qualunque stato il dare, promettere, e far'obblighi per le sudette cause sotto pene corporali gravi ad arbitrio nsotro, e quanto à quelli che faranno, è procureranno tali estorsioni la pena sarà della Galera per anni 10 & altre à nostro arbitrio, e per venire à più precisa cognitione di simili delitti si promette l'impunità perdono, e premio à quelli che spontaneamente rivelaranno, e denuntieranno li Contraventori del presente Capitolo per il quale sarà la prova privilegiata con ammettersi ancora quelli che dalle leggi communi fussero à ciò stimati inhabili."

218 Ibid.: "25. E per ovviare coll'accrescimento delle pene alle fughe dei forzati, à altri, si notifiche che se alcun forzato fuggirà dalla Galera, ò altri luoghi dove stasse, ò fosse mandato per ordine de Commananti incorrerà ipso facto nella pena della ridupplicatione del tempo per il quale fusse stato condannato per altro, e della vita se sarà per altro forzato in perpetuo, ò schiavo benche fusse ripreso per strada, ò altrove nell'atto, & attentato, ò procuramento di fuga benche non perfetta."

219 ASR, Serie Bandi, b. 100, 1770, I semester, f. V.

the final one on 8 August 1715, included 77 articles. Over time, not only were new regulations added, but several existing articles were supplemented with more precise provisions and harsher penalties—particularly in response to the apparent rise in escape attempts. One notable example is the edict of 15 January 1709 issued by the Navy Commissioner General Carlo de Marini, specifically to address the rampant issues of sodomy and blasphemy, which had reportedly plagued galley crews since the Battle of Lepanto.[220] Whereas the 1668 proclamation imposed life at the oars for blasphemers—or one hundred lashes for those serving life[221]—the 1709 edict introduced a more extensive system of denunciation and surveillance. This included the threat of flogging for uncooperative rowers and demotion for negligent officers.[222] The escalating volume of regulations and the severity of the associated punishments not only reflected heightened vigilance over disruptive behavior, but also betrayed the authorities' inability to enforce discipline. The frequent re-issuance of regulations aimed at curbing specific practices and offenses serves as indirect evidence of their persistence.

6.4 Spiritual care

In addition to military discipline, there was an urgent need to enforce spiritual discipline aboard the galleys, whose objective was to "serve God through the proper administration of the sacraments, and to reform morals whenever necessary."[223] It was essential not only to ensure that the crew attended confession and received the sacraments, but also to re-educate the *galeotti* in the principles of morality and the Christian faith.

Catechesis began with the chaplain's obligation to celebrate Mass and have the convicts recite prayers according to a fixed schedule.[224] Galley time was thus structured around a rhythm of liturgical time: even before navigation or military training, spiritual exercises shaped the daily routine. Mass signaled the start of each day, while reciting the Hail Mary signified its conclusion. Saturdays were reserved for reciting the litanies. Pastoral care, however, also legitimized tighter control over the crew—evident in the introduction of special registers for Confes-

---

220 AAV, Misc. Arm., II, b. 110, ff. 394–396.
221 AAV, Misc. Arm. IV/V, b. 54, f. 85.
222 Ibid., f. 105.
223 ASVR, Atti della segreteria, b.74, *Giursiditione dell'e.mo vicario sopra le galere pontificie ed, in Civitavecchia, sopra l'ospedale di S. Barbara e su di alcune chiese (1722—1773)*, f. 49.
224 AAV, Misc. Arm. IV/V, b. 54, f. 85.

sions and Communions from the pontificate of Alexander VII onward.[225] This was hardly a novel development, but rather the maritime extension of long-established practices in daily life throughout society at large. Following the Fourth Lateran Council (1215), parish priests were required to maintain lists of the names of those who had fulfilled their annual obligation to confess—a prerequisite for Easter Communion. From the late thirteenth century onward, the use of *schedulae confessionis* [confession slips] became widespread. These personal documents confirmed whether an individual had attended confession and was eligible for communion. Over time, the Church's drive to monitor the faithful extended to recording excommunications, baptisms (especially after the Council of Trent), marriages, and funerals, following the adoption of the Roman ritual in 1614. The *status animarum*—parish style registrars listing baptized individuals—also existed aboard the galleys.[226] The meticulous recording of convicts' participation in the sacraments became a key tool in reinforcing moral discipline.

The vigilance exercised by the religious aboard went well beyond spiritual matters. In the aftermath of the Battle of Lepanto, their responsibilities broadened to include the monitoring of criminal or immoral behavior. Chaplains were not only expected to denounce the faintest suspicion of unethical and heretical conduct, but also to serve as models of rectitude. According to the instructions issued to chaplains during the Cyprus War, they were tasked with teaching Christian doctrine to the crew—referred to as a form of spiritual medicine, and regarded as the only true remedy capable of reforming and converting souls.[227]

This drive to convert galley crews became increasingly evident during the 18th century when the Jesuits—who replaced the Capuchins as spiritual directors of the crews—embarked upon serious missionary activities on Civitavecchia's galleys. These religious missions were organized down to every last detail, from the Jesuits' arrival in the city—they were supposed to be four in number—to their departure. Efforts were made to align these missions with Easter, ensuring that the two ecclesiastical commandments—annual Confession and Communion at Easter—could be fulfilled simultaneously. The overriding concern was to

---

[225] Ibid., f. 106; ASVR, Atti della segreteria, b. 74, *Giurisditione dell'e.mo vicario sopra le galere pontificie ed, in Civitavecchia, sopra l'ospedale di S. Barbara e su di alcune chiese (1722—1773)*, ff. 273–278.
[226] Groebner, *Identità*, p. 67f.
[227] AAV, Misc. Arm. II., b. 110, ff. 370–373, *Instruttioni del molto Ill.re Sig. D. Marniese à gli Preti, frati, e persone religiose che anderanno nell'armata del ser.mo S.V. Don Gio. D'Austria et princi-palm.te agli Preti [..] delle Galere*. As Michele Camaioni has aptly observed in *Pulpiti*, p. XIII: the expression "to convert souls" denoted not only the conversion of non-Christians to Christianity, but also the transition from one Christian confession to another.

have convicts repent for their sins. To achieve this, they needed to be convinced of the gravity of their crimes through preaching about the shamefulness of sin, purgatory and its punishments, and by insisting that God patiently await the repentant sinner. Yet, should the sinner fail to repent, God would punish him more severely on the Day of Judgment. For this reason, severe penances were not to be imposed—no matter how much they were deserved. Meekness was seen as the best means to guide the sinner to mend their ways; the process had to appear voluntary, not imposed from outside.[228] The entire population of Civitavecchia participated in the celebration of Masses, processions, and sacraments, engaging actively with the significance of the events—especially during the 1716 mission which culminated in the conversion of four Protestants.[229]

6.5 Control of the body, control of the soul: the *Governatore di Civitavecchia* and the *Vicario di Roma*

As in Livorno, the ambiguous legal status of the *galeotti* in Civitavecchia often led to misunderstandings—if not outright conflict—among the legal authorities. The situation was further complicated by the dual nature of oversight exercised aboard the galleys. Chaplains, for example, frequently intervened in matters beyond pastoral care and exercised authority similar to that of captains, which only exacerbated tensions.

The temporal governance of Civitavecchia was entrusted to the Governor, who himself was subordinate to the Governor of Rome.[230] Under the pontificate of Innocent XII, it was decided that the Governor would be assisted by a prelate from the curia, holding the rank of protonotary, who was to be appointed in accordance with the instructions of the pope.[231] In theory, the Governor's authority extended to almost all legal matters. In practice, however, exceptions were frequent. Certain groups—most notably the clergy—remained subject to other tribunals, such as that of the diocesan bishop residing in Viterbo. The *galeotti* were formally under the authority of the Governor of Civitavecchia. Initially, however, galley crews were overseen by a separate official—the *Auditore delle Galere*. To resolve the numerous disputes arising from this overlap of authorities, the of-

---

[228] ARSI, Fondo gesuitico, manoscritti, n.10, *Breve instruttione Per quelli che vanno Alla Missione delle Galere*.
[229] ARSI, Rom. 138. ff. 61–63, *Breve relazione delle missioni fatte in Civitavecchia il 1716 per ordine di nostro signor papa clemente XI. Missione alle galere pontificie*.
[230] On the Governor of Rome see Fosi, *La giustizia*, pp. 21–23.
[231] Calisse, *Civitavecchia*, p. 445.

fice of *Auditore* was abolished in the early 17th century, and responsibility for convicts was transferred to the Governor of Civitavecchia, who then also assumed the title of "General of the Papal Galleys."²³²

Conversely, spiritual oversight of the *galeotti* fell under the ecclesiastical authority of the Roman Curia.²³³ In Civitavecchia, which was part of the diocese of Viterbo, the Vicar of Rome exercised authority over the galleys whenever they were docked in port—and even at sea, so long as they were within a hundred miles of Civitavecchia. His authority also extended to the Ospedale di Santa Barbara in Civitavecchia, which was designated for the care of convicts and slaves. Elsewhere in the city, galley rowers and personnel fell under the ecclesiastical authority of the Bishop of Viterbo.²³⁴ Both the Vicar and the Bishop were responsible for approving and licensing confessors to serve aboard the vessels, and ensuring that chaplains properly fulfilled their duties, especially in administering the sacraments.²³⁵

---

232 Ibid., pp. 451–452.
233 See Rocciolo, *Della giurisdittione del Vicario*, pp. 35–71.
234 ASVR, Raccolta di notizie di vario genere sui Diritti, giurisdizione e prerogative del vicariato di Roma, 1650—1740, tomo 55, ff. 678–679.
235 ASVR, Atti della segreteria, b. 74, *Giurisditione dell'e.mo vicario sopra le galere pontificie ed, in Civitavecchia, sopra l'ospedale di S. Barbara e su di alcune chiese (1722—1773)*. At the opening of the tome there is a *Breve relazione di quel ch'è seguito nell'esercizio della giurisdizione spirituale dell'E.mo Vicario di Roma sopra le galere pontificie, l'Ospedale di S. Barbara ed alcune Chiese di Civita Vecchia* which has been transcribed in Rocciolo, *Della giurisdittione del Vicario*, pp. 350–352.

Chapter 3
Medicine and the Early Modern Galleys

As early as the 16th century, Leonardo Fioravanti (1517–1583), protophysician of the Spanish Imperial Army, sent a letter to the Papal court concerning the health of armies both at sea and on land, in which he argued that their most formidable enemy was undoubtedly disease, which, by killing men and weakening their strength, would cause armies to lose their opportunities for glory.[1] The presence of medical experts aboard vessels was, therefore, essential. When it came to providing medical care for naval forces, these experts had to be capable of addressing any kind of infirmity that might occur while at sea. Contributions by physicians and surgeons were equally significant. Furthermore, Fioravanti asserted that the inability to medically treat the crew was not only a cause of a high mortality rate among soldiers and oarsmen, but also served as a deterrent to enlisting. Knowing that in the case of illness or injury they would not receive the necessary medical care discouraged soldiers and rowers, and further diminished their motivation. Having an experienced and competent medical team was supposed to benefit the soldiers' and rowers' morale, as they became more motivated to fight, knowing that someone was there to care for them if needed.[2] Clearly, this focus on the ship's physician appears partly propagandistic and self-celebratory, for Fioravanti himself had served as a doctor on a Spanish vessel sailing to Africa in 1551.[3] It is therefore reasonable to assume that the letter was written to improve his position in the eyes of the papal authorities, perhaps with the hope of obtaining a place of honor in the papal fleet. This does not, however, detract from its value as a direct testimony to the high regard in which medical care was held, and to the level of despair soldiers must have felt when it was lacking.

Doctors—especially surgeons—have been a constant presence aboard ships. As Fioravanti's testimony reveals though often overlooked by historians, doctors played a central role not only aboard early modern ships but also on land in dock hospitals, and were increasingly held in high esteem by naval authorities.

---

1 AAV, Fondo Pio, b.112, f.333r.
2 Ibid.: "perché quei poveri soldati amalati et feriti che non hanno de quibus non si trova medici che li voglia guardare non che medicare et cosi li convien morire al suo dispetto et questo è cose che mette tanto spavento al mondo che non si ritrova huomo che ardisca andare a servire in armata, ma quando vi fossi tal hordine ognuno andaria alegramente, sapendo di essere aiutati nelle loro calamità."
3 On the figure of Leonardo Fioravanti see Camporesi, *Camminare il mondo*; Furfaro, *Leonardo Fioravanti*; Gadebusch Bondio, Verità e menzogna.

Open Access. © 2025 the author(s), published by De Gruyter. This work is licensed under the Creative Commons Attribution 4.0 International License. https://doi.org/10.1515/9783111654133-007

This chapter will, therefore, aim to partially fill this gap by attempting to present, as exhaustively as possible, a defined profile of those physicians involved in caring for the *galeotti*, both on land and at sea. First, I will reconstruct their biographical and professional profiles to understand who these medical practitioners were and what their career trajectories entailed. I will then analyze the tasks they were required to perform. Essentially, doctors performed three functions. The first was therapeutic—medicine's primary objective—aimed at maintaining the crew in good health. The second could be described as "inspective," consisting of observing and examining the rowers' physiques for technical, military, but also economic purposes. And, finally, there was an "expert judgmental" intervention, executed *a posteriori*, such as when physicians were asked to ascertain the causes of death or the severity of injuries in the case of murders and fights, to decide on the aggressor's degree of guilt, and thus determine the most appropriate punishment.

1 The profile of a galley doctor

Unfortunately, sources provide little information about the profiles of galley physicians, and generally, all that can be gleaned is the individual doctor's name and, sometimes, their years of service as evidenced by the reading of the crew rolls, or by finding some reports signed by them. In his treatise on the Papal navy published in 1856, Alberto Guglielmotti complained that reconstructing the biographies of these professionals was quite a task, and that no sources on the subject were to be found before the 1550s. Guglielmotti's disapproval is evident: despite the fact that "at all times physicians and surgeons, whether voluntarily or forced, followed the armies on land and at sea: histories and documents only speak of them in a general way."[4]

In the case of the Papal galleys, Guglielmotti was unfortunately right: the archives do not reveal any specific information, apart from a few names. In general, we know that each galley had its own surgeon aboard. Alongside the surgeons, one physician was in charge of the entire fleet. According to Pantero Pantera, this physician had to be "intelligent, capable, and experienced in his field." He could only be employed on the basis of his proven reputation, backed up by certificates testifying to his skills and education. Apparently, he also had to be trained

---

4 Guglielmotti, *La guerra dei pirati*, Vol. II, p. 202f: "In ogni tempo i medici e i chirurghi hanno seguito, o volontari o condotti, gli eserciti di terra e le armate di mare: le storie e i documenti ne parlano solo per le generali."

in surgery so that he could give instructions to surgeons in the most appropriate way.[5] Ultimately, however, sources testify that surgeons almost always acted without instructions from above. This was especially true when they were on the high seas or in battle, when they had to react as quickly as possible, and were often the sole medical professionals aboard. The lack of control from above should not have posed a problem, since surgeons had to prove their expertise in practice before being recruited. Moreover, in the absence of a physician, the surgeon had to handle illnesses that were theoretically the physician's responsibility. Therefore, it was good practice for the surgeon to have a barber as an assistant, who often happened to be a slave.[6]

Compared with the Papal context, the Tuscan one is much richer in information, and provides a more precise idea of this social category's biographical-professional profile—one that was probably similar, or even identical, in other Italian regions. However, even here, information is fragmentary, as the most detailed records date to the late 17th century, and for the preceding years we have just a few scattered names. It is not entirely clear whether this is due to a gap in the archives, or to less systematic recording in the previous century. Both factors likely influenced the situation. As in the Papal fleets, each Tuscan galley had a surgeon, and a physician was in charge of the entire fleet, serving aboard the *Capitana* during the sailing season. As reported in the rolls of the crews, in 1680,[7] the physician on the *Capitana* was an individual called Clemente di Salvatore Cosci from Livorno, who was in his thirteenth year of service. The surgeon was the 60-year-old Agostino Jacopo Amiconi from Palermo, who has been a medical practitioner for 29 years and previously worked at Florence's Hospital of Santa Maria Nuova, and subsequently in the city of Livorno. Aboard the *Santo Stefano*, the surgeon was Giovan Battista di Lorenzo Barizeni, aged 45. Although he had only been in that position for four years, he was highly experienced, having had the opportunity to train at Rome's Santo Spirito Hospital—a privileged institution for training Roman surgeons—and having already held the position of surgeon at sea aboard merchant ships. Two surgeons were aboard the *Santa Maddalena de' Pazzi*: Carlo Antonio di Gasparo Franceschin from Volterra, aged 43, and Gaetano di Zanobi Pantalino from Livorno, assistant surgeon, aged 24. Franceschin had previously served as a surgeon for 14 years, after having studied at the Hospital Santa

---

5 Pantera, *L'armata navale*, p. 110.
6 Ibid., p. 126. ASF, MP, 2131, dossier 4, Chapter IV, c.29: It seems that by the late 16th century, surgeons' assistants were always slaves or forced rowers. Since assisting a surgeon was considered a low-skilled job, they did not need to be educated and, as convicts and slaves, they were not paid. For more on the use of slaves in healthcare, see Bono, *Slaves*, p. 167.
7 ASF, MP, 2130, cc. n. n.

Maria Nuova, worked as a town physician [*medico condotto*] in the Maremma, and later in Livorno. Pantalino, on the other hand, began his naval career at a very young age, being the son of one of the Tuscan Navy's boatswains. The surgeon on the *Santa Margherita* was Jacopo d'Antonio Roccatagliata, aged 33. He was trained at the hospitals in Pisa and in Florence and was already in his eighteenth year of service—initially as an assistant, and later as a surgeon.[8] Although this information is scant, it nonetheless allows for some more general assumptions. First, we know that the *Capitana* alone had both a physician and a surgeon aboard, while the other vessels had only one or two surgeons each. Surgeons and physicians must have had a good reputation and been highly experienced, given the importance attached to their role in ensuring the crew's good health. Not only were sick galley slaves and convicts considerably expensive to treat, but they also represented a loss of much-needed manpower, so it was imperative that they remained healthy and fit enough to row. It remains unclear who was responsible for appointing the medical staff. According to archival sources—such as letters and licenses issued to galley doctors—an initial decision was likely taken by the Captain of the Galleys, which was then officially confirmed by the Grand Duke in the form of a license countersigned by him and the Secretary of War.[9]

Despite the challenges involved, the position of fleet doctor was highly coveted, as any physician holding this position was entitled to a fixed salary—which could vary from eight to 25 *scudi* in the 17th century, even reaching 30 to 40 *scudi* in the 18th century. This variation in salaries did not follow a linear pattern, but changed according to contingent factors, such as the urgency of finding new medical professionals in times of military emergency, economic conjunctures, and other circumstances.[10] In addition, holders of this post enjoyed "special honors, privileges and prerogatives," as stated in the fleet doctor's license granted to Dr. Don Diego Galletti in February 1716. He had distinguished himself by his zeal and skill in caring for the Tuscan crew during their stay in Messina on their return from the Levant. As a sign of gratitude and benevolence, Galletti was rewarded with the title of supernumerary physician on the Medici galleys in the city of

---

8 Ibid.
9 See the manuscript licenses conserved in ASF, MP, 2131, dossier 6, cc. n. n.
10 One *scudo* was equal to one *ducato*, i.e., 7 Florentine *lire*, and weighed 21.231 grams. See Martini, *Manuale di metrologia*, p. 209. The typical monthly salary for a soldier was 3 *scudi*. On this topic, see Goldtwhwaite and Cipolla. For galley doctor's salary, ASF, MP, 2132, cc. n. n. in 1639, a surgeon received 11 *scudi* at month. ASF, MP, 2130, cc. n. n. In 1650, a chaplain received a monthly salary of 3 *soldi*, a captain 29.5 *scudi* and a surgeon 8 scudi. After1680, a surgeon's salary increased to 15.5.9 *soldi*.

Messina.¹¹ The appointment of galley surgeon was, therefore, viewed as both a privilege and a reward in the eyes of the authorities, as is evident in the following cases. In 1716, as an acknowledgment of his medical skills in Portoferraio, it was proposed to designate Filippo Sampieri, brother of the Sampieri surgeon at the *Bagno*, as galley surgeon. In the same year, the naval authorities also recommended appointing Agostino Giustiniani as assistant to the *Capitana's* surgeon; the latter had been registered in the city's medical college for more than ten years, and who had proven himself as a *medico condotto* in Campiglia¹² and at the Hospital of Santa Maria Nuova. Should Giustiniani not accept the post, the name of Donato Ercolani was put forward as an alternative. Although both surgeons possessed comparable professional experience, Ercolani was regarded as the second choice due to his younger age and his having practiced for only one year, which was perceived as a disadvantage.¹³

Some petitions addressed to the Grand Duke reveal just how prestigious the positions of physician, surgeon, and assistant surgeon aboard the naval fleet were. For example, in 1694, Luigi Montorsi, a physician, pleaded for the position of galley doctor, and was even willing to accept a monthly salary of six *scudi*, compared to the usual twelve.¹⁴ An essential requirement for the position was to demonstrate competence. Thus, in May 1749, following the death of Carlo Springipill, who had served as first surgeon on the *Capitana* for a monthly salary of 33 *scudi*, two applications were submitted to replace him. The first came from Francesco Corona, who had already served aboard the galleys in August 1747, without pay, replacing one of the enlisted surgeons at the time because he was at sea aboard warships. Corona requested to be officially hired by the Tuscan navy. Given his outstanding performance, his certificate of practice in Livorno, and endorsements from other hospital doctors attesting to his abilities [*fede*], he was deemed suitable and duly approved.

However, even more intriguing is the second of the two applications, this one submitted by Dr. Giuseppe Carlesi, who had studied surgery for five years at Pisa's Hospital of Santa Chiara and had practiced as a surgeon in Livorno for three years. He presented all his certifications and declared his willingness to be publicly examined in Florence by any master surgeon lest his credentials appear insufficient. Furthermore, in a bid to substantiate his good reputation, he brought to their attention that one of his older brothers, Tommaso di Francesco Carlesi, had

---

11 ASF, MP, 2131, c.n.n.
12 On the *medico condotto* [town physician] see Russell, *State Physician*; Mendelsohn/Kinzelbach/Schilling, *Civic Medicine*.
13 ASF, MP, 2110, c.n.n.
14 ASF, MP, 2103, c.n.n.

also previously served on the fleet for seven years—initially as an assistant and later as a surgeon, from 1735 to 1742. Carlesi produced no fewer than five documents in support of his claim. The first was from the Commissioner of the Galleys, certifying his abilities and good health. The second was from the physician and surgeons of the city of Livorno, attesting to his years of service and the success of all his operations. Carlesi also enclosed two letters from his erstwhile professors, one of whom was Giovan Pietro Bernardini, master of theoretical and practical surgery at Pisa's new municipal hospital, in which he stated that Carlesi had always excelled throughout his 39 months as a regular student and his subsequent 20 months as a surgical trainee. The other was signed by Domenico Baofanti, master of theoretical surgery in the same hospital, in which he asserted that Carlesi had consistently been skillful and diligent with the sick. Finally, Carlesi presented a certificate of merit awarded to him by Francesco Maggio, Knight of Santo Stefano, and Rector of Florence's Hospital of Santa Maria Nuova, Pisa's Ospedale Nuovo, and the Spedale del Ceppo in Pistoia.[15]

Notwithstanding his impeccable credentials, Carlesi had no maritime experience, so one might surmise that this was why he was not selected. In any case, the documents accompanying his application show just how demanding the selection criteria for galley doctors were. Not only did they have to be registered and licensed to practice in Livorno, but they also had to train at a prestigious institution, such as the hospitals of Santa Maria Nuova or Santa Chiara. Judging from Carlesi's records, he trained at Santa Chiara's hospital surgical school. The available statements from his professors confirm that he had completed his studies, though there is no mention of a university or a doctorate—the title conferred upon those who finished their university education.[16] The fact that none of those surgeons had a university education, but had all been trained in hospitals or similar institutions underscores how hospitals had become established as educational centers of the highest caliber.[17] After all, hospital practice had long been a critical form of training for surgeons. In these settings, future surgeons not only learned their trade through direct experience but also gained the professional credibility essential for their careers.[18] Over time, this form of training evolved, becoming progressively more disciplined and structured following the creation of dedicated surgical schools within medical institutions. Unfortunately, to date,

---

15 ASF, MP, 2132, dossier 5, c.n.n.
16 Siraisi, *Medieval Renaissance Medicine*, p.55.
17 On Roman hospital schools, see Conforti/De Renzi, Sapere anatomico.
18 Cavallo, *Artisans of the Body*, p. 146 f.

these schools have not been examined in detail, and their foundation dates and internal organization remain unclear.

Recent studies have highlighted the development of institutionalized medical training in Tuscan hospitals during the Renaissance.[19] This evolution was closely linked to the gradual decline of surgical instruction in the *Studia*, especially following the closure of the Florentine *Studium* in 1472 by order of Lorenzo the Magnificent,[20] and the abolition of the chair of surgery at Pisa after the death of its last holder in 1699: Carlo Vasoli. In Pisa, the transition from university-based to hospital-based surgical training was relatively seamless, as Vasoli also served as a master surgeon at Santa Chiara. After his appointment as a lecturer in 1692, many surgical students followed him to the hospital's operating theatres after his lectures, with some even choosing to reside there. Thus, hospital-based schools made up for the gap in surgical training at universities, offering aspiring surgeons the highest level of both practical and theoretical training.[21] However, scholars such as John Henderson have questioned whether such schools existed before the late 18th century. Informal links between Tuscan hospitals and the medical faculties at Pisa and Florence were certainly evident; many hospital doctors also served as university professors or lecturers.[22] Furthermore, the statutes of Italian universities explicitly stated that medical training required an internship in hospitals, where students worked alongside senior medical staff to acquire hands-on experience. Despite this, there is no definitive evidence of formal collaboration during the 17th century.[23] However, the absence of institutionalization does not necessarily negate the existence of such schools. On the contrary, it underscores their independence from universities. As demonstrated by Carlesi's case, even if these schools were not officially recognized, they were publicly acknowledged by the 18th century as effective training grounds for competent, practically-skilled surgeons. These institutions specialized in preparing surgeons for serving the public.[24] Thus, it is reasonable to conclude that, alongside university education and an apprenticeship with senior surgeons—often taking the form of knowledge transmitted from father to son or master to trainee—there existed a viable alternative

---

19 In general, see Agrimi/Crisciani, *Edocere Medicos*; Siraisi, *Medieval Renaissance Medicine*; Nardi, Statuti e documenti, pp. 245–248; Sani/Zurlini, *La formazione del medico*.
20 Baldanzi, Nell'Ospedale di Santa Maria, p. 287.
21 Ibid., p. 281f. See also Coturri, Le scuole ospedaliere, pp. 3–8.
22 Ciuti, Il medico e l'ospedale, pp. 63–88; Henderson, *The Renaissance Hospital*, p. 246f.
23 Ibid.
24 On the early modern hospital as a space for educating surgeons, see Baldanzi, Nell'Ospedale di Santa Maria; Cavallo, *Artisans of the Body*; Ciuti, Il medico e l'ospedale; Conforti/De Renzi, Sapere anatomico; Henderson, *Renaissance Hospital*.

that served as a middle path between these two institutional forms of training. As Paolo Savoia observed, the practical reality was more complex, making it possible to identify five distinct profiles for the early modern surgeon.[25]

First, the graduated physician who also devotes himself to surgery—a category into which Savoia includes graduate surgeons; second, one who received some formal teaching, typically in a surgeon's workshop, but who had not graduated; third, a barber-surgeon, who lacked any academic credentials; fourth, one specializing in specific procedures; and, finally, the broad group of professionals who could be categorized under the generic label of "quacks."[26] Galley surgeons primarily fell into the second category, though some could also be classified under a hypothetical sixth category of fully trained surgeons—both practically and theoretically, through hospital-based education. Indeed, hospital training was often preferred over university training when it came to recruiting surgeons for the galleys. At least a year's practical training under a more experienced surgeon was a key requirement for obtaining a license to practice.[27] That all professional training took place in hospitals served as a stronger guarantee of a surgeon's practical skills, further highlighting the critical importance of hospitals in the hands-on education of medical professionals.

Despite the privileges associated with the position, being a galley doctor was a difficult and demanding task. First, the doctors themselves had to be in good health. As stated in the doctors' licenses and petitions, it was crucial that they not suffer from seasickness, as medical operations were more likely to be performed during stormy conditions.[28] In cases where a doctor or surgeon suffered from ailments such as ulcers, poor eyesight, sciatica, and other similar afflictions, he would be deemed unfit to set foot aboard ship, and, as a result, would be removed from his position.[29] Moreover, many were the instances of galley physicians and surgeons who, as a reward for their years of service, petitioned to continue serving the fleet on dry land, in the *Bagno*, away from the risks and hardships of constant sea duty. For instance, the aforementioned Luigi Montorsi, who had pleaded for his position aboard the galleys in 1694, requested a waiver from the Grand Duke after 22 years of service, citing severe asthma, which had plagued him for four years and left him unable to sail.[30] Similarly, the case of the surgeon Filippo di Giovanni Sampieri from Livorno, aged 52 and in service

---

25 Savoia, Early Modern Italian Surgeon, p. 32f.
26 Ibid.
27 Cavallo, *Artisans of the Body*.
28 ASF, MP, 2132, dossier 5, c.n.n.
29 See, for example, the list of surgeons unable to board ship in ASF, MP, 2113, c.n.n.
30 ASF, MP, 2109, c.n.n.

for 24 years—seven as an assistant surgeon on the galleys, nine as surgeon in the city of Portoferraio, and eight as a galley surgeon—illustrates the challenges faced by medical personnel. In February 1722, no longer fit for sea service due to frequent bladder stones, cramps, and blood in his urine, he requested to be relieved of duty and sent ashore. He was replaced by Agostino Giustiniani, who had accepted the position of assistant surgeon in 1716.[31] Sampieri was then appointed surgeon at the Hospital of Sant'Antonio—a medical facility reserved for soldiers and free rowers.[32]

Doctors and surgeons working in the *Bagno*'s infirmary shared similar professional and biographical profiles; many of them had previously served aboard the fleets. In 1680, for example, the hospital's doctor was Antonio Francesco Tossi from Livorno. Aged 45, he had served for 20 years and had previously worked as a naval doctor. The *Bagno*'s surgeon was Salvator Clemente Cosci from Pisa, aged 69. He had been active for 52 years—38 years as an assistant surgeon, then as a galley surgeon, followed by 15 years on *terra firma*. Unable to set foot aboard a ship due to gout, Cosci petitioned to be appointed as surgeon of the *Bagno* in 1655. As a reward for past services, his request was granted—albeit not without some regret from the authorities at losing such a skilled physician.[33] The position of physician in the *Bagno* was, therefore, even more coveted than that of a galley doctor, not only because it offered better working and living conditions but also entailed the benefit of residing within the *Bagno* compound with one's family.[34] In addition, the doctor of the *Bagno* was often attached to the *Ufficio di sanità* [Health Office] in Livorno, making him responsible for any sanitary measures to be taken within the city in the event of contagion or suspicion of an epidemic.[35]

Among the names of physicians appearing in the sources, two stand out: Francesco Redi and Giovanni Cosimo Bonomo. In 1690, when the physician of the *Bagno*, Romanello Romanelli, fell seriously ill, both Redi and Bonomo were summoned to treat him.[36] Redi (1626–1697) was a physician, naturalist, and a man of letters of great fame: arch-consul of the *Accademia della Crusca*, co-founder of the *Accademia del Cimento*, and first physician to Grand Dukes Ferdinando II (1621–1670) and Cosimo III (1670–1723). His most important works include *Esperienze intorno alla generazione degl'insetti* (1668), in which he disproved

---

31 ASF, MP, 2112, c.n.n.
32 ASF, MP, 2113, c.n.n.
33 ASF, MP, 2130, c.n.n.
34 Santus, "Il turco", p. 38.
35 Ciano, *La Sanità Marittima*, p. 43.
36 ASF, MP, 2101, c.n.n.

the theory of the spontaneous generation of insects, and *Osservazioni intorno agli animali viventi che si trovano negli animali viventi* (1684), the earliest known extensive and methodical study of human and animal parasites, which is considered the foundation of modern parasitology.[37] Redi had already been in Livorno since 1687, having been summoned to treat Cosimo III, who was recovering from a fever and bouts of vomiting at the Fortress Palace.[38] That a physician of Redi's caliber was summoned to personally cure Romanelli likely indicates how highly he was regarded. Indeed, the two may have been closely connected, for Romanelli's name appears in several of Redi's letters,[39] further underscoring Romanelli's professional and social status.

The figure of Bonomo (1666–1696) is even more intriguing. A native of Livorno, he studied medicine in Pisa, and was licensed to practice in Florence, where he encountered Redi, who soon became his main patron. In 1684, Bonomo began to frequent the pharmacy of the famous naturalist and close collaborator of Redi, Giacinto Cestoni (1637–1718). In May 1684, thanks to Redi's intercession, Bonomo—then in dire financial straits—obtained the position of galley doctor from the Grand Duke, following the naval expedition against the Turks ordered by Pope Innocent XI, an operation that culminated in the landing at Santa Maura (present-day Lefkada). The account of this venture—written by Bonomo and sent to Redi in 1685—confirms the severity of the assignment, as the expedition was plagued by a series of illnesses, infections, and deaths, ultimately decimating the crew, whose numbers fell from 370 to 160 men. Bonomo himself was twice taken ill, with life-long consequences. The order to return to port at Livorno was a godsend, bringing an end to the extreme hardships of the voyage, which Bonomo thereafter sought in vain to avoid.[40]

Bonomo is still best remembered for discovering the mite-like nature of scabies, based on observations made in the *Bagno*. This discovery—which contradicted the traditional explanation based on humoral theory—sparked a fierce academic controversy, particularly involving physician Giovanni Maria Lancisi (1654–1720). Despite Redi's patronage, Bonomo struggled to establish a practice on land. In May 1690, he set sail again aboard the *Santo Stefano*, bound for Spain, as the ship's physician. According to the report on the selection of the galley doctor, Bonomo was regarded as Livorno's top physician, and his presence aboard

---

[37] Altieri Biagi/Basile, *Scienziati*, pp. 555–561; Bernardi/Guerrini, *Francesco Redi*; Bernardi, Uno scienziato aretino, pp. 17–36; Di Tommaso, The Erudite Pratictioner.
[38] Ciano, *La Sanità Marittima*, p. 123.
[39] Some correspondence between Redi and Romanelli can be found in Redi, *Opere di Francesco Redi*, p. 103; Redi, *Lettere di Francesco Redi*, p. 182.
[40] The letter was published by Pera, *Curiosità Livornesi*, p. 111f.

was deemed essential.[41] Finally, in March 1691, thanks to the fame he had acquired and Redi's continued support, he was chosen as physician to Cosimo III's daughter, Anna Maria, who was to marry the Elector of the Palatinate.[42]

The overall professional profile that emerges is that of doctors—both physicians and surgeons—who were highly specialized and competent. They had all received advanced education, either at universities or in hospitals (surgical schools), and were licensed by a *Protomedicato*. The need for a healthy crew with skilled rowers was critical for seafaring, military, and economic purposes. This necessity made it crucial to rely on an experienced medical team. The significance of their role was reflected in their monthly salary, which was relatively high for the time. Galley doctors enjoyed respect, honors, and privileges in exchange for providing excellent medical care. It was a highly coveted position, undoubtedly more prestigious than that of the *medico condotto*. However, it should not be overlooked that this role also involved numerous risks and discomforts due to the constant life at sea. When aboard the galleys, doctors were exposed at all times to hardships similar to those faced by the rowers—enduring the elements and being crammed into narrow spaces with poor hygiene. Clearly, the prospect of a stable, well-regarded, and well-compensated job, along with the expectation of professional and social advancement in a highly competitive medical marketplace, likely made these challenging conditions more tolerable.

2 Health and manpower at sea

A healthy crew was crucial for successful seafaring; vessels could not sail without manpower. As Pantera observed, the crew was the "soul of the galley," and it was necessary for oarmen to be at full strength at all times.[43] In this regard, the Battle of Lepanto undoubtedly represented a turning point, as it forced the authorities to confront the inadequate levels of medical care aboard the Italian fleet. A report on the Papal galleys commissioned by Pope Pius V in 1571—written either by a Capuchin friar, or by Domenico Grimaldi, the Papal general commissioner for the galleys[44]—denounced the utterly precarious and inadequate sanitary condi-

---

41 ASF, MP, f. 2101, c. n.n.
42 Altieri Biagi/Basile, *Scienziati*, pp. 709–712.
43 Pantera, *L'Armata Navale*, p. 130.
44 Civale, *Guerrieri*, p. 113.

tions aboard the ships which, combined with the chaos engendered by sea battles, led to widespread despair.[45]

Space aboard was extremely limited. For example, the dimensions of a typical Venetian galley—which can be regarded as standard for early modern Italian galleys—measured, in the 1550s, approximately 42 meters in length, just over 5 meters in width, and 1,75 meters in height, with between 25 and 30 benches.[46] This space, barely sufficient to accommodate 300 to 350 healthy rowers, made it impossible to provide a separate bay for the sick. Rowers were constantly exposed to the elements, the fatigue of long hours at the oar, and the beatings inflicted by officers. Physically and mentally exhausted, they often fell ill. Common ailments included fever and lung diseases caused by the cold and the sea water, from which they could barely protect themselves, with only a wool shirt and a cloak.[47] Crammed together in dangerously unsanitary conditions, contagious diseases spread rapidly and were practically impossible to avoid.

It is noteworthy that in the literature—however dated—on the galleys, one of the primary causes of illness is attributed to the presence of slaves, as they were believed to have brought contagious diseases with them from across the Levant.[48] In early modern imagery, epidemics invariably came from "outside," and the Ottoman Empire was often identified as one of the major sources of contagion. Slaves were seen as the primary vector of its transmission among the crews.[49] However, the sources I have reviewed provide little confirmation of this accusation, as the diseases recorded were already widespread across the Italian mainland. Consequently, it is plausible that these illnesses could have been introduced aboard not only by slaves, but by convicts as well. Furthermore, when slaves were purchased or acquired from regions impacted by epidemics, they were quarantined in lazarettos before being sent to the galleys.

In any case, the most common illnesses were not of an infectious nature. Rather, they resulted from inclement weather and the grueling physical demands

---

[45] AAV, Misc. Arm. II, b.110, *Secondo Avvertimenti sopra I disordini delle galere di S. Santità occorsi nell'anno passato 1571 dati da certe religiose persone et da bene con I rimedij necessari et opportuni per emendargli*, f. 385: "Primo disordine fu circa gli infermi, però che molti ne morsero di necessità per non essere sovvenuti pur di cose minime come di pan cotto et c'è di più molti ne morivano disperati, vedendosi così abbandonati, et pregavano d'esser gettati in mare."
[46] Aymard, Chiourmes et galères, p. 73 f.
[47] Lo Basso, *Uomini da remo*, p. 353. As Lo Basso recalls, the rowers' clothing consisted of a coat, a wool shirt, two light shirts and a pair of trousers made of hemp.
[48] Calisse, *Civitavecchia*, p. 143.
[49] Harrison, *Contagion*, p. 2 f.

of rowing-conditions that led to ailments such as pneumonia, leg ulcers, blindness, among others.[50] Moreover, sources record that convicts on land fell ill more often, and more easily, than slaves. While confined to port, slaves enjoyed better treatment, as they were unchained and assigned less strenuous work, such as running taverns or managing shops at the docks. Convicts, however, were used for forced labor and endured levels of fatigue and ill-treatment similar to those they faced aboard the galleys.[51]

2.1 Hygiene

One of the most pressing problems encountered aboard the vessels was the poor sanitary conditions. The crews—typically numbering about 300 men—were confined to seafaring crafts usually about 40 meters long and five meters wide. To enhance the galley's speed and balance, the hull was raised only one meter above water level, which meant the galley had virtually no lower deck. Most crew members had to sleep in the open—rowers on their benches and soldiers on the floor. In good weather, an awning could be stretched over the entire length of the vessel to provide some shelter from the sun or rain. However, this was not possible during open sea voyages or in strong winds. Additionally, oarsmen were not permitted to leave the oar room and were forced to live amidst their own excrement.[52]

Disease, therefore, ran rampant due to poor hygiene and the practical impossibility of effectively separating the healthy from the sick, which, in turn, facilitated the spread of epidemics.[53] Fleas and lice were especially prevalent, as rowers were chained to their benches, and the scarcity of fresh water aboard meant that they seldom had the opportunity to wash.[54] On land, conditions were scarcely better. Inside the *Bagno*, for example, the presence of goats and rams meant to feed the rowers only worsened the situation, as the animals' excrement increased the amount of dirt and potential for infection.[55] This might seem surprising, given that one of the reasons for constructing the *Bagno* compound was to improve sanitary

---

50 See, for example, the records of diseases among rowers hospitalized in Livorno in 1684 in ASF, MP, cc. 639, 644, 708,736,744, etc.
51 See ASF, MP, 2101, 2107, 2115, c.n.n.
52 Civale, *Guerrieri di Cristo*, p. 90 f.
53 Ramazzini, *De morbis artificum*.
54 Civale, *Guerrieri di Cristo*, p. 91.
55 ASF, MP, 2101, c.n.n.

conditions for the rowers, who otherwise were compelled to spend their days and nights crammed into the galleys, even when not sailing.[56]

Among the most widespread contagious diseases were those transmitted by insects—especially fleas and lice, with scabies being particularly prevalent. Bonomo and Cestoni had developed their theory about the etiology of scabies by observing galley slaves and convicts confined in the *Bagno*. Paradoxically, the cramped conditions endured by rowers provided a unique opportunity to develop a medical approach. It allowed physicians to conduct experiments on galley slaves, who, momentarily deprived of their humanity, were observed with what could be called a "clinical gaze."[57] This detailed observation of oarsmen formed part of a wider project championed by Redi and his circle to study contagious diseases, which were common in Livorno due to its role as a commercial and military port. More generally, it was an effort to expand scientific knowledge through first-hand experience and the new "experimental" method of enquiry.[58] The *Bagno*, like the lazarettos in Livorno, represented an exceptional setting for such work, as these were restricted spaces, isolated from the rest of society, where patients could be carefully examined. Although physicians in the lazaretto were strictly controlled by the city's health authorities—leaving them little room for experimentation—Bonomo enjoyed the necessary autonomy to try innovative remedies, supported by Romanelli, who was also part of Redi's scientific circle.[59]

Contrary to Galen's theory—which attributed scabies to a humoral imbalance, particularly due to melancholia—Bonomo argued that it was caused by tiny animals called *pellicelli* that lived under human skin. By closely observing the bodies of those afflicted with scabies, he discovered that the cause of their itching was erythematous vesicles. When one of the vesicles was squeezed and studied under a microscope, Bonomo found "a tiny white body" identified as a small insect—later recognized as a mite. Scabies' contagiousness was thus explained by the transfer of these mites from one person to another, and could even be spread through clothing. The remedy, he argued, was to eliminate these mites through various treatments such as washings, baths, the application of salts, sulfur, vitriol, mercury, and other substances. Ultimately, scabies was recognized as an "external disease," and any remedies dispensed were, therefore, not to

---

56 Bernardi, *Relazione*, p. 13.
57 This term is taken from Foucault's *Naissance de la clinique*. Without delving into the merits of the criticism of anachronism, I find that the term—coined to denote the dehumanizing medical separation between the patient's body and their identity—serves the concept well here.
58 Altieri Biagi/Basile, *Scienziati*, pp. 55–556.
59 Ciano, *La Sanità Marittima*, pp. 122–124.

be taken orally, as prescribed by Galenic medicine, but rather to be applied externally, directly to the skin.[60]

In June 1687, Bonomo shared his observations with Redi who, despite not being fully convinced by his protégé's discovery, nonetheless decided to publish the findings. The same year, the *Osservazioni intorno a' pellicelli del corpo umano* [Observations on the Hair of the Human Body] appeared in the form of a letter to Redi dated 18 July 1687, in which, however, the discovery of the mite as the cause of scabies and its effective treatment was relegated to a secondary note. Despite Redi's decision to downplay his protégé's hypothesis, the treatise was a model of the experimental "new science." As such, it circulated not only in Italy but also across Europe, aided by its Latin translation in 1692.[61] Far from being a mere academic exercise—often dismissed as incapable of making a meaningful contribution to naturalistic research—Bonomo's discovery had tangible practical applications. By 1717, Romanelli had instructed practitioners to apply sulfur boiled in oil to the skin of inmates with scabies. According to Romanelli's records, those who underwent the treatment recovered.[62]

2.2 Epidemics at sea

During an epidemic, medical vigilance was considered of the utmost importance, not only to ensure the mariners' wellbeing, but, above all, to protect public health. This was particularly true in times of plague. Since time immemorial, the sea has been the primary route through which epidemics entered new territories. The connection between contagion and the sea—or more specifically between contagion and maritime trade routes—had been widely accepted since late antiquity.

After the Justinian Plague of 541–762, the plague was thought to have vanished from the European continent until the Black Death struck in 1347. Tradition has it that this outbreak originated in Asia in 1346. While the exact location re-

---

60 Bonomo, *Osservazioni intorno a' pellicelli*, p. 3: "Trovammo con facilità il rognoso, ed interrogatolo, dove egli più acuto, e più grande provasse il prurito, ci additò moltissime piccole bolluzze, e non ancora marciose, le quali volgarmente son chiamate Bollicelle acquaiuole. Mi misi intono con la punta d'un sottilissimo spillo ad una di queste acquaiuole, e dopo averne fatta uscire, con lo spremerla, una certa acquerugiola, ebbi fortuna di cavarne un minutissimo globetto bianco, appena appena visibile, e questo globetto osservato col Microscopio, ravvisammo con certezza indubitata, che egli era u minutissimo Bacherozzolino […] Non ci fermammo a credere, ne ci contentammo di questa prima veduta, ma ne facemmo molte, e diverse altre esperienze in diversi corpi rognosi."
61 Altieri Biagi/Basile, *Scienziati*, p. 710 f.
62 ASF, MP, 2110, c.n.n.

mains uncertain, the hypothesis proposed by McNeill in 1976, which identifies the Khanate of Mongolia as the most likely source, is considered credible. From China, the disease spread along the caravan routes controlled by the Mongols and the Tartars, travelling north to the Caspian Sea coast, south to Azerbaijan, and west to the Black Sea. From there, it reached Europe, following the siege of the emporium of Kaffa by the Mongols of Kipchak Khan Janibeg. The Tartars' siege of the Genoese emporium is often cited as the earliest example of bacteriological warfare *ante litteram*. The Tartars, decimated by the plague themselves, were unable to continue this siege. Before surrendering, they made one final, desperate attempt to infect their enemies by catapulting the corpses of plague victims over the ramparts.[63]

Plague thus entered Europe by sea, introduced by Genoese merchant ships fleeing the Tatars. In the aftermath of this new and devastating epidemic—estimated to have reduced Europe's population by between 33% and 60%—plague remained endemic on the continent practically until the mid-18th century, returning cyclically to urban centers, primarily following merchants along trade routes and armies on their campaigns.[64]

The earliest measure taken when a city was declared "infected" was to isolate it geopolitically, leading to a ban or suspension of any kind of relations—primarily commercial—with neighboring cities.[65] At the same time, city gates were closed, lazarettos opened, and quarantines imposed on people, ships, and goods. Numerous epidemics erupted during the early modern period, which, according to the sources, were often caused by ships failing to comply with quarantine regulations or naval officials lying about their contacts with infected ports en route. For example, the plague outbreak that struck Marseille in 1649 was introduced by a ship from the Levant, which presented a forged certificate that falsely indicated that it had not come from an infected region.[66] Similarly, the plague outbreak that struck Naples in 1656 was likely caused by a ship from Sardinia, another vassal kingdom of the Spanish empire—probably a ship carrying troops destined for the Spanish territory of Milan, an area of conflict between Spain and France. Contemporary reports suggest the vessel had evaded quarantine requirements in the

---

[63] Harrison, *Contagion*, p. 2f.
[64] Alfani/Melegaro, *Pandemie d'Italia*, p. 12f.
[65] Cipolla, Crisi di mortalità, p. 198: "Bando si intendeva quando il blocco era decretato dopo che si era accertata la presenza della peste nella città o nel territorio bandito e perciò il bando poteva essere tolto solo dopo che si fosse accertato la fine dell'epidemia. Sospensione si intendeva quando il blocco era decretato solo sulla base di presunzioni o sospetti."
[66] Calvi, La peste napoletana, pp. 418–421.

port of Civitavecchia and had falsified the official documents certifying that it had undergone quarantine—known as "health licenses" [*bollettini di sanità*].[67]

In this context, doctors' vigilance in recognizing any signs or symptoms of plague, and reporting even the slightest suspicion of infection, was crucial to safeguarding public health. Seafaring ships and coastlines thus became the first line of defense, where vigilance and prompt reporting were essential.

As Bernardo Ramazzini wrote in 1700 in *De morbis artificum*, in the event of an epidemic nothing could be done for mariners due to the logistical impossibility of separating the healthy from the sick. As the proverb goes, they were "all in the same boat."[68] Ramazzini was not entirely correct, however, and there is evidence that that this problem could, at times, be overcome. Whenever a vessel was at sea and was suspected of having come into contact with an infected site, traditional remedies against contagion were employed. Medically, plague was believed to be caused by purely "natural" factors, in line with traditional humoral theory. For the Hippocratic-Galenic tradition, plague was considered an "epidemic"—a disease with "universal causes," particularly one linked to the air. As Hippocrates wrote: "When many men are stricken with a single disease at the same time, the cause must be imputed to that which is most common and which we all use first of all; and this is what we breathe."[69]

The prevailing belief was that plague arose from poisonous atoms exuding from decaying bodies or infected individuals. These atoms, once dispersed in the air, would then infect the atmosphere, rendering it "miasmatic" or poisonous. If inhaled, this bad air would cause a general corruption within the body, defined by Galen as "heat against nature," generating symptoms like the appearance of buboes. Plague was understood as a disease of heat and dampness, and in line with the principle of attraction, individuals naturally predisposed to these elements were considered the most vulnerable. This predisposition was referred to as *aptitudo patientis* [the patient's susceptibility].[70] Indeed, while not everyone struck by plague contracted the disease, it had been observed since antiquity that this disease—which theoretically should have affected both sexes and various social strata indiscriminately—often had a greater impact upon certain categories of the population. In an attempt to explain this, Galen hypothesized the presence

---

[67] Fusco, *La grande epidemia*, p. 1. On the *bollettini di sanità*, see Bamji, *Health Passes*.
[68] Ramazzini, *De morbis*, p. 388: "Sovente accade che qualche malattia Epidemica s'introduca nelle navi [...] In tal caso non vi è scampo alcuno trovandosi tutti, come dicesi per proverbio, nella medesima nave."
[69] Cited in Cosmacini, *Storia della medicina*, p. 20.
[70] Stevens Crawshaw, *Plague Hospitals*, pp. 27–29.

of λοιμοῦ σπέρματα [pestilential seeds] in the air, which did not affect every individual indiscriminately, but only those predisposed to them.[71]

While the miasmatic theory remained dominant throughout the learned medical world, it was eventually challenged in the 16th century by Girolamo Fracastoro's "contagionist" theory, which significantly influenced how subsequent plague epidemics were managed. Fracastoro (1476–1553), a Veronese physician who studied medicine in Padua, published *De contagionibus et contagiosis morbis et eorum curatione—Libri III* in Venice in 1546. According to Fracastoro, the nature of contagions did not lie in occult properties, as ancient authors believed, but rather in the consensus and dissensus of things [*consensus et dissensus rerum*], expressed as sympathy and antipathy.[72]

The principle of infection was thus linked to putrefaction, the dissolution of the material composition of the body due to heat and moisture from outside the body. The putrefying body would emit imperceptible particles, known as *seminaria* [seeds], which, when hot and moist, could cause decomposition upon contact with a second body. These *seminaria* were the agents of infection, but this process only occurred when the second body began to undergo decay after being infected by the first. However, the contagious process was not indiscriminate. For the *seminaria* to act on a second body, it had to share similar qualities with the body from which the contagion originated.[73] If the two bodies were analogous, contagion could occur through direct contact. If the bodies were not so, however, the object touched would not directly receive the infection, but rather retain the *seminaria*, which could later lead to an infectious outbreak known as *fomes*. If infection occurred at a distance—via the air—it happened because the inherent qualities of the *seminaria* allowed them to move through the air and survive for varying periods even far from the original source of infection.

Thus, sympathy—the principle of contagion—was understood as a purely physical process of transmission, operating according to the elemental qualities of bodies. Contagion could occur in three ways: through direct contact between an infected body and a healthy one; through an intermediary *fomes*—an object carrying the seeds of infection; or through transmission at a distance, as with plague. Fracastoro offered an alternative to the Galenic idea of a "patient's suscepti-

---

71 Pennuto, *Simpatia*, pp. 425–427.
72 Pennuto, La natura dei contagi, p. 57, Nutton, The Seeds of Disease, p. 22: he suggests that contagion resulted from a corruption in the substance of one body, which then passed to another body via the transmission of imperceptible particles.
73 Pennuto, *Simpatia*, p. 407f.

bility": the concept of the "attitude of the physical substance" [*aptitudo materiae*], which was grounded in sympathy.[74]

Thus, according to the official miasmatic theory, plague was believed to be caused by atoms that rendered the surrounding air "putrid and corrupt." These atoms were not only highly toxic, but also incredibly viscous. Plague was often referred to as "the sticky disease" [*male appiccicaticcio*], and because of this, anything a sick person touched was deemed "infected" and capable of transmitting the contagion. Not all objects were considered equally dangerous, however; some items, such as bedding and clothing, were, by their very nature, seen as more susceptible to becoming infected, while others, like metals, were not. When confronted with the presence of "infected" objects, two primary solutions were recommended: disinfection, known as "purging," typically carried out through fumigation, or destruction by fire.

Hippocrates recommended lighting fires in city squares at night during the winter, using *herbe calide* such as sage, rue, and rosemary, while in summer, herbs such as yellow sandalwood, roses, cardamom, and camphor were used.[75] It was also considered good practice to wash one's hands, wrists, and face with vinegar as a disinfectant, and to sprinkle clothes with aromatic herbs.[76]

Similar precautions were taken aboard the galleys. For example, during the plague outbreak of 1591–1592, the crew of the Tuscan galley *Santo Stefano*, who had just returned from Marseilles and were quarantining in Portoferraio,[77] were instructed to bathe their wrists once or twice with anti-infection oil and inhale vinegar fumes. In addition, they were ordered to burn rosemary and juniper and frequently wash the vessel and the crew's clothes with seawater.[78] The decision to have the ships take refuge in Portoferraio can be explained as an attempt to avoid any potential contact with Livorno, which was also infected. According to the miasma theory, the only effective remedy against plague was to stay as far away as possible from the infected area.[79]

Thus, whenever a galley with infected or suspected crews docked, it was quarantined like any other seafaring vessel. If anyone aboard was found to be infected, they were immediately locked up in the city's lazaretto. To illustrate this procedure, let us examine the quarantine of the Tuscan galleys in Portoferraio in 1679.

---

[74] Ibid., pp. 423–428.
[75] Benvenuto, *La peste*, p. 107.
[76] Ibid., pp. 97–126.
[77] Ciano, *La Sanità Marittima*, p. 28.
[78] ASF, MP, 2131, dossier 3, *Instrution a voi Pietro Dini di quello che havrete à fare nello sciorinare, et purgare la robba della presa delle Galere*, c.n.n.
[79] Bergdolt, The Discourse, p. 378.

As outlined by Livorno's *Ufficio di sanità*, the Grand Duke's galleys, which had just set sail from Civitavecchia, were instructed to stop at Portoferraio before returning to Livorno because they had recently taken aboard a number of slaves from the Barbary Coast, an area then suffering from a plague outbreak. During the quarantines, the galley physicians were relieved of their duties, and it was the lazaret doctors who looked after the crews. The galley physician remained responsible for reporting all details to the authorities in Livorno. According to his reports, plague victims were transferred to the city lazaretto, which was guarded throughout the quarantine period. At the same time, the galleys were searched to ensure that belongings of infected individuals—potential vectors of contagion—were removed and disinfected ashore. The rest of the crew were also placed under quarantine, but remained aboard the vessels, anchored at a safe distance from the shore. No one was allowed to disembark without the doctor's permission.[80] Similarly, on 15 July 1697, Livorno's health authorities ordered the quarantine of the Tuscan galleys in Portoferraio. During their sea voyage, the galleys had attacked two ships from Algiers, where plague was rampant. This precautionary quarantine was further justified because, a week earlier, a slave had died of what appeared to be plague.[81]

The case of Civitavecchia's galleys during the plague outbreaks of 1656–1657 is unique and remarkable for its efficiency in managing galley crews during such an epidemic. The wave of plague that hit the Italian peninsula in 1656 is believed to have originated in Sardinia, where the disease had already been raging since 1652. From Sardinia, it made its way to the Italian peninsula, first surfacing in Naples and Genoa.[82] According to the chronicles, the plague likely arrived in Sardinia via a ship from the Levant or the Berber Coast; it initially erupted in Alghero before spreading to Sassari and Oristano. The earliest reported plague deaths in Naples were recorded as early as March 1656, though the contagion was not immediately recognized as such. The cause of these "sudden deaths," which primarily affected the lower classes, was attributed to the poor quality of food consumed by the population during the previous Lent, particularly cheap salted cod. However, as the mortality rates rose over the following months, and symptoms such as boils and carbuncles—classic signs of bubonic plague—appeared, any remaining doubts were soon dispelled. On 12 May 1656, plague was officially declared in

---

80 ASL, Magistrato poi Dipartimento di Sanità (1606–1806), b. 71, ff. 389–394.
81 ASF, MP, 2128, c. n.n.
82 Benvenuto, *La peste*, p. 25.

Naples. It then spread throughout southern Italy, sparing only parts of Calabria and southern Puglia, particularly the area around Otranto.[83]

Rome was not unprepared: the *Ufficio di sanità* was immediately reactivated and, on 20 May 1656, an edict was issued prohibiting trade with Naples and its suburbs.[84] Multiple accounts exist regarding how plague arrived in the Papal States. As reported in Cardinal Sforza Pallavicino's memoir, the outbreak was likely caused either by Sardinian or Neapolitan vessels, as they had allegedly evaded controls and quarantine and continued to trade despite the bans.[85] Another version of events is provided by the *Commissario Generale alla Sanità*, Girolamo Gastaldi, who claimed that the contagion had entered the territories of the Papal States via a Roman galley that had docked in Naples without realizing that it was infected.[86] Both these reports converge on one key point: they identify the deaths from plague that occurred in late May 1656 in Civitavecchia as the first outbreak of epidemic in the region.

Regardless of how the plague actually arrived in Civitavecchia, one thing was certain: it came by sea. The first person to contract the disease was a soldier from the Papal galleys who, upon falling ill, was transported to the Hospital of San Giovanni di Dio, only to die five days later. In the typical fashion of the time, the disease was not immediately recognized, and the hospital doctor attributed his death to a combination of venereal disease and malignant fever. His opinion changed some days later, however, when the disease again struck one of the hospital's nurses, Angelo Ferrugio from Sicily, who had developed several boils in the groin area. Nevertheless, the authorities sought to downplay the gravity of the situation in order to prevent public alarm, presenting it as an isolated incident. During the early stages of plague epidemics, physicians usually sought to minimize the issue, fearing that hasty judgments could trigger unnecessary disorder. It was widely believed—not only by medical professionals but also by learned people in general—that fear of the plague could be as dangerous as the disease itself, if not more so.[87] This perspective had its roots in Aristotle, who argued that the imagination's great suggestive power could influence reality, and in Avicenna's belief that fear itself attracted the plague in the first place. While certainly an extreme case, the episode of citizens murdering a doctor who confirmed the presence of the contagion in Busto Arsizio in 1630 is undoubtedly significant.[88] Eight

---

83 Corradi, *Annali delle epidemie*, p. 184.
84 Gastaldi, *Tractatus politico-legalis*, p. 271 f.
85 Pallavicino, *Descrizione del contagio*, p. 1.
86 Gastaldi, *Tractatus politico-legalis*, p. 25 f.
87 Cipolla, *Il pestifero e contagioso morbo*, p. 113.
88 Ibid.

days later, a lay friar named Stanislaus the German fell ill. He presented similar symptoms to the aforementioned nurse: boils under his left armpit and on his right groin; he died two days later. Realizing the gravity of the situation, the authorities sent a nuncio to Rome from where, on 29 May, an edict was issued banning all trade with Civitavecchia and mandating the disinfection of the galleys and the adjacent coastlines.[89]

A report compiled by one of the galley surgeons of the time, Francesco Casella, provided information on how the penal fleet in Civitavecchia was managed during the epidemic. Written in the form of a letter addressed to the city's protophysician, Giovan Battista Bindi, it is contained in Bindi's *Loemographia Centumcellensis* (1658)—a treatise on plague. A widely established literary genre from the late 15th century onward, the aim of these treatises was to offer patients remedies for achieving and maintaining bodily integrity and good health, usually by prescribing a particular diet and promoting a healthy lifestyle.

Medical treatises on the plague are typically divided into five or six sections: a dedication or preface; an overview of the hypothetical causes of the disease; a section devoted to the signs of the plague (both in terms of symptoms and precursors); a section describing the illness; and a concluding part explaining the methods of purification.[90] In addition to this standard structure found in such texts, Bindi's treatise presents a reconstruction of the plague epidemic in Civitavecchia, where the need to prevent the galley crews from falling ill was considered paramount. The governance of the ships during the epidemic is detailed in Francesco Casella's letter, which stated that after the initial reports of contagion, the galleys sailed away from the coast, maintaining a safe distance in an effort to avoid infection. Isolation proved to be a highly effective strategy; among the five galleys in the Papal squadron, four remained unaffected by the contagion.

The *Capitana*, by contrast, experienced a much higher death toll: within a three-week period 80 cases of plague were reported aboard, 52 of which were fatal. As reported by Casella, the outbreak on the *Capitana* was caused by two *buonavoglia* who had served as undertakers to bury Stanislaus' corpse before setting sail. Despite clear instructions not to touch anything, the two men, tempted by greed, supposedly stole the deceased's personal belongings and sold these items to a fellow shipmate, who hid a belt and a hat aboard the vessel, thus inadvertently allowing the contagion to spread. Faced with the ever-increasing number of deaths among the Papal crews, they resorted to a strategy that was often imprac-

---

89 Gastaldi, *Tractatus politico-legalis*, p. 25 f.
90 Jones, The Plague, p. 102.

tical to implement in the galleys—separating the healthy from the infected. The *Capitana* returned to dock and the infected rowers were taken ashore and hospitalized in a storeroom—which later became the dock's infirmary. The healthy remained aboard, while those convalescing were isolated on a *galea polmonare*—a hospital galley named *Santa Caterina*.

The remaining four non-contaminated galleys continued their sea-passage, keeping as far away as possible from the infected regions. In mid-September, once signs of infection had subsided, the warehouse was vacated. The convalescents were placed on the *Santa Caterina*, while the rest were accommodated ashore. The *Capitana* was then thoroughly scrubbed and disinfected with fire, sulfur, and lime, and repainted in a bid to remove any traces of infection. The convalescents were then taken ashore and compelled to wash naked in hot vinegar—considered one of the most effective remedies against the plague.[91] A similar account is found in the letters to the Roman Curia written by the lieutenant Stefano Lomellini, who personally intervened to contain the contagion on the penal fleet. It is noteworthy that, while Casella's letter extols his own achievements and those of his medical colleagues, Lomellini's correspondence paints quite a different picture. It describes the reluctance of galley doctors to treat those infected, fearing that they would contract the disease. Lomellini notes how the naval authorities had to constantly monitor medical staff to ensure they provided proper care for the sick rowers.[92]

2.2.1 Floating hospitals: the *galee polmonari* during the 1656 plague epidemic

What, then, was this mysterious *galea polmonare* that played such a crucial role in managing the 1656 epidemic? The cramped conditions aboard the galleys were not only conducive to causing illness, but also symptomatic of a poor medical infrastructure. There was no designated space for treating the sick, making it difficult to provide effective healthcare. Additionally, the vessels often lacked the medicines and utensils needed to treat the sick, with the apothecary being located only aboard the *Capitana*. As noted in Grimaldi's report, the lack of medical staff and tools posed significant problems for the other galleys, as, in emergencies

---

[91] Bindi, *Loemographiae Centumcellensis*, pp. 12–16.
[92] AAV, Segr. Stato, Particolari, b. 34, ff. 254, 269, 271, 304, 327, 398.

such as sea battles or bad weather, the medical team could not move easily from one vessel to the next to obtain medicines or access surgical instruments.[93]

To address this pressing issue, the naval authorities of the Holy League, in the aftermath of the Battle of Lepanto, proposed to the Pope the idea of converting a galley into a floating hospital. These *galee polmonari* [hospital-galleys] were to be unarmed, lacking both oars and weapons. Two planks would be laid across the galley from stern to bow, with only a narrow passageway left open. Mattresses were placed between the planks to accommodate up to 400 sick men. The galley was to be fully equipped with everything necessary to care for and feed the sick: an apothecary, a storeroom, and a kitchen. Its medical staff would consist of a physician, a surgeon, and someone to prepare the required medicines. Additionally, ten *buonavoglia* could be deployed as servants to provide for the galley's needs, such as water, wood, and so on. A crew of forty was deemed sufficient to operate this hospital galley, and they could be recruited from among the oarsmen, without burdening the navy with additional costs.

A galley was chosen over a regular ship for several key reasons. Economic considerations were foremost. A ship was simply more expensive to maintain. The risk of infection also played a decisive role. On a ship, the sick would have to be accommodated below deck, in enclosed and likely cramped spaces, heightening the risk of contagion. Additionally, logistical factors favored the galley. Unlike a ship, which would have struggled to keep up with the fleet, a galley could easily navigate along the coasts or function without wind.

The use of these galley hospitals proved to be invaluable to the army. By providing a designated space for caring for the sick, they allowed for more effective treatment and helped prevent the spread of illness among the troops. Furthermore, ensuring an efficient medical system aboard was seen as a way to boost morale and motivate the soldiers to continue fighting—or, at least, that was how the authorities viewed it.[94] The term *galea polmonare* or *pulmonare* was derived from

---

[93] AAV, Misc. Arm. II, b. 110, *Secondo Avvertimenti sopra i disordini delle galere di S. Santità occorsi nell'anno passato 1571 dati da certe religiose persone et da bene con i remedij necessari et oportuni per emendargli*, ff. 385–386.

[94] AAV, Misc. Arm. II, b. 110, *Secondo Avvertimenti sopra i disordini delle galere di S. Santità occorsi nell'anno passato 1571 dati da certe religiose persone et da bene con i remedij necessari et oportuni per emendargli; Rimedio facile et utile à sani et infermi e di poca spesa pare saria questo*, ff. 387–389.

the Italian word *polmoni* or *pulmoni* used to refer to those unfit for labor, thus aptly naming the galley designed to take care of the sick and the infirm.[95]

The project was not realized in 1572, likely due to difficulties encountered in arming a new naval squadron. The pontiff thus opted for a more traditional solution, financing the Venetian initiative to construct a hospital for each member of the League in Corfu.[96] However, some *galee polmonari* were established at a later stage, as confirmed by the case analyzed above, as well as by contemporary naval treaties, such as the *Armata Navale* by the naval captain Pantero Pantera. The pages the captain devotes to these galleys are noteworthy, as these vessels' hybrid nature—part hospital and part prison—is clearly evident. As the captain observes, these galleys required both trained medical staff and guards "not only to help them [the galley rowers] with their infirmities, but also to prevent their escape."[97] Despite the undeniable utility of the *galee polmonari*, I could not find any traces of their systematic use after 1660, the year the Hospital of Santa Barbara was opened in Civitavecchia's harbor dock. It is likely that this type of galley was no longer used because, by the late 17th century, all major ports had hospitals where seafaring crews could be treated. In any case, the absence of specific documents on the subject does not rule out the hypothesis that a galley could still have been used as a hospital in the event of an emergency.

Throughout the early modern period, the systematic presence of the hospital galley was only observed in the Papal States. However, it was not an early modern Roman invention, as evidence exists of the presence of special hospital ships dating back to the Peloponnesian War, during which a trireme called *Therapia* operated in the Athenian fleet. Similarly, in Roman naval squadrons during the imperial era, ships called *Aesculapius* or *Asclepius*, evidently intended for medical purposes, are often mentioned.[98] As for the Tuscan case, I found only two references to the practice—the first, dated October 1556, describes how, during a sea voyage to Naples, it was decided to cast off hundreds of sick rowers onto an unarmed galley in the port of Genoa. Judging by the polemic debate it provoked among naval officers about its effectiveness, this practice was clearly not the

---

95 Guglielmotti, *La guerra dei pirati*, Volume Secondo, p. 149: "Pulmonara è la galera che serve per infermeria: ed è detta così, come si dicono pulmoni gli uomini inetti alla fatica: perché è galera dimessa e poco atta alla navigazione."
96 Civale, *Guerrieri di Cristo*, p. 96 f. AAV, Brevi, Registri, b. 20, f. 320r–v, *Breve per la costruzione di un grande ospedale a Corfù per tutti i membri della Lega, Roma, 12 febbraio 1572*.
97 Pantera, *L'armata navale*, p. 111 : "sarà necessario deputare alla cura loro persone, che non solamente gl'aiutino nelle infermità, ma gli custodiscano ancora, accòche non possano fuggire."
98 Aymard, Chiourmes et galères, p. 73 f.

norm.⁹⁹ According to Aurelio Scetti's diary, we learn that, in 1575, the author was aboard a "galley for the sick crew" anchored in the Arno in Pisa, possibly near the dock of Porta a Mare.¹⁰⁰ This appears to contradict the regulations, which explicitly stated that unarmed galleys should not be used as hospitals. However, we can infer that these regulations referred to a later period.¹⁰¹ For less serious infirmities, sick oarsmen had to be treated aboard the galleys while at sea, where they could rely on an experienced surgeon.¹⁰² Salvatore Bono also recalls that the practice of using an unarmed galley as a hospital was attested even in Genoa in 1559.¹⁰³

2.3 Dock hospitals

The lack of space to care for rowers conflicted with the need to ensure their good health. For this reason, in the 17th century, hospitals were constructed in the docks of both Civitavecchia and Livorno, designated exclusively for treating slaves and convicts. These medical facilities served a dual purpose: providing effective medical care while ensuring that rowers did not escape under the pretext of illness. It is noteworthy that while every rower was guaranteed medical and hospital care, the cost for such treatment was borne by the rowers themselves,¹⁰⁴ leading to a situation where they accumulated debts they were unlikely to repay, at least in the short term. The establishment of medical facilities and the high level of care they provided not only had the merit of keeping the crew healthy, but also served to provide a place where they could be contained and controlled in the event of illness. Additionally, by fostering a state of indebtedness, these institutions ensured workforce continuity.¹⁰⁵

In Livorno, references to a dock hospital for treating galley oarsmen can be found in the reports and letters on galley management as early as the late 16th century.¹⁰⁶ No further details are available, nor do we know the specific type of hospital it was, or whether it was solely used to treat galley crews. Considering that

---

99 ASF, MP, 2078, c.n.n.
100 Monga, *Aurelio Scetti*, p. 152.
101 ASF, MP, 2082, c.n.n.
102 ASF, MP, 2100, c.n.n.
103 Bono, *Schiavi*, p. 218
104 Lo Basso, Condannati alle galere, p. 124. This is also supported by the receipts of expenses for each rower conserved in ASF, MP.
105 As mentioned in Chapter 2, in the case of debts to the navy, convicted rowers were not released at the end of their sentence, but were usually employed as *buonavoglia* aboard galleys until they fully paid off their debts.
106 ASF, MP, 2131, dossier 3, c.n.n.

the earliest medical facility at the *Bagno* compound was located outside the walls, we can assume that this hospital mentioned in the 16th century was its original nucleus. Perhaps it was a garrison hospital, or a warehouse repurposed as a medical unit, as happened in Civitavecchia during the 1656 epidemic. The first hypothesis seems more plausible. In any case, whether the same hospital or a different one, we may turn to the hospital designated for crews after the construction of the *Bagno* between 1598 and 1604, as recorded by Father Filippo Bernardi.

One of the primary reasons for constructing the *Bagno* was to ensure the hygiene and care of the convicts and slaves. Initially located outside the *Bagno*, Cosimo III relocated the medical facility to a site beyond the *Bagno*'s warehouses. Ultimately, in 1697, to address the growing religious "promiscuity" within the hospital, the Capuchins decided to build a special, similarly-sized medical unit to care for Muslim slaves, adjacent to the Immaculate Conception of Mary Hospital, which was designated for Christians. Both facilities were equipped with their own apothecary, along with all the essentials to treat the sick as efficiently as possible. The attending doctors were mandated to visit the sick every morning, ensuring they received the care, medicines, and provisions necessary for their recovery.[107] The Commissioner of the Galleys was responsible for overseeing that doctors, surgeons, and their assistants treated the sick with diligence and charity, and that the necessary medicines were always available and properly administered. Medical staff were also required to send the Commissioner of the Galleys a daily updated list of rowers admitted to the infirmary, detailing the nature of their illnesses and the treatments they were receiving. This directive to register all the sick, as well as all those entering and leaving the hospital, had a twofold purpose. First, therapeutically, to acquire a better understanding of the illnesses suffered and the treatments given. Second, for monitoring purposes, to ensure no one could escape unnoticed, as might happen if the exact number of patients in the facility were not known.[108]

There were also twelve medical attendants, all of whom were forced *galeotti*. The most experienced worked as nurses, while the others served as cooks or orderlies. They were required to stay in the medical unit at night, both to care for the sick and ensure they did not escape. While the hospitalized rowers were under constant supervision, cases of successful escapes were not unknown. Two or three attendants worked in the Turkish infirmary, where they were joined by one or two *vigilanti di Maria*, who were responsible for checking that no criminal or blasphemous behavior occurred and, if it did, for notifying the nurse or

---

[107] Bernardi, *Relazione*, p. 21.
[108] See the aforementioned lists of hospitalized rowers preserved in ASF, MP, 2099–2118.

medical warden so that appropriate action could be taken. Naturally, working as a servant in the hospital provided an improvement in the convicts' living conditions. If they failed to perform their duties properly, these privileges could be revoked and the offenders returned to the *Bagno*'s dormitories.[109]

While designed as a highly rationalized space, sanitary conditions were extremely poor even within the *Bagno*. This was particularly evident in the Muslim medical unit, as noted by Father Filippo of Florence. In contrast, he praised the cleanliness of the Christian hospital, which appears to have been spacious and airy, with sixty beds whose sheets were regularly changed.[110] Despite the large number of beds, conditions within the Christian hospital were far from ideal. Records reveal that the number of inpatients frequently ran into the hundreds, especially during the winter months. The fact that there was only one medical facility for forced rowers often led to overcrowding, as no one could be refused access if they had been prescribed treatment by a doctor. Despite the directive that slaves and convicts were to receive care indiscriminately, there were clearly practical differences in their treatment. The lesser zeal applied to the care of the Turks is further evidenced by the fact that, when space had to be made for foreign troops, it was precisely the Muslims' medical unit that had to be vacated. For instance, in 1736, imperial soldiers were housed in a wing of the Turkish hospital while the Turks themselves were temporarily accommodated in a room next to the hospital.[111]

This comparatively limited—albeit minimal—care for the slaves may seem surprising, given that the official directives discouraged mistreatment and prescribed charitable treatment. Yet the Grand Duke had a vested interest in ensuring the well-being of the Muslim captives, in line with the rationale of reciprocity that governed Mediterranean captivity. News of any abuse could result in the immediate retaliation against Christian captives held in the Levant.[112] Although Bono's theory of Mediterranean "reciprocity" remains compelling, the sources suggest that in practice, the logic was far from absolute: slaves were often neither released nor repatriated.

A medical facility to care for convicts and galley slaves in Civitavecchia was established relatively late. Although the idea of establishing a hospital to maintain healthy crews had been considered several times, the fear that galley slaves might view hospitalization as an opportunity to escape the drudgery of rowing had prevented the project from being implemented. Numerous cases were reported of

---

[109] Bernardi, *Relazione*, p. 42.
[110] Ibid., p. 21.
[111] ASF, MP, 2118, c.n.n.
[112] Fiume, *Schiavitù*, p. IX; Santus, *"Il turco"*, pp. 49–51.

convicts feigning illness or infirmity to obtain better treatment or to be excused from oar duty. Their complaints were frequently not taken seriously, as they often pretended to be crippled or ill. When they fell ill for real, they were often left untreated, and many died as a result.[113] The high mortality rate, exacerbated by the plague outbreak of 1656, made it necessary to address the issue. In 1660, the Hospital of Santa Barbara was opened, with the stated goal to "cure, assist, and control them in every possible way."[114]

To prevent fraud, no rower could be admitted to the hospital without prior evaluation by the galley physician and surgeon. Once the severity of the illness had been confirmed, the rower was registered on a special list of the sick. The chain that shackled him to the bench was removed, and another was placed on his feet to prevent escape. He was then taken to the infirmary for as long as necessary. Once in the infirmary, the chain was removed to avoid further injury. Armed guards were stationed at the entrance day and night to restrain patients from escaping.

As in Livorno in May 1684—when the Capuchins were appointed to oversee the hospital and care for the sick "with devotion and sacrifice"—Christians were housed in a separate sickbay from Muslims, even on separate floors.[115] This segregation aimed to avoid religious promiscuity, although the therapeutic needs of both groups were similar. For rowers sentenced for a fixed duration, their hospital stay had to be recorded to prevent it from being counted as part of their sentence—which otherwise would have been reduced. This measure was introduced to avoid fraud, with severe consequences, especially in cases involving slaves.[116]

Doctors were instructed to visit the sick daily, or multiple times in cases of serious illness, and ensure proper care and treatment. The Papal authorities insisted that physicians, rather than medical assistants such as barbers, look after the

---

[113] AAV, Misc. Arm. IV/V, b.54, f. 38: *Ordini per la nuova infermeria di S. Barbara nella Darsena di Civitavecchia:* "Quante volte s'é pensato, è parlato sopra al curare in terra l'infermità della Gente di catena delle Galere Pontificie, altrettante se n'è dismesso il pensiero, e discorso per le frequenti invenzioni, che ella è stata sempre mai solita usare à fine d'haver trattamento migliore, & esimersi dal Remo talmente; che altri apparire impediti de' membri, ò destituiti da forze sono col tempo riusciti stroppiati da vero, ò per altro inabili, altri per fingersi febbricitanti sono appresso ammalati effettivamente, e morti ancora."

[114] Ibid.

[115] Angioni, Leone, *La Guerra di Morea*, p. 19.

[116] AAV, Misc. Arm. IV/V, b.54, f. 38: *Ordini per la nuova infermeria di S. Barbara nella Darsena di Civitavecchia.*

sick, as the latter lacked proper training and the required medical skill set.[117] Additionally, assistant barbers were often forced rowers or slaves, raising concerns about their reliability.[118] The physician was assisted by a surgeon and a chaplain.[119] The five galley surgeons helped the physician during daily rounds, remaining in the infirmary for a week, even overnight, to assist the surgeon.

One of the key issues was how to feed the sick, as the authorities knew that the nutrition aboard galleys was neither healthy nor balanced, often contributing to various illnesses. This was particularly true of the water, which was frequently left to stagnate. Although it was impossible to provide fresh and plentiful food, especially meat, for the entire crew, such foodstuffs were still necessary for the recovery of the sick. Their diet consisted of fresh meat, eggs, and bread, as well as hot soups made with rice and noodles, since a well-balanced, rich diet was essential to stimulate the healing process.[120]

Despite the hospital's establishment, sources indicate that hygiene and sanitary conditions improved only slightly, remaining largely unchanged from those in Livorno—too many sick people in a space that, while larger than the galleys, was still cramped. At times, the infirmary was so overcrowded that many could not be admitted, leaving many to suffer in agony on the decks of ships. Writing about their missions, the Jesuits even described the galley hospital as "hell on earth."[121] This lack of space became an urgent problem whenever an epidemic struck. In 1716, during a suspected contagious diarrheal epidemic,[122] it became necessary to separate the healthy from the sick, who were housed in a warehouse adjacent to the hospital. As we learn from Governor of Civitavecchia Niccolò Maria Lercari's letters to Cardinal Paulucci, the convalescent were later sent back to the galleys, where they were served a soup with sheep's blood [*sangue ircino*] to help

---

[117] Ibid., f. 85; BCR, 34B13, *Raccolta di notizie e scritture diverse sopra le galere pontificie, armamento di vascelli, fatto dal papa Alessandro VII per soccorso di veneziani contro il turco, fortezza e porto di Civitavecchia*, f. 337f.
[118] ASR, Camerale III–Comuni, b. 846, *Capitoli per l'assento dello spedale delle galere*, f. n.n.
[119] ASVR, Atti della segreteria, b. 74, *Giurisditione dell'e.mo vicario sopra le galere pontificie ed, in Civitavecchia, sopra l'ospedale di S. Barbara e su di alcune chiese (1722—1773), Breve relazione di quel ch'è seguito nell'esercizio della giurisdizione spirituale dell'E.mo Vicario di Roma sopra le galere pontificie, l'Ospedale di S. Barbara ed alcune Chiese di Civita Vecchia*, ff. 57v, 68v–69r.
[120] Ibid., ff. 67r–68r.
[121] ARSI, Rom. 132.I, *Breve relatione della Missione fatta alle Galere Pontificie in Civitavecchia da quattro padri della compagnia di Gesù l'anno 1649*, ff. 248–249.
[122] It is possible that it was a typhus outbreak. The directive to find and dispose of all bad salami suggests an infectious disease like typhoid. However, the high number of hospitalizations and the fact that it coincided with the cold season suggest that another viral disease migh have been circulating alongside typhoid.

them regain their strength. According to Dioscorides, an ancient medical authority, consuming fried sheep's blood was an effective remedy for dysentery and abdominal pain.[123]

In addition to traditional medicinal remedies and pharmacopoeia, physicians also relied on the authority of prominent contemporary writers. As Lercari explicitly noted, the strategies for maintaining good health were grounded in the "physical-political reflections" proposed by Giovanni Maria Lancisi.[124] Serving as the first physician to Popes Innocent XI and Clement XI, Lancisi authored several works on how to contain and eliminate epidemic diseases, focusing on street cleaning, combating stagnant water, and separating the sick from the healthy. As the pope's personal physician, Lancisi held the highest medical authority of his time, and his influence was especially strong, particularly in the territories of the Papal States. While it is unclear which of Lancisi's works Lercari was referring to, it could have been either *De subitaneis mortibus*, published in 1707 after the epidemic that struck Rome in 1706, or *Dissertatio de Nativis*, published in 1711, which analyzed the issues of air quality after the catarrhal influenza outbreak that struck Rome in 1709[125]—or perhaps both. Regardless, the ongoing high number of sick people, which showed no signs of abating, made it necessary to repurpose the *Annona* warehouse—typically reserved for storing public grain—into a hospital.[126]

In 1716, Jesuits documenting their galley mission described hospital wards as filthy, desperate spaces.[127] This situation persisted well into the 1730s, when there was still a severe shortage of beds for sick rowers, forcing them to share the limited space. This overcrowding not only caused great discomfort, but it likely also facilitated the spread of diseases.[128] By 1770, the infirmary had not yet been expanded, and the solution to overcrowding was to quarter the sick in adjacent rooms. These additional sickbays, however, had to be outfitted with the full provisions required for proper treatment.[129]

---

123 Dioscoride, *Della materia medicinale*.
124 AAV, Segreteria Stato e Prelati, b. 129, f. 48
125 Donato, *Morti improvvise*, pp. 64–68.
126 AAV, Segreteria Stato e Prelati, b. 129, f. 71.
127 ARSI, Rom.138, *Breve relazione delle missioni fatte in Civitavecchia il 1716 per ordine di nostro signor papa Clemente XI. Missione alle galere pontificie*, ff. 61–63.
128 BCR, 34D18, *Memorie e scritture diverse appartenenti alle galere Pontificie e condannati nelle medesimo Principalmente in tempo del tesorierato di mons. Lorenzo Corsini poi cardinale, e papa col nome di Clemente XII*, f.33.
129 ASR, Serie Bandi, b. 100 (1770, I semester), *Bando e Ordinazioni Pel buon regolamento delle Galere Pontificie nel tempo che sono in Darsena, e del loro Spedale, Ordini e provvedimenti per lo Spedale*, f. XIV.

2.4 Curing the body, purifying the soul

The coexistence of lay and religious professionals in the oversight of galleys could foster collaboration and tension. In general, whenever a chaplain intervened, it was often perceived as an intrusion into a sphere beyond his competence. In particular, serious disputes arose with the galley officers. As noted in a directive concerning the administration of the papal galley crews and land-based hospitals in the 18th century, chaplains were even accused of meddling in shipboard operations—punishing or ordering the punishment of criminals as though they themselves were naval commanders, while neglecting their assigned pastoral duties.[130] In contrast, within the context of Livorno, there is no explicit criticism of the chaplains' work, although several letters to the Grand Duke reveal a certain irritation toward Friar Ginepro's managerial choices regarding the inmates in the *Bagno*—at times deemed excessively harsh, at others overly lenient.[131]

During the early modern period, physical and spiritual care were closely intertwined, making collaboration between lay and religious personnel in hospitals inevitable.[132] Since their establishment in late antiquity and the early Middle Ages as places of refuge for travelers and pilgrims, hospitals had primarily been centers of religious devotion.[133] Within their walls, fluctuating dynamics of cooperation, competition, and conflict between medicine and religion are evident—both in the overlapping care of body and soul, and in the coexistence of lay and clerical medical and administrative staff.[134] Caring for the body and caring for the soul were seen as two sides of the same coin, even if the spiritual aspect usually took precedence over the physical aspect. Although the early modern period represented a moment of progressive secularization for the medical profession, its independence from religion was still a long way off.[135] One need only consider how certain areas of life remained off-limits for doctors, as evidenced by their obligation to yield to the priest when a prognosis of certain—or even presumed death—was given, so that the patient could receive Extreme Unction.

During the Counter-Reformation, religious oversight over medicine grew even stronger. Not only were medical staff expected to follow Catholic doctrine and

---

130 BCR, 34D18, *Memorie e scritture diverse appartenenti alle galere Pontificie e condannati nelle medesimo Principalmente in tempo del tesorierato di mons. Lorenzo Corsini poi cardinale, e papa col nome di Clemente XII*, f.33.
131 ASF, MP, 2107 etc.
132 See Henderson, *Renaissance Hospital*.
133 Nutton, *Late Antiquity and the Early Middle Ages*, pp. 77–79.
134 Donato, *Medicina e religione*, p. 27; Tomassetti, Dentro e fuori l'ospedale, pp. 118–123.
135 Ibid.

morality,[136] but hospitals themselves were conceived as spaces designated not only for physical recovery but also for the moral re-education of the sick. Illness was viewed as divine punishment, and the body could not be healed unless the soul was first purified.[137] Medicine itself was believed to be a gift from God, and doctors were seen merely as His instruments. In particular, St. Augustine promoted the image of *Christus medicus*—Christ as a divine physician concerned with healing humanity's spiritual illnesses. Unlike the human physician, the Almighty Physician was infallible[138]; thus, medical practitioners were expected to recognize their inherent limitations, especially in the face of death, where their power ended and they had no choice but to invoke divine help and defer to priestly authority.[139]

Pastoral care became even more urgent when it came to institutions dedicated to the care of slaves and convicts, who were highly immoral and often blasphemous individuals. The report on the disorder aboard the galleys during the Battle of Lepanto emphasized the urgent need to send eight or ten Capuchin friars to oversee the infirmary and to rely on a devout lay medical staff.[140] Once the Hospital of Santa Barbara was established, Pope Innocent XI entrusted both the management of the hospital and the sick's spiritual care to the Capuchins of the Roman Province with the Bull *Cum nos ad spiritualem militiam,* issued on 15 May 1684.[141] As a result, the hospital fell under the jurisdiction of the Vicar of Rome.[142] That it was entrusted to the Capuchins is unsurprising, given the Order's strong tradition in hospital care. The initiators of the Capuchins' Reform, Matteo da Bascio (1495–1552) and Ludovico di Fossombrone (c. 1490–1560), had distinguished themselves in caring for plague victims during the epidemic that struck Camerino in 1523. This influence was so strong that the earliest testimonies regarding the Capuchins outside the Marche region were recorded within the *Ospedali degli incurabili* [hospitals for the incurables] in Rome (1529), Naples, and Genoa (1530). Finally, in the Order's Constitutions of 1535–1536, it was stated

---

136 Pastore, Errori e peccati, pp. 775–797.
137 Minois, *Il prete e il medico*, (translation) p. 22 f.
138 Henderson, *Renaissance Hospital*, pp. 113–117.
139 See, for example, Zerbi, *De cautelis medicorum*, p. 61 f.
140 AAV, Misc. Arm. II, b. 110, *Secondo Avvertimenti sopra i disordini delle galere di S. Santità occorsi nell'anno passato 1571 dati da certe religiose persone et da bene con i remedij necessari et oportuni per emendargli, Rimedio facile et utile à sani et infermi e di poca spesa pare saria questo,* ff. 387—388.
141 Calisse, *Civitavecchia*, p. 412.
142 ASVR, *Raccolta di notizie di vario genere sui Diritti, giurisdizione e prerogative del vicariato di Roma, 1650—1740,* tomo 55, f. 652.

that the Capuchins must be at the service of the sick, even if this meant risking their lives, particularly during outbreaks of plague.[143]

By order of the Vicar of Rome, pastoral care was provided in Civitavecchia once the sick were admitted to the Hospital of Santa Barbara, where they had to confess prior to receiving treatment. The chaplains visited them daily—in the morning for Communion and in the evening for Confession. In addition to these regular sacraments, a chaplain always slept in the hospital in case of an emergency. This ensured that the sacraments could be administered even at night, if necessary.[144] There was also a chapel within the hospital, taking up the entire ground floor of the building, as can be seen from the map preserved in the archives of the Vicariate of Rome.[145]

Spiritual care was also provided in the Turkish infirmary by a Turkish minister called the *Papasso*, a name typically used to refer to mosque workers. The hospital's religious staff vehemently opposed the presence of this figure, as he was perceived as an obstacle to converting slaves to Christianity. However, based on the logic of reciprocity that characterized Mediterranean captivity, they were also aware that preventing "Turks" from practicing their religion would have meant that Christian slaves in the East would suffer a similar fate. Thus, the *Papasso* was tolerated.[146] Despite the great attention paid to the pastoral care of convicts and slaves, both on land and at sea, reports indicate negligence by both doctors and chaplains. In particular, the sacraments were often not administered, especially at the time of death. This occurred, for example, with Francesco Loggietti, a galley convict on the *San Benedetto*. Delirious from his illness, he was brought to the hospital for treatment. After being diagnosed as insane, he was refused admission, as no medical treatment could help him. He was then taken back to the galley, where he died without the assistance of a priest and without receiving Extreme Unction.[147] It is likely that such poor pastoral and medical care stemmed from the sheer volume of sick individuals and the scarcity of those able to tend to them.

---

143 Criscuolo, *I Cappuccini*, p. 71.
144 ASVR, Atti della segreteria, b. 74, *Giurisditione dell'e.mo vicario sopra le galere pontificie ed, in Civitavecchia, sopra l'ospedale di S. Barbara e su di alcune chiese (1722–1773)*, Breve relazione di quel ch'è seguito nell'esercizio della giurisdizione spirituale dell'E.mo Vicario di Roma sopra le galere pontificie, l'Ospedale di S. Barbara ed alcune Chiese di Civita Vecchia, f. 68r.
145 Ibid., f. 76v–r.
146 Ibid., ff. 57v–58r.
147 Ibid., f. 178.

With regard to ministry to the sick in the *Bagno's* medical unit, records refer to a relatively late period, as the arrival of the Capuchins marked a significant development. In 1697, the hospital for galley crews was divided into two sections: one for Christians and one for Muslims. The reason for this separation was not medical: it was religious. As Father Bernardi indicates, it was blasphemous to Father Ginepro that "a priest should commend the soul of a dying Christian while, at the same time, blasphemous words from the Koran were being pronounced by a Turkish minister over a dying Muslim next to him."[148] Consequently, if a slave converted to Christianity, he was likely to be sent to the Christian hospital. However, if he disrespected the Christian religion, he was immediately transferred to the Turkish medical unit, as occurred in 1699, when a baptized "Turk" admitted to the Christian hospital was heard blaspheming the Pope.[149]

The following anecdote illustrates religion's key role in the spatial organization of medical care and the importance of providing space for Muslims to practice their faith unhindered. Apparently, in 1698, a scene depicting the Grand Duke's army was painted on the entrance door to the Turkish hospital, and a crucifix was also placed there. This fresco must have caused quite a stir. The Grand Duke was asked not only to remove the cross to avoid offending the "Turks" in the Barbary States, but also to officially state that the cross was not meant to mock the slaves who had been hospitalized.[150]

As in Civitavecchia, chaplains were specially appointed for pastoral care in the *Bagno* and, by extension, the hospital. Three in total, they lived in the hospice next to the hospital. The Muslims also had a religious minister, known in Tuscany as the *Coggia*, who was responsible for comforting sick slaves and assisting them at the moment of death.[151] In Livorno, the chapel in the *Bagno* was not adjacent to the hospital and, since it was either impossible or extremely difficult to move patients, Mass was celebrated both in the chapel and in the hospital.[152] For the same reason, the oil used in Extreme Unction was kept in a shrine above a tabernacle in the hospital, next to the convicts' beds, so that it would be readily available when needed. This sacred oil remained there for a long time, until, in 1749, a request was made to transfer it to the chapel. Keeping it within reach of the convicts

---

148 Bernardi, *Relazione*, p. 21: "un abuso del tutto medesimo tempo, un Sacerdote raccomandasse l'anima di un moribondo Cristiano, ed all'altra parte il Ministro Turco borbottasse tra denti parole scomunicate dall'Alcorano per aiuto di qualche Maomettiano agonizzante."
149 ASF, MP, 2104, c.n.n.
150 Ibid.
151 Bernardi, *Relazione*, p. 22.
152 Ibid., pp. 42–43.

seemed disrespectful, as they were criminals who could easily be blasphemous, or might even dare to steal it.[153]

2.4.1 The exorcism of Volumnio Maria Merucci in Livorno

The complex relationship of part-cooperation and part-competition between medicine and religion is most evident in supernatural cases such as demonic afflictions and exorcisms. Cases of demonic possession among galley rowers were not common, or at least, not many records have survived. However, in the correspondence of Romanello Romanelli, Vittorio Vitolini, Scribe of the *Bagno*, and Giuseppe Prini, Minister of the *Bagno*, several letters concerning one particular case of exorcism have been preserved. Why it was kept is unclear. Perhaps it was due to the length of the exorcism, which lasted almost a year; or because of the involvement of prominent figures, such as the Archbishop of Pisa. Alternatively, and perhaps more likely, it was because the authorities mistrusted the credibility of the affliction, which necessitated increased vigilance in managing the situation.[154]

On 11 May 1712, Dr. Romanelli sent a letter to Prini informing him that a Sienese convict, Volumnio Maria Merucci, who had been hospitalized at the Christian Hospital for tuberculosis, was suspected of being demonically possessed.[155] The suspicion, the doctor wrote, arose from various reported signs, considered by the hospital's chaplains to be "obvious" indications of possession. While the specific signs remain unclear, it is recorded that although Merucci was physically weakened, he showed no signs of spiritual submission. He fiercely resisted and refused the chaplains' exorcisms. It appears that the Capuchins were only temporarily able to confine the evil spirit from relocating from the body's upper organ—perhaps the heart or the brain—to its lower regions such as the feet. During these brief moments of freedom from the demonic spirit, the captive experienced some relief and regained a limited ability to move or respond. He apparently declared to the chaplains that he had entered into a pact with the devil, selling both his body and soul until his death. The exorcisms were most likely conducted at the behest of the Capuchins, who would have recognized Merucci's signs of demonic possession during their daily visits to the sick.

---

153 ASF, MP, 2132, dossier 5, c.n.n.
154 All the letters by Vitolini, Prini, and Romanelli analyzed here are conserved in ASF, MP, 2108, cc. n.n.
155 See *"possessione demoniaca"* in Lavenia, *Dizionario Inquisizione*, pp. 549–554.

The question arises as to whether Romanelli's opinion was sought on the matter. While Romanelli provides a direct account, he appears to have been a passive observer, recording what he had witnessed and reporting it to his superiors, as expected of someone in his position. He did mention that, by nature, he was not inclined to believe in diabolic possessions, but he seemed to accept the Capuchins' judgment without much questioning, as they were convinced that the possession was genuine. The Capuchins continued to exorcise Merucci every morning and night.[156]

The *Bagno*'s authorities, however, were less certain and took precautionary measures to prevent the patient from escaping. Alongside preparations for the exorcism, an inquisitorial style investigation was launched to gather information about Merucci's past, character, and any potential motives for faking his condition to avoid his sentence at the oars. In fact, the chaplain at the Stinche prison, where Merucci had been held before his transfer to Livorno, confirmed that the convict had a reputation for swindling, and described the "very bad qualities of such a man, including deceiving others."[157] He also revealed the prisoner had previously sought to have his galley sentence commuted. His questionable nature led some to believe that he might have succumbed to the devil's temptations. The chaplain advised that exorcism should be delayed until the prisoner had been properly converted, confessed, and received communion, noting that "exorcisms have little effect on those with a guilty conscience who fail to perform the good and holy acts prescribed by the authors for the liberation of spirits."[158]

Despite these concerns, the initial exorcisms apparently had some effect, with Merucci appearing to be freed from the demon. Five days later, however, Romanelli informed Prini that the demon had once again possessed him. The second possession was preceded by what initially seemed to be a series of accidents but, in the eyes of the Capuchins, constituted unmistakable signs of demonic presence. Upon arriving at the hospital, the monks discovered not only an apple tree, which they had planted in a pot, smashed on the floor, but also a cage containing a live goldfinch—its wings and tail missing.

As in his first letter, Romanelli does not explicitly articulate his opinions, though a note of skepticism can be detected, especially in light of the aforementioned omens, which, as he observed, could have been caused haphazardly by

---

[156] ASF, MP, 2108, cc.n.n, letter by Romanelli to Prini: "Io sono di natura di non credere cosi facciano queste cose, ma essendovi puntualmente mattina e sera, mi convince crederlo tale."
[157] Ibid., letter by the Stinche's chaplain to Prini: "delle pessime qualità di un tal huomo, tra le quali una del rigirare e aggirare le genti."
[158] Ivi: "Gl'esorcismi poco vagliono in quegli che hanno cattiva coscienza e non esercitano quegli atti buoni e santi richiesti dagli autori per la liberazione de i spiriti."

the wind, or even a cat. In one revealing sentence, which cautiously asserts a sense of professional caution, he tentatively takes a stance by arguing that Merucci's apparent demonic possession was at least partly faked, and at least partly due to the devil's intervention.[159] Romanelli, therefore, holds belief and doubt in suspension, demonstrating that credulity and skepticism are not necessarily incompatible, and that fiction and truth are not merely opposites. Conversely, throughout history, there have been frequent oscillations between these two positions, such as the outright rejection of the miraculous, on one hand, and the belief that it can never be called into question, on the other.[160] After all, to entertain doubt surrounding demonic affliction did not equate with denying the devil's existence, and like any other doubt, it could always be dispelled. In any case, given the involvement of the supernatural and the supposed possession that took place in the hospital, Romanelli took a step back, declaring that the case would have been better entrusted to the Inquisition.[161] Days passed, and while the Capuchins became increasingly convinced of the devil's presence, Prini remained equally skeptical and, so as not to be deceived, had Merucci wear the ring and iron sock he had removed upon admission to the hospital. The ring and the iron sock were the most common forms of punishment inside the *Bagno* compound. The ring [*anello tondo*] was the chain attached to the foot, while the iron sock [*calzetta di ferro*] could be described as a type of shin guard made of iron, with a varying number of meshes, generally between twelve and eighteen, depending on the prisoner's constitution. It was placed on the calf under the ring to limit the physical damage caused by the chain.

The use of chains as a punishment is unsurprising, given that oarsmen were usually not chained inside the *Bagno*. This penalty was even harsher when applied to slaves, as it meant depriving them of their privilege to step outside the *Bagno* to carry out their commercial activities in the city. As Romanelli noted in another letter, the exorcist—the parish priest from the church of Santa Maria del Giudice in Pisa—arrived at the *Bagno* on 9 June. The exorcism took place in front of the altar in the hospital, with great theatricality: the spirit inside Merucci began to defy the priest's incantations, causing the convict to fall to the ground, groaning like an animal. Within half an hour, he was reportedly free of the demon. The hospital chaplains claimed, however, that the exorcism had been so effective due to

---

[159] Ibid., letter by Romanelli to Prini: "Quanto a me crederei ci fosse del briccone, molto del furbo, e parte del demonio."
[160] For more this subject, consult Veyne, *Les grecs ont-ils cru à leur mythes?*
[161] ASF, MP, 2108, c.n.n, letter by Romanelli to Prini: "e dalla Santa Inquisizione potrebbesi credo io assicurarsene meglio."

their preliminary work, which had led Merucci to confess twice and receive communion three times.

As Prini wrote, the exorcism's successful outcome was interpreted as confirmation of its veracity. Further validation came in the form of some "obvious signs" which, it was claimed, had gone unnoticed—such as a scar on the convict's left arm from which blood had allegedly been drawn to seal a pact with the devil. It is difficult to believe that these scars appeared suddenly, and even harder to accept that they had gone unnoticed, given that doctors were required to thoroughly inspect the patient's body during their evaluations. More likely, the medical staff had not been consulted in advance about such signs and scars, or the convict had acquired them later. Prini, too, admitted that this could be a genuine case of diabolic possession, yet suspicion lingered that Merucci was feigning affliction in his bid to avoid being sent back to sea. The most suspicious aspect, in Prini's view, was that the convict persistently requested alternative duties rather than a return to the galleys.[162]

Volumnio Merucci's exorcism appeared to be a closed chapter until February 1713, when the Archbishop of Pisa visited the *Bagno*. After celebrating Mass and distributing communion to nearly all the convicts, the Archbishop was taken to the sacristy to speak with Merucci, who—once again allegedly possessed—underwent two exhausting exorcisms, one lasting an hour and the other three and a half hours. Finally, on 3 March, Romanelli noted, Volumnio was officially and definitively freed—by chance, it seems—after being threatened with chaining and flogging.[163] Though Romanelli offered no commentary on the matter, it is reasonable to suspect that the convenient timing of this deliverance may have fueled lingering doubts about the authenticity of this possession. In any case, the affair closed on a positive note for all concerned: the chaplains could claim to have vanquished the devil, while Merucci was assigned the role of caretaker in the *Bagno's* main chapel, in the hope that proximity to a sacred space would protect him from future diabolic interference.

---

[162] Ibid., letter by Prini to Vitolini: "due però sono i miei sospetti, e forse gran cosa lontani uno che abbia pattuito con il demonio e che vi sia qualche promessa di cavarlo dove si ritrova, e l'altro sia per astenersi dalla navigazione, e quest'ultimo me lo fa credere perché ora va cercando di qualche impiego per levarsi d'ozio, e da potersi esercitare nella pena, o vero in qualche cosa altro pure che resti franco dal remo."

[163] Ibid., letter by Romanelli to Prini: "si crede liberato dopo averlo minacciato di catena e bastone."

2.4.2 The physician and the exorcist

Volumnio Maria Merucci's exorcism is noteworthy for our analysis of the relationship between medicine and religion because it illustrates how a doctor's opinion, which in theory should have been sought in hospitals, was not always deemed necessary. This was especially true when "spiritual diseases" were involved.

Diabolic possession—the presence of malignant forces within a body—was solely within the authority of the Church.[164] Nevertheless, inquisitorial practice was required to be extremely skeptical toward those claiming possession, often relying on medical consultation before determining its authenticity. In practice, with the affirmation of the doctrine of *discretio spirituum*—the ability to distinguish between someone possessed and someone ecstatic, or between the possessed and sainthood—recourse to medical advice became secondary, primarily serving to validate the theological diagnoses.[165]

The likelihood that certain diseases could be attributed to the devil had been discussed since ancient times: Galen claimed that Hippocrates rejected the possibility that some afflictions might have non-natural causes. The term "divine" was often understood not as supernatural but as celestial, implying that sudden changes in the air could affect temperatures and the humoral balance. Throughout the 15th and 16th centuries, this controversy gained renewed momentum, with the medical, theological, and legal worlds divided between those who accepted the possibility of the existence of "spiritual" diseases—and, by extension, medicine's limits—and those who categorically rejected the idea that diabolic forces could influence the natural world.

A notable example of supernatural causes of occult diseases and miracles is found in *De incantationibus* (1520) by the philosopher Pietro Pomponazzi. In this work, Pomponazzi attributes occult diseases and miracles to the power of the stars and the *vis imaginationis* [the suggestive power of the imagination]. Drawing on Galenic medicine, he argued that phenomena attributed to the devil were, in fact, natural effects caused by an imbalance of black bile. This imbalance, when it became melancholic, could affect individuals to the point of altering their imagination.[166]

Despite—or perhaps because of—the dissenting voices of many physicians, since the papacy of Pius V, the Church sought to regulate medicine, not only by

---

[164] Lavenia, I diavoli di Carpi, p. 94.
[165] Brambilla, *Corpi invasi*, p. 95.
[166] Lavenia, La medicina dei diavoli, pp. 163–169.

requiring physicians to report suspected heretics before treating them, but also by fostering the convergence of official medicine and exorcism.

One of the most influential treatises advocating for this convergence was Codronchi's *De Christiana ac tuta medendi ratione* (1591). Initially a skeptic, Codrinchi began to believe in sinister influences after one of his daughters became possessed. Even doctors who supported the belief in diabolic possessions did not advocate for excluding medicine in such cases, as the body of the possessed person was still viewed as a natural body. The theory of humoral imbalance was not rejected, but instead linked to the devil's involvement. Medicine's role remained crucial, as the physician had to determine the reality of demonic intervention.[167]

Even Zacchia, though fundamentally skeptical of the numerous possessions and miracles documented at the time, acknowledged their likelihood by mediating between natural and supernatural causes. According to the Roman physician, those possessed were individuals suffering from melancholia, a true *instrumentum diaboli* [devil's instrument] that enabled possession. Zacchia's skepticism was not about the existence of the devil and the possibility of curses, but rather about how to recognize them, criticizing the excessive credulity and superstition of the masses.[168] According to the Candido Brugnoli's definition in *Alexicacon, hoc est opus de Maleficiis et Morbis maleficis* (1668), the *maleficium* [curse] was defined as an "an evil operation on the body, carried out by diabolical power as a result of a tacit or explicit pact between man and the Devil, involving the use of natural substances and the individual's collaboration, allowing the expression of his inclination toward evil."[169] While this curse had physical consequences, in the case of demonic possession, the treatment had to begin with the spiritual: the soul needed saving before the body could be healed.

As demonstrated by the case of Merucci's possession, the hospital doctor emerges as a largely passive observer. While he recorded events as they unfolded, he offered no explicit opinion. Yet the issue is not whether Romanelli believed in the possession, but that his input was never solicited—precisely because this was an instance where spiritual purification was considered a prerequisite to physical healing. Ultimately, the episode reveals a broader institutional tension: even if the

---

[167] Ibid., pp. 180–182.
[168] Brambilla, *Corpi invasi*, p. 99 f.
[169] Biondi, *Tra corpo ed anima*, p. 407: "maleficio è operazione malvagia sul corpo, realizzata per potere diabolico, a seguito di un patto tacito o espresso intervenuto tra l'uomo e il diavolo, tramite l'applicazione di cose naturali e collaborazione dell'uomo stesso a sfogo della propria inclinazione al male."

doctor appeared to play a marginal role, his presence and documentation were essential to preserving the memory—and perhaps the legitimacy—of the event.

2.5 Unmasking those who feign infirmity

As indirectly evidenced by the numerous provisions instructing doctors and surgeons to exercise extreme caution in ensuring that only seriously ill rowers were exempt from duty and allowed to recuperate in the hospital, the practice of simulating diseases must have been quite common among galley crews. In March 1703, a letter arrived in Livorno from the *guardiano* of the crew working at Portoferraio, stating that six forced convicts had attempted to escape during the night by digging a tunnel beneath the dormitory floor. The leader of the escape, Agostino Bianchi from Rome, had feigned a severe toothache to remain in the dormitory, and while the others were out working, he dug a hole in the floor and prepared the escape. The six were ultimately caught and punished by having the ring placed on their feet with the iron sock, effectively shackling them as they would have been aboard the galleys.[170]

While feigning illness was prevalent in both the Tuscan and Papal fleets, more extensive legislation can be found regarding this issue for Civitavecchia's galleys. As evidenced by the edict for the construction of the Hospital of Santa Barbara, the project was deferred for as long as possible due to concerns that hospitalization would be used as an excuse to avoid serving in the galleys, resulting in an excessive number of inpatients, only a small proportion of whom were actually ill.[171] Once the situation became unmanageable for the opposite reason—too few patients and an alarmingly high mortality rate among oarsmen—it was finally decided to open the facility. However, access was only granted after the ship's doctor or surgeon had conducted a thorough examination and issued a medical authorization. Anyone who left the galleys without permission or on false pretenses was punished with a ten-year sentence in the galleys.[172]

Despite the fact that these orders and their associated punishments were renewed annually, the situation did not improve. Even in the late 18th century, action was still required in this regard. As late as 1770, it was still decreed that no convict could be unchained on account of being ill without first being seen by

---

170 ASF, MP, 2113, c.n.n.
171 AAV, Misc. Arm. IV/V, b. 54, f. 38.
172 Ibid., f. 85.

the doctor or surgeon assigned to the galley in question, and without a written report validating the illness and its specific characteristics. No rower could be allowed to enter the medical facility without such a certificate. Upon recovery, a discharge certificate was required and had to be submitted to the sailors on watch or the sub-guard, who were also required to inspect the recovered person to ensure that he had not stolen any items or food from the hospital.[173] Such directives were once again renewed in 1784 because, according to the Commissioner of the Galleys, rowers continued to exploit the status of "infirm" with extreme malice to avoid the drudgery of rowing and remain in the hospital for two or three extra days at the institution's expense.[174]

Even though all the edicts treat the simulation of disease as a matter of urgency, there are not many accounts of rowers feigning illness. This suggests that it was a common practice, and the authorities did not consider it important to record every single case. Likewise, it is plausible that most attempts to simulate illness occurred during the rowers' initial medical assessment, and therefore did not interfere with naval operations. More common, however, than feigned illness, are the documented cases of another form of deception: disguise.

In 1714, for example, reports surfaced of two escape attempts by a convict and a slave, who were found by the guards in possession of clothes, hats, and wigs. Particularly interesting is the case of the slave Ametto di Alì, sentenced to serve in the house of a Jew, who was caught stealing jewelry, gold, and silver worth 250 *pezze*. The slave had made a deal with two German soldiers who, in exchange for the loot, promised to help him escape and gave him a skirt and a woman's girdle to wear so that he would not be noticed. Unfortunately for them, all three were discovered, and the slave was sentenced to the gallows.[175]

Feigning illness was a pressing issue for early modern medicine and jurisprudence. In particular, the authors of the *Methodus Testificandi* noted how often convicted or suspected criminals simulated illnesses in court to avoid punishment. Aboard galleys, given the danger and harshness of sea voyages, there was a constant fear that rowers might feign illness to escape an uncertain and painful fate. In this regard, Chapter IX of the fourth book of the *Medicus-Politicus* by the Portuguese physician Rodrigo de Castro is noteworthy.[176] This chapter is entirely dedi-

---

[173] ASR, Serie Bandi, b. 100 (1770, I semestre), *Bando e Ordinazioni Pel buon regolamento delle Galere Pontificie nel tempo che sono in Darsena, e del loro Spedale, Ordini e provvedimenti per lo Spedale*, f. XII.
[174] ASR, Camerale III—Comuni, Civitavecchia, b.846, *Editto concernente il governo delle galere pontificie*, 1784, f. VIII; *Capitoli per l'assento dell'ospedale delle galere*, f. n.n.
[175] ASF, MP, 2109, cc. n.n.
[176] De Castro, *Medicus Politicus*, p. 251.

cated to the theme of simulating diseases and describes how, in 1588, de Castro himself was instructed to visit several Portuguese sailors accused of feigning illness to avoid the order to sail with the Spanish Armada and fight the English fleet. These sailors' desperation was so great that some of them were even willing to injure themselves in a manner that involved bloodshed to make their deception more credible.[177]

The naval captain Pantera, in his writings on galley management, also advises his readers to remain alert to the possibility that sick convicts were feigning illnesses to avoid active duty—or worse, escape. He recalls how many convicts were vagrants, condemned to serve in the galleys because they were accused of simulating diseases to shirk work. In particular, physicians were instructed to be extremely vigilant in this regard, for they were the only ones who, thanks to their knowledge, could ascertain whether an illness was real or not.[178]

3 The rower's physical examination

The galley doctor's responsibilities extended beyond treatment; he also had an oversight role, being tasked with examining the physical condition of the rowers to assess their fitness for duty. Rowing was an arduous activity, demanding good health prior to embarkation and throughout the voyage. Ultimately, only a physician could determine whether someone was healthy enough, and, therefore, capable of manning the oars.

A medical assessment was requested at the time of conviction to determine the condemned person's suitability for galley service. This involved a thorough physical examination—including observation, auscultation, and palpation—typically carried out by surgeons, who were, at least in theory, the sole medical practitioners permitted to perform hands-on procedures. For convicts, this moment was pivotal: if declared unfit, an alternative punishment had to be arranged. In such cases—due to missing limbs, lameness, blindness, or old age—the sentence would specify a substitute for galley service: imprisonment, forced labor, banishment from the city or for the gravest crimes, execution.

In general, a sentence was commuted to imprisonment for a period considered equivalent to that spent at sea, or to forced labor either aboard the galleys,

---

[177] Ibid.: "Anno 1588. Quo in gens illa classis ad versus Angliam Olisippone parabatur, naturae & milites plerique data opera lecto decumbentes testimonium ex me petiere, ut à praefectis veniam impetrarent domi manendi, aut maris aut belli taedio, aut quia exitum praeviderant, & ad majorem fidem sanguinem sibi mitti nonnulli curarunt."
[178] Pantera, *L'armata navale*, p. 111.

for example as a cook, or on land, in the city's dockyard. For instance, in 1756, Giuseppe Puleggi from Soriano was sentenced to seven years in the galleys for theft. Due to his advanced age, he was not assigned to manning the oars. However, given that he was still in good health and fairly robust, it was decided to send him to work in the Annona warehouses, transporting stones to the masons working there. The following year, in May 1757, Puleggi managed to escape from Civitavecchia by exploiting the very infirmity that had spared him from rowing duty. His condition—a source of suffering and humiliation—now became his unexpected ally. Due to his advanced age, Puleggi lagged behind the others and was always the last to return from the stores to the galleys. One day, taking advantage of being left behind, he escaped. Five months later, he was found by the guards on one of his farms outside Soriano, arrested, and taken back to Civitavecchia, where he was tried by the City Governor. When asked how he managed to escape unobserved, Puleggi explained how he had concealed the iron chainmail on his foot by loosening his trousers and had covered his shaved head with a handkerchief, as travelers were wont to do. Disguised in this way, he pretended to be chasing some pack animals and managed to slip out of the city through Porta Corento.[179]

Turning to the broader practices established in the late 17th century in the Papal States, following the establishment of the *Carceri Nuove* [new prisons], the convict's preliminary medical examination was conducted in the *galeotta*—the room where convicted rowers were held before transport to Civitavecchia.[180] According to Aurelio Scetti's diary, there was likely a similar room with the same function in the prison in Pisa during the late 16th century. However, no mention is made of a preliminary examination, suggesting that such an assessment may not have been required in the early years of galley punishment.[181]

Enslaved rowers underwent a thorough physical inspection at the time of their capture, during which they had to be "inventoried with details such as their names, surnames, homeland, age, height, hair, face, and body markings."[182] Similar to convicts, the decision whether to employ a slave aboard was predicated upon the surgeon's medical opinion, as he was tasked with determining whether they were "useful" or "useless" for rowing duty. Those deemed healthy and robust

---

[179] ASR, Tribunale poi Governatore di Civitavecchia (1589–1913), b. 640, dossier 9, f. n.n.
[180] Calzolari/Di Sivo/Grantaliano, *Giustizia e criminalità*, pp. 31–32.
[181] Monga, *Aurelio Scetti*, pp. 35–36.
[182] ASF, MP, 2131, dossier 3, *Ordine, che si ha da tenersi dal commissario delle Galere, e da gl'altri offitiali di esse da qui avanti, non ostante qual'si voglia altro uso in contrario et prima*, c. 25r.

were assigned as oarsmen, while others were either put to forced labor or sold off, along with women and children.[183]

Thus, for example, in the inventories of the galleys in 1555, we find the entry: "Saim Granatino of Algiers, aged 35, is blind and cannot row."[184] In another case, in a 1620 letter addressed to the captain of the galleys, it is noted that the young 20-year-old slave Abdi, captured while fleeing Naples, had never served in the galleys "for having always been ill, and is now convalescing in the *Bagno*." Abdi's infirmity was marked by visible signs on his body, particularly his face, which was "pitted by smallpox."[185]

Likewise, a surgeon's assessment was required when it was necessary to ascertain an infirmity incurred during duty. If a rower was certified as unfit due to an incident while in service, he was relieved from galley duty. If his inability only prevented him from rowing but did not hinder him from performing other duties or sea voyages, he would still be assigned to work either on land or on the galleys, until his sentence was completed.

For example, in 1621, Candio di Horato Bellieri from Genoa served the last two years of his sentence as a deckhand in the stern of the *Santa Maria*, after completing his forced labor sentence—begun in April 1615—as a rower in the Tuscan galleys. Master Antonio, the hospital surgeon, reported that Bellieri was "forty-and-a-half-years old, of good stature, and crippled in the legs. He has a blunderbuss wound that passes across his body from one side to the other, causing discomfort to his soul."[186]

In the same year, Girolamo Ferrini, surgeon of the *San Carlo* galley, recounts how Lorenzo di Bastiano Mazzieri from Florence, bound in chains since 16 May 1600, aged about 37, after having rowed for ten years on the *Santo Stefano*, had spent two years as a cabin boy in the bow chamber due to liver pain.[187] Similarly, in 1728, there is a record of a forced oarsman Lorenzo Stefanini, who, because he was "broken on both sides"—though the exact ailment is unclear—was no longer able to row and thus served as a nurse at the *Bagno*'s hospital.[188] Those who were totally incapacitated—especially the elderly—were often confined to prisons. In

---

183 Ibid., c. 30r–v.
184 ASF, MP, 627, c. 63r.
185 ASF, MP, 2083, c.n.n.
186 Ibid., c. n.n: "d'età d'anni 40 mezzo, statura giusta di buona vita e rotto dalla parte da basso, et ha una archibusata nel corpo a banda dritta che passa da l'altra banda, che li da fastidio al'Anima, che tanto referisce mastro Antonio cerusico dello spedale."
187 Ibid., c.n.n.
188 ASF, MP, 2114, n.n

the Papal States, this meant the prisons of Rome,[189] while in Livorno, they were sent to the *Bagno*. In some cases, declining health could even result in freedom. In the end, an unfit rower became a financial burden, and it was considered more economical to release or reassign him.

In 1697, 120 convicts were released from the Papal galleys as part of a clemency act; all were deemed too old or otherwise no longer viable for service and were either freed or reassigned to public works. Among them was Marc'Aurelio Benedetto Fabri, whose life sentence was commuted to exile. As he was considered unfit due to his failing eyesight and advancing age—69 years old—Fabri was not sentenced to the galleys but instead was banished from the region where he had committed the crime. Bernardino di Giovanni Paolo from Rieti, who had been sentenced to perpetual galley service, was released due to his disability. Bernardino d'Antonio Colarroni, alias Tuccio from Terni, had his ten-year galley sentence for theft and escape—which commenced on 12 May 1697—commuted to forced labor until his sentence was completed, due to a debilitating injury to his left hand.[190]

Despite the harsh living and poor sanitation conditions to which they were subjected, many rowers lived to a ripe old age. Clearly, a 50-year-old man was considered elderly at the time. Yet, the sources reveal a surprising number of rowers in their 60s, 70s, and even 80s, often describing them as "decrepit old men."[191] Medical care for galley rowers was relatively good, especially compared with the situation in public hospitals. In Livorno, for instance, on 3 August 1710, some 30 slaves were freed by an act of clemency because they were old and infirm; most were blind and frail, however, and rarely left the *Bagno* except to go to the hospital.[192] More than an act of mercy, this decision to free these slaves was motivated by practicality. As long as they remained in the *Bagno*—where they were entitled to daily treatment, food, and clothing—they were a financial burden. Furthermore, given that they would never recover, admitting them to hospital only meant that they took up space and beds that could instead be allocated to younger slaves who might recover. Finally, given that these elderly patients and

---

189 See, for example, ASR, Tribunale Criminale del Governatore di Roma (1505—1871), Congregazione della visita alle carceri (1528–1870), b. 140, f.4.

190 BCR, 34D18, *Memorie e scritture diverse appartenenti alle galere Pontificie e condannati nelle medesimo Principalmente in tempo del tesorierato di mons. Lorenzo Corsini poi cardinale, e papa col nome di Clemente XII, De' forzati Liberati da Innocenzo XII—e aggraziati di commutazione, e diminuzione di pena*, ff.129–197.

191 Bono, *Schiavi*, p. 122. As Bono reminds us, age was a determining factor in setting the price of a slave, and individuals were considered "old" at the age of 60.

192 ASF, MP, 2107, c.n.n.

inmates were unable to move, they apparently washed themselves even less than others, contributing to the accumulation of filth and the spread of disease.[193]

Of note among these freed slaves are Alì of Ametto from Biserta, Ametto of Califfa from Giggieri, and Said of Alì from Tripoli. The first two were 76 and 80 years old, respectively, and had been slaves for 35 and 40 years, never leaving the *Bagno* because they were crippled and blind in both eyes. The 80-year-old Said was also blind.[194]

That there are more references to old, frail slaves than to forced convicts can be explained by the fact that the slaves were, in many ways, better suited to rowing. The assumption that Turks were more suited to life in the galleys is not to be understood as implying a difference based on the anachronistic concept of "race" as theorized in the 19th century.[195] Indeed, as outlined in Chapter 2, Pantera's naval treatise distinguishes between various categories of slaves—Turks, Moors, and Blacks—suggesting that a concept of race may have already existed. This racialization, however, would appear to be based on geographical and religious grounds rather than genotypic.[196] Pantera's position on the supposed superior suitability of Turkish slaves for maritime labor is, in fact, the opposite. He argued that Turks were unsuited to life at sea and were employed as oarsmen not for their nautical skills but for their presumed docility, which made them more likely to obey orders.[197] Most Turks aboard the galleys, however, had been captured from Ottoman ships, and so, if not sailors, were at least familiar with life at sea.[198]

The *forzati*, by contrast—many of whom came from the Italian mainland—were convicted of various crimes and may never even have set their eyes on

---

[193] Ibid.

[194] Ibid.

[195] For a provocative discussion on the significance of "race" in the early modern times, see Heng's *The Invention of Race*.

[196] We must not forget that, according to Galenic medicine, the human body was composed of a specific temperament, often influenced by climate and habitat. This personal temperament not only shaped the personality of each individual but also determined their susceptibility to certain diseases.

[197] Pantera, *L'Armata navale*, p. 131.

[198] See, for example: BAV, Stamp.Ferr.IV.8532 (int.4), *Fedel relazione mandata dall'Illustrissimo Signor Balio di Cremona C. Bernardo Vecchietti generale delle Galere della Sacra Religione Gerosolimitana Del viaggio, e presa delle tre Galeotte, Fuste, e vascelli d'Infedeli. Fatta dalle medesime Galere in Levante*, in Roma, appresso Ludovico Grignani, 1641: "Si rinforzerà la Religione, ché oltre a 36 ebrei assai ricchi, ci sono parimenti molti Turchi mercanti, conseguentemente di buon riscatto [...] si rinforzeranno anco le Galere, per esse molti dell'altri assuefatti al remo."

the sea.[199] Many were sentenced to the galley for fixed periods, and when they continued rowing after completing their sentence, they did so as *buonavoglia*, until they had repaid their debts. It is difficult to determine how long they lived after release, or how the harsh conditions they endured as galley oarsmen affected their health. Yet, a comparison of the death records of convicts and slaves preserved in the archives gives the impression that the slaves lived longer. Furthermore, daily life for convicts on land was far harsher than for slaves—who, when not at sea, were unshackled and were allowed to run workshops and taverns in the port districts. They could also learn trades such as cooking and surgery, the income from which was put toward their ransom and their eventual emancipation. Convicts, on the other hand, remained chained at all times and were assigned to strenuous labor, requiring greater physical exertion, exposing them more acutely to disease and bodily decline.[200]

A more thorough physical examination took place during what was known as the *taglio degli schiavi*—the formal appraisal of a slave's condition used to determine his market or ransom value.[201] If slaves were treated as commodities, then their state of health had to be assessed accordingly. The use of medical expertise to evaluate a slave's health was an ancient and integral practice in buying and selling—or redeeming—slaves. Like any other transaction, the exchange of slaves was not only an economic act involving a calculated price, but also the transfer of rights and property. To guarantee fairness and confirm the validity of such transfers, medical practitioners were frequently called upon to examine the slaves before the sale.[202] The same applied in cases of ransom, to ensure that captives were not released for a sum below their assessed worth.

Undoubtedly, this physical inspection was one of the most discriminatory moments for slaves, as it further deprived them of their humanity and underscored their status as commercial commodities.[203] While the slave trade was practiced in both the Christian and the Muslim Mediterranean, practical manuals on how to physically examine the enslaved body were initially drafted by doctors of Muslim origin. One such example is the *al-Kitāb al-Malikī* by the Persian physician Haly Abbas (930–994). Thanks to the Latin translation by Constantine the African, the work was known in the West by the 12th century under the title *Liber Pantegni*,

---

199 Lo Basso, *Uomini da remo*, pp. 344, 355 f. The same applied to contracted rowers, many of whom were poor people from the countryside who had signed up in search of an income, see p. 20 f. This thesis is also accepted by Bono, *Schiavi*, p. 219.
200 ASF, MP, 2108, c.n.n.
201 Chizzolini, *The "taglio degli schiavi"*.
202 Barker, *Precious Merchandise*, pp. 98–104.
203 Ibid., p. 92.

though it enjoyed little success.²⁰⁴ An apparent exception to this Arab monopoly was the *Cirugia Universal* (1597) by the Spanish royal surgeon Juan Fragoso and the aforementioned *Medicus Politicus* by Rodrigo de Castro.

These two treatises are notable for several reasons. First, they are unique for the period in which they were written. To our knowledge, there are no other medical treatises produced by Europeans during the 16th and 17th centuries that address the topic of physically examining slaves—even if only in a specific chapter. Furthermore, Fragoso and de Castro, while both drawing on earlier Arab authors, differ from them in that the Arab manuals were not intended for surgeons, but rather for those buying slaves, who were expected to perform the physical examination. One possible hypothesis is that these two authors wrote their treatises to systematize a body of knowledge that had been passed down from the Arab world and that had become established in the Mediterranean at that time. This is particularly significant in the case of de Castro. Although originally from Portugal, de Castro published his work while in Hamburg, and the intended medical audience was likely that of the Hanseatic port, which was probably not yet familiar with the practice of examining slaves as was common in the Mediterranean. Indeed, from the Thirty Years' War (1618–1648) onward, Germany, along with France and England, became involved in the slave trade between Africa and Brazil.²⁰⁵ De Castro was familiar with Fragoso's work, as both hailed from the Iberian Peninsula, and the texts are nearly identical in content, with both simply repeating the teachings of Arab authors.

In the second book, entirely devoted to the different types of declarations surgeons had to make after their examinations, Fragoso laid out instructions on how to determine the sale price for a slave in the paragraph *Como se ha de aver el Cirujano en la declaracion y examen de un esclavo que se vende* [How the surgeon should proceed in the declaration and examination of a slave being sold]. As the author explicitly states, he primarily drew upon the teachings of the Arab physician al-Razi (865–925).²⁰⁶ First, the slave's skin color had to be examined to ensure

---

204 Ferragud, Role of Doctors, p. 146.
205 Rissel, Hamburg in the Atlantic Slave Trade, pp. 75–96; Mallinckrodt/Lentz/Köstlbauer, *Beyond Exceptionalism*. The sources I analyzed also mention slaves sold by German merchants. In 1710, for example, there is a reference in the *Bagno* to a slave named Macameto di Calilla di Marasci, aged 66, who arrived from Germany in 1694 (ASF, MP, 2107, c.n.n.). The presence of the British and French in the Mediterranean slave trade has been well documented by Massimo Bomboni in his paper *Northern Experiences of Mediterranean Slavery in the Tuscan Sources*, presented at the international workshop *Captivities: Experiences and Institutions of Slavery in the Early Modern Mediterranean*, Bologna, 18–19 May 2023.
206 Fragoso, *Cirugia universal*, p. 570 f.

that there were no signs of disease, such as white spots [aluarazos] or ulcers, especially in areas that could be concealed, such as the armpits and groin. A poor skin color or paleness could suggest liver or stomach disease, or very bloody hemorrhoids. The examination continued with a hearing test. The slave's speech and behavior were acutely observed to assess his overall health and character. What followed was an eye test to evaluate the size and condition of the eyes. Special attention was given to the color of the sclera; brown suggested leprosy, yellow pointed to liver disease, and red indicated inflammation. If the eyelashes moved heavily and with difficulty, or if they were thick or rough, it was more likely a sign of leprosy.

From the eyes, the examination moved to the nose and mouth. The surgeon had to ensure that the slave's breath did not emit a foul odor. The nostrils were then examined in sunlight to detect fistulas, and the mouth was opened to inspect the teeth, which needed to be straight, strong, and clean. Small, fragile teeth that could easily be lost were considered a sign of bodily weakness.

The neck and throat were then inspected for any signs of swelling or previous swelling. If dry, these areas could easily develop into ulcers. The chest was then examined to determine whether it was large and fleshy, as a small, thin chest with protruding spinal bones could suggest that the slave might eventually suffer from typhus. Thin, excessively protruding hips, and frequent bleeding were also symptomatic of typhus.

The surgeon then inspected the abdomen by touching or lightly squeezing it with his fingers, palpating for any tumors or pain, especially in the area of the liver and spleen, as well as the pylorus. The slave was asked to walk to assess the strength of his steps. He was instructed to squeeze something so that the surgeon could determine how much force he could exert. This would help assess the state of his nervous system. The slave was then asked to run so that the surgeon could observe whether he was coughing or breathless at the end, possibly due to asthma or choking. In addition, the hands and feet were observed to see whether they had good proportions, with neither being larger than the other. Finally, the surgeon examined the legs to check for thick and wide veins, as these were often symptoms of varicose veins or leprosy.[207]

In the fourth book of de Castro's *Medicus Politicus*, devoted to various types of medical examinations, Chapter XIII focusses on the physical assessment of slaves to determine their sale price: *Declarandi ratio circa emptitios servos* [The procedure for declaring the purchase of slaves].[208] A slave's physical evaluation was

---

207 Ibid.
208 De Castro, *Medicus Politicus*, p. 263 f.

akin to an ordinary medical check-up. The doctor's first step was to carefully examine the slave's complexion in the light to assess his overall condition. Next, the skin color and condition had to be analyzed to rule out the presence of diseases such as scabies, leprosy, or elephantiasis, as well as other infectious diseases. Any deviations from "normal" skin color indicated poor nutrition and could be divided into three types: white, black, and reddish, depending on the predominant hue.

Although the procedures described by de Castro often mirror those found in Fragoso's manual, they are presented with greater clinical specificity and applied in a broader diagnostic framework. Where Fragoso's approach centers on legal declarations, de Castro's text integrates more active testing-posture, muscular exertion, and ocular responsiveness—suggesting a shift toward functional evaluation.

This shift is particularly evident in the detailed examination of physical symptoms where de Castro looks beyond mere observations to diagnose specific conditions. For example, the presence of varicose veins on the back might indicate "melancholic blood." Sparse and loose hair, especially in the eyelashes and eyebrows, combined with a hoarse voice and a flushed face, could suggest the onset of leprosy. Venereal diseases might be diagnosed through hair loss and weakness in the extremities and joints. To evaluate a slave's overall strength, de Castro recommended doctors monitor their posture and movements, as well as the way they held objects. The next step was to test their hearing and eyesight. Doctors had to ensure their teeth were both healthy and complete. The eyes, in particular, required close examination, for blindness would render a slave completely incapable of working. The pupils had to be equal in size, the eyeball had to be white, free from secretions or redness, and the movement of both eye and eyelid had to be normal. Breathing should be unobstructed, and no discharge should be expelled from the mouth or nose. The assessment concluded with palpation of the slave's abdomen and chest, whereby he was positioned on his back to detect any tumors or other irregularities.[209]

Unfortunately, no evidence of price fixing or negotiations for the ransom of slaves from the Papal galleys have been found. There are several sources for the Tuscan context, however. Here are just a few examples: In October 1682, it was decided to ransom the slave Asano of Usaino from Biserta, who had served for six years. He was described as having brown hair, of strong build, two scars from a stab wound on his left arm and a small burn on his wrist, and a "large flower"—perhaps a birthmark or a hemangioma—on his right arm extending to his shoulder. Despite a ruptured intestine on his right side, he was relatively

---

209 Ibid.

young at 35 and strong enough to row. More importantly, since he had been a Janissary at the time of his capture on the Turkish galley, his ransom was set at 500 gold *scudi*.[210]

An exchange that took place in July 1696 illustrates the extent to which the ransoming of slaves was primarily driven by economic motives. At that time, a ransom had been offered for the slave Amor Muccio, registered in the *Bagno* under the name of Amore of Abdallà from Tripoli, commander of the galley on which he had been captured off Capo Spartivento on 10 July 1688. He was 65 years old, and blind, as confirmed by his pupils' dilation and opacity, according to Romanelli's account. Muccio had never undergone the aforementioned *taglio*, but it was clear that, given his state of health, his ransom price would have been very low.

At the same time, however, a slave named Macametto of Amore from Tripoli submitted a memorial in which he claimed to be Amor Muccio and offered 250 pieces of 8 *reali* for his freedom.[211] This relatively low price was due to the fact that, although he was only 35 years old, he had been unable to man the oars for three years due to sciatica in his left leg, as confirmed by the galley surgeons' reports.

The *Bagno* authorities decided to accept the exchange and finally released him upon payment of 290 *pezze*. Given that the nerves in his leg were damaged, he was unable to row and, by extension, was considered worthless as a laborer. However, Macametto's better health and younger age allowed for a margin of profit, as the price paid for Amor Muccio would have been even lower due to his poor health.[212] Such considerations continued to determine ransom decisions over the following decades. Thus, in August 1702, the slave Durach of Amida from Cavalla was freed for a mere 40 *pezze*. It was impossible to ask for more for a man who

---

[210] ASF, MP, 2099, c. 441. The Janissaries were slaves of Christian origin who were forcibly conscripted as young men and taught the Muslim religion. As a result, many of them became the most fanatical followers of Islam and were considered the best military corps in the Ottoman army. See the positions expressed in BAV, Stamp.Cappon.V.683 (int.93), *Breve relatione dell'imprese fatte contro Turchi delle galere di Malta della Religione di S. Giovanni. Et dalli Galeoni dell'illustruissimo & Eccellentissimo Sig Duca d'Ossuna Vice Rè di Sicilia. Dove s'intende la presa fatta nell'Arcipelago di due Vascelli Turcheschi, e d'altri quattro nel porto della Goletta, con la presa, e morte di molti Turchi, e liberazione di molti schiavi Christiani*, in Viterbo, 1616: "360 schiavi, la maggior parte sono Giannizzeri, che sono li migliori, & più bravi soldati che habbia il Gran Turco."

[211] The price of the ransom in Livorno was usually fixed in *pezze da 8 reali*, a silver coin weighing 4.032 grams, minted by Duke Ferdinand II in 1656. See Martini, *Manuale di metrologia*, pp. 209, 283f.

[212] ASF, MP, 2103, c.n.n.

had been enslaved for 43 years, had no teeth, and had spent all his savings on alcohol.²¹³ Similarly, in April 1728, the ransom demand for the slave Mustafà of Abdallà from Stanchio was accepted for 80 *pezze*. Given that he was 74 years old and in imminent danger of dying, the Tuscan authorities hastened to accept the ransom to avoid losing even that modest sum.²¹⁴

Paying as low a price as possible was a priority when buying slaves. There were slaves aboard the galleys who were not captives of war, but had been purchased to man the oars. In 1549, the Grand Duke ordered the purchase of slaves for his galleys. They had to be between 18 and 35 years old, not older than 40, free from incurable diseases, and in good health. They could be of Turkish or Morlach origin, provided the cost of buying and transporting them did not exceed 35 gold *scudi*.²¹⁵

In August 1702, a French vessel arrived in Livorno from Cagliari with some slaves for sale. One of them, Ali of Salem from Morocco, was 25 years old, and described by the surgeon Franceschini as "of olive complexion, rather tall, and handsome." While the ring finger on his left hand had been crippled by a bullet wound, he could still row well, according to the surgeon, who had found him strong and robust, with no other physical defects. He had been bought in Cagliari for the sum of 105 *pezze* of 8 *reali*, and sold to Livorno's navy for 110 *pezze* of 8 *reali*.

Another slave, whose name is unknown, was young and strong, though he had a blemish on his cheek. Fearing it might be a sign of disease, the galley surgeon suggested that they wait before acquiring him and re-examine him a few days later. As the blemish had not changed, the surgeon concluded that it was a birthmark, recommending the purchase of the slave. This recommendation was particularly compelling, as even the French must have believed the birthmark indicated a disease, and had set their price at just over 40 *pezze*. After all, the surgeon wrote, it was better to acquire a slave for 42 *pezze* than to pay a salary to a free rower.²¹⁶

4 Controlling the crew: crime, punishment, and medical expertise

In Chapter 1, I discussed the central role of medical action *a posteriori*—not aimed at providing treatment, but rather at assessing the health status and judging

---

213 ASF, MP, 2105, c.n.n.
214 ASF, MP, 2115, c.n.n.
215 ASF, MP, 2077, c. 333r.
216 ASF, MP, 2105, c.n.n.

whether a disease or injury had already occurred and, in particular, emphasizing the importance of medical expertise in court proceedings, the obligation to report suspicious cases, and the function of post-mortem examinations.[217] While the galleys and the docks were distinct social settings—somewhat removed from official society—similar dynamics in judicial prosecution and in the roles assigned to doctors in cases involving prosecuted *galeotti* were observed. As previously mentioned, these tasks centered on the belief that the doctors could reveal conclusive truths by examining physical evidence, particularly in post-mortem cases. In instances of murder or violent injuries, medical examinations aimed to assess the severity of the injuries and determine the causes of death, establish the accused's innocence or guilt, and evaluate the degree of their culpability to decide on punishment. Punishing transgressions was essential for maintaining discipline among slaves and convicts, serving as a deterrent against excess and disorder. While the primary purpose of medicine was therapeutic, one of its greatest potentials emerged when healing was no longer possible, and the focus shifted to evaluation.

4.1 Injuries as evidence

The directives governing the management of crews on Papal and Tuscan galleys prohibited various forms of behavior aboard, including altercations and other aggressive acts. Regardless of whether the violence was in response to an offense or in self-defense, any form of violent assault was strictly forbidden, and anyone committing a violent act was to be severely punished—either through exemplary fines, corporal punishment, or both.[218] Those who witnessed a confrontation and failed to notify naval officers were subject to punishment. The threat faced by trained medical officers onboard was even more severe. Barbers aboard the Papal galleys could be sentenced to up to five years at the oars if they treated rowers showing signs of violence on their bodies without reporting the injuries to the authorities.[219]

---

217 De Ceglia, *Body of Evidence*; De Renzi, La natura in tribunale; Pastore, *Il medico in tribunale*; Watson, *Forensic Medicine*.
218 ASF, MP, 2131, dossier 3, *Instruction a voi Pietro Dini di quello che havrete à fare nello sciorinare, et purgare la robba della presa delle Galere*, c.n.n.; AAV, Misc. Arm. IV/V, b. 54, f. 85.
219 AAV, Misc. Arm. IV/V, b. 54, f. 85, *Frà Vincenzo Rospigliosi generale delle galere pontificie, governatore di Civitavecchia, Sovraintendente generale delle fortezza e Torri marittime di tutto lo Stato Ecclesiastico*: "7. Li Signori Capitani, Comiti & altri Officiali siano tenuti, e debano far mettere in catena qualsivoglia persona, che rissasse, ò tumultuasse, ò alterasse di parole in Galera alla presenza loro con darne à Noi subito parte afinche li Delinquenti venghino castigati sotto

Before analyzing some trials initiated for altercations between enslaved and convicted rowers, it is important to note that most of these trials involve fights that took place ashore, rather than at sea. Even when violent skirmishes occurred aboard, they always took place during breaks from sailing on the high seas. This is hardly surprising, for rowers—free or enslaved—were chained to their decks during sea voyages. Consequently, it would have been more difficult to commit acts of violence at sea, given their severely restricted movement. If such acts did occur, there would likely have been no means to properly prosecute them during the voyage. Furthermore, even if such prosecutions had been possible, there was neither the time nor the interest. For example, in 1600, a notary aboard the Tuscan galleys was specifically tasked with prosecuting knights, soldiers, and officers while at sea. None of the recorded trials mention slaves or convicts, however.[220] In contrast, many trials took place on land, especially in Livorno, where slaves and convicts were not shackled and could move about freely, particularly within the *Bagno*. Furthermore, slaves had even greater freedom of movement, as they were unshackled during non-sailing periods, and could work in the city's dockyard. As a result, they had access to and from the *Bagno* compound, which increased opportunities to get involved in violent altercations, even with the general population of Livorno.

Let us analyze a few cases of violent assaults and murders where medical advice was required, starting with the Tuscan context. The sources examined fall into two categories. The first consists of reports from trials for assault and murders committed by galley rowers, found in a legal booklet titled *Liburnensis Iurisdictionis*, dated 1733, and signed by the lawyer Giovanni Fei. The pamphlet, designed as a tool for the Commissioner of the Galleys to assert his authority over the Governor's intervention in cases of violent behavior between rowers, first presents a series of examples in which the two jurisdictions intersected, followed by a summary in which the surgeons' reports are documented for each case.[221] The second category consists of a series of criminal trials held by the Governor. In some instances, it has been possible to locate the criminal trial corresponding to the cases described in *Liburnensis Iurisdictionis*.

---

pena à chi trasgredirà di scudi 100 & altre ad arbitrio &c | 8. Li Comiti, Sottocomiti, Agozzini & altri respettivamente debano, succedendo alcun delitto ò scandalo, denunciarlo alla nostra Corte subito sotto pena di scudi 50. E di anni cinque di galera, & altro ad arbitrio &c nelle quali pene incorreranno li Barbieri delle Galere, loro Barbierotti & altri che cureranno feriti, stroppiati, e altri offesi non li denunciando nel modo suddetto." For the Tuscan galleys, see ASF, MP, 2131, dossier 3, n.n, *Ordini da osservarsi nelle Galere del Ser.mo G.Duca di Toscana*.
220 ASF, MP, 2082, c.n.n.
221 ASF, MP, 2132, part II, dossier 8, *Liburnen. Iurisdictionis*, cc. n.n.

4 Controlling the crew: crime, punishment, and medical expertise — 181

On 1 August 1650, Casone of Gimillo, a slave on the *Santa Vittoria*, was wounded in the throat while in the bow room. The surgeon reported that he had severed an artery. Due to the severity of the injury, the aggressor faced the maximum penalty. As Casone was unable to speak, he could not identify his attacker. A search was conducted among the rowers, and Bastiano Petrini, a forced rower, was found with a knife smeared with blood. The discovery of the weapon, combined with witness testimony stating they had seen Petrini with bloody hands shortly after the attack, was considered evidence of his guilt. Petrini's crime was deemed extremely grave, not only because it involved attempted murder, but also because it occurred within the galleys. As punishment, Petrini was sentenced to imprisonment with the "iron sock" in the *Bagno*, plus five years of galley service.[222]

In March 1651, a murder trial was held before the court of the Governor of Livorno, following the death of the slave Amet of Persia—known as Ciuff—who, on his return to the *Bagno* at night after a day working in the dockyard, encountered Orazio di Domenico Vitali, a woodcutter from the town.[223] The two quarreled and Ciuff, who insulted the boy by calling him *bardassone*—a depraved sodomite—was wounded below the rim of his right eye. Although initially the attack appeared to have caused a minor bruise, four days later, the slave began to suffer from "acute fever and chest pains" and died.

To determine whether death had occurred as a result of the beating, and hence whether Vitali should be considered guilty of murder, the court relied on the report by the doctor who had examined and treated the slave: the surgeon of the *Bagno*, Salvatore Cosci. While the surgeon argued that death had been caused by chest pains, the fever, which certainly worsened the condition, was probably linked to the head wound. Since Vitali was a legally considered a minor—he was 25 years old—the trial concluded with him being sentenced to six months in exile. Should he refuse to go into exile, he would be sent to Florence's Stinche prison.[224]

This trial is notable because it triggers reflection on another duty of doctors, inextricably linked to assessing the severity of wounds or performing autopsies: the duty to notify. In fact, during the trial, the Governor, who had intervened as the attack took place outside the compound, claimed to rule not only on the woodcutter's presumed culpability, but also on that of the surgeon, who was accused of "not having made the usual report" to the Governor, as required by

---

222 Ibid., *sommario* 7.
223 Chizzolini, Medici a Livorno, pp. 86–88.
224 ASLi, Capitano poi Governatore poi Auditore Vicario, 1550–1808, b. 3082, f. 112 r–v; b. 3233, f. 14 r–v.

law. Medical practitioners were obliged to report any case of violence they treated so that controls could be tightened and violence reduced both at sea and on land. Ultimately, however, the surgeon was acquitted because the slave's death was not directly caused by the wound he had sustained. Although he had not reported the incident to the Governor, he had informed "those who administer justice in this *Bagno*," specifically the Scribe, as was required.[225]

As reported in *Liburnensis Iurisdictionis*, in February 1652, a forced convict by the name of Francesco di Lorenzo was recovering in the *Bagno* infirmary. According to the surgeon's report, he had been "wounded in the lower abdomen below the ribs with a penetrating wound" after having been stabbed by another forced rower during an argument. Given that di Lorenzo was "in danger of [losing] his life," the aggressor—the friar Giovanni Battista Allegrini de Servi—had to be punished in an exemplary manner and was subsequently sentenced to an additional five years in the galleys.[226]

In 1657, Macometto of Ali from Tituano was sentenced to 200 lashes for having wounded Rais Mostafà of Macometto of Malvagio—a slave on the *Capitana*—by cutting the nerves in his neck. The attack was reportedly motivated by Mostafà's doctrinal beliefs, which did not seem to be entirely in line with the precepts of the Muslim religion.[227]

In another case, brought in 1657, Barca of Suleiman from Algieri, a "black" slave in the *Bagno*, was accused of having wounded a certain Ametto of Saide from Gerba with a knife. Ametto's injuries left him with a crippled middle finger, rendering him unable to row. Barca was therefore condemned to have his nose and ears severed.[228] Amputation of ears was a punishment generally reserved for thieves. Thus, to deprive the fleet of a skilled oarsman might be considered a form of theft. The reference to *denasatio* [cutting off the nose] is notable, especially when considering the particular symbolic meaning, closely linked to the sphere of personal identity, that this practice carried in the Middle Ages. As early as the 13th century, Albertus Magnus claimed that cutting off someone's nose symbolized the unmasking of moral deformity, a corruption that persisted even without its physical material form.

As the Venetian Penal Code of 1443 postulated, a man's face was the mirror of his honor: to damage it was to dishonor his person. In a society such as the European one—where honor and reputation were paramount, anyone without a nose

---

[225] Ibid., f. 14 r–v.
[226] ASF, MP, 2132, part II, dossier 8, *Liburnen. Iurisdictionis*, n. 16, *sommario* 8.
[227] Ibid., n. 17, *sommario* 9.
[228] Ibid., n. 18, *sommario* 10.

was instantly recognizable as belonging to a specific social category: criminals. Losing one's nose meant a loss of face, and therefore, a loss of honor.[229]

Such was the gravity of the offense that, though there were legal provisions for it, amputating someone's nose was used more as a threat and only executed in exceptional cases—unlike cutting off someone's ears, which held less symbolic significance. Amputation was an extremely grisly punishment, likely reserved for the gravest offenses as an alternative to capital punishment, in order to prevent the navy from being deprived of able-bodied men.

Thus, in August 1701, it was decided to cut off the nose and ears of the slave Macometto of Fesso Manetta because he had wounded Cosimo of Alì from Mes, the *spalliere* of the *Padrona* galley,[230] by stabbing him in the arm. Macometto had previously wounded a slave and killed the chief slave trader, for which he had already been sentenced to ten years in prison. Wounding Cosimo only worsened his predicament. Cosimo was considered the best of all available slaves, and the loss of such a skilled oarsman was regarded as a great blow to the fleet.[231]

Similarly, in October 1674, Mustafà of Ibraino from Tunis was sentenced to 160 lashes, and had his nose and ears severed for stabbing Giuseppe—the son of the Captain of the *Bagno*—in the back.[232] Evidently, Mustafà had come to work in the biscuit shop drunk. After repeated disturbances and bouts of insolent behavior, he was slapped by Giuseppe, who told him to get back to his cell and sleep off his drunkenness. As Giuseppe was about to leave the *Bagno*, Mustafà jumped on him. Striking him on the back, Giuseppe managed to shake off the slave, but when he started walking again, the slave pounced on him, stabbing him. He would have struck again had the other slaves not intervened. Once again, the slave blamed his actions on alcohol and begged for forgiveness. Drunkenness, however, was clearly not a mitigating factor.[233]

Notwithstanding, it is reasonable to assume that, in addition to alcohol, a voluntary element of hatred and opposition to superiors played a role in such violent

---

229 As written by Gadebusch Bondio, I denasati e i medici, p. 159: "In quanto parte prominente del viso, il naso è strettamente connesso all'individualità e alla dignità della persona [...] La *scissio nasi* o *denasatio* appare però anche tra le punizioni fisiche previste dalla legge tra il Quattrocento ed il Cinquecento proprio perché si prestava meglio di altre a distruggere l'immagine della persona segnandola per sempre." For a general reflection on physical marks as identifying signs —such as cutting off ears and nose—see the works of Groebner, *Defaced* (translation), pp. 68–86; *Storia dell'identità personale* (translation.), pp. 92–100.
230 Lo Basso, *Uomini da remo*, p. 31. These rowers, positioned on the first two stern benches, were tasked with timing the rest of the crew.
231 ASF, MP, 2105, c.n.n.
232 ASF, MP, 2132, part II, dossier 8, *Liburnen. Iurisdictionis*, cc.n.n.
233 ASLi, Capitano poi Governatore poi Auditore Vicario, 1550–1808, b. 3082, trial n. 335.

outbursts. Altercations between slaves, forced oarsmen, and officers were common. For instance, in July 1705, Francesco Mosti, a forced rower, was tried for injuring Zaccaria Bertelli, another convict and *vigilante*, after preventing him from entering the *Bagno di San Giuseppe* a week earlier. Although Francesco Mosti was not housed in that wing of the *Bagno*, it did house Raffaello Braccini, with whom he had been caught sodomizing at Christmas—an offense for which both had already been flogged. In revenge, Mosti struck Bertelli seven times, inflicting six life-threatening wounds, including fractures and bleeding, as reported by the surgeon Giuliano Cini.[234]

As already noted by Cesare Santus, alcohol seemed to be the primary cause of violence within the *Bagno*.[235] Given that wine was part of their daily diet, it is not difficult to imagine that many of the *Bagno*'s inmates abused alcohol in an attempt to render their appalling living conditions more bearable.[236] Inside the compound, desperation was so prevalent that even cases of self-inflicted violence were common. From the sources, it appears that it was primarily slaves, rather than convicts, who attempted suicide.[237]

The higher percentage of suicides—an extreme act of courage and desperation to escape slavery—among Muslim slaves, as compared with Christians, had been noted by Salvatore Bono.[238] This can be explained by the stricter condemnation of suicide in Christian teachings. In *Summa Theologiae* (1265–1274), Thomas Aquinas places suicide among the gravest sins an individual could commit. In contrast, in Roman culture—thanks to the influence of Stoicism—suicide was forgivable, tolerable, and even recommended as an act of great nobility. According to Christian theology, however, the act was considered contrary to the work of the Creator. Suicide was stringently condemned as a violation of natural law, and thus to the principle of self-preservation, of civil law, insofar as the individual was part of the community; and of divine law, as human life was a gift from God to humankind.[239]

Among the trials analyzed, one particular murder-suicide case is notable. In 1674, the slave Casagno of Ametto from Tunis—a cabin boy on the *Capitana*—was tried for the murder of Abassisi of Ametto from Tripoli, also a slave. On the pretext of wishing to shave his private parts, Casagno, managed to obtain a razor, with which he slashed Abassisi's throat, and then attempted in vain to

---

[234] ASLi, Capitano poi Governatore poi Auditore Vicario, 1550–1808, b. 3087, c. n.n
[235] Santus, *"Il turco"*, pp. 122–123.
[236] Bono, *Schiavi*, p. 185.
[237] Some cases of suicide are reported in ASF, MP, 2102, 2113, and 2114.
[238] Bono, *Schiavi*, p. 291.
[239] On suicide see Barbagli, *Congedarsi dal mondo*.

kill himself by slashing his own throat twice. Due to the cruelty of the act, Casagno was sentenced to death, ironically achieving the goal for which he was punished.

What is noteworthy about this case is the absence of any apparent motive for such a brutal act. The authorities could not fathom what had led Casagno to commit such violence, since both he and Abassisi were comrades who ate together and had never been seen fighting.[240] The murder was neither caused by enmity nor by alcohol. We cannot offer a definite answer, but we can speculate—even indulge in some conjecture. The trial records suggest that Casagno's actions stemmed from sheer desperation. He may have either realized the gravity of his act in a moment of madness or premeditated it as a means of escaping the painful and inhumane conditions of slavery.

In Civitavecchia, due to the lack of onshore facilities for confining slaves and convicts when not at sea, all the trials related to riots and murders took place aboard the galleys. Unfortunately, the surviving trial records for the Papal fleet are sparse and mainly pertain the mid-18th century. Nevertheless, the procedures followed were similar to those used on land under normal conditions. The key difference from the Tuscan case, however, lies not only in the smaller number of extant trials, but also in the reportedly less rigorous manner in which offenders were prosecuted. Most trials held by the court under the authority of the Governor of Civitavecchia did not result in custodial sentences. It remains unclear whether this indicates that the trials were left unresolved, or whether the judgments are preserved in an as-yet unidentified archive. Given that some of the surviving trials do involve a sentence, the former hypothesis seems more likely. Furthermore, the alleged leniency of the punishments seems to be corroborated by the testimonies of those involved. In 1745, for instance, galley captains complained that armed fights were still breaking out too frequently aboard ship. To address this issue, they asked the Governor to replace the ineffective beatings—punishments that clearly failed to instill fear—with harsher measures such as the *strappado*—a method of suspended torture that dislocated the arms.[241]

In any case, what is of particular interest here is that, in the event of a brawl or murder, a surgeon's expertise was always required. In March 1760, Nicola Cuchiarelli from Benevento and Francesco Savero Ciolli, both life-sentenced rowers on neighboring benches, were prosecuted by the Governor of Civitavecchia. Allegedly, after an argument, Cucchiarelli struck Savero on the right ear with an axe, and Savero retaliated by stabbing him in his left side with a knife. Seriously injured, both were taken to the Hospital of Santa Barbara. According to the gal-

---

240 ASLi, Capitano poi Governatore poi Auditore Vicario, 1550–1808, b. 3082, trial n. 47.
241 ASR, Tribunale poi Governatore di Civitavecchia (1589–1913), b. 650, dossier 1, f. n.n.

ley surgeon, Domenico Siri, both men were in mortal danger. Cucchiarelli had a wound three fingers long and half a finger wide in the lower ribs on his left side, while Savero had a wound two fingers long, extending from his temporal bone along the length of his ear and down to the bone, with torn muscles and nerves.

When asked what had happened, Cucchiarelli claimed that Savero had started the fight and that he had simply acted in self-defense, though he could not recall whether he had struck Savero with a wooden or an iron stick. He insisted that he had been the one to receive the first blow. On the other hand, Savero claimed that Cuchiarelli had attacked him out of anger because Savero had sided with a certain Guerrino, whom Cucchiarelli had previously insulted. The two men began to fight and when Cucchiarelli realized he could not win with his bare hands, he grabbed the axe.[242]

While the exact sentence is unknown, the court records suggest that the Governor believed Savero's account. If the absence of a verdict indicates an unresolved trial, it can be surmised that no punishment was deemed necessary, as neither man died. Furthermore, both convicts were already serving life sentences in the galleys, and imposing corporal punishment would have risked losing valuable rowers.

In July 1763, a trial was initiated against Francesco Verzelli, a cabin boy on the *San Prospero*. He was accused of fatally wounding the convict Giuseppe di Marco in the left arm with a gunshot from a harquebus while the vessel was entering the port of Anzio. The bullet shattered bones, tore muscles, veins and arteries, and the galley surgeon was forced to amputate the arm in an attempt to control bleeding and prevent infection.

Upon admission to Santa Barbara, di Marco was asked who had injured him and why. He identified Verzelli as the assailant, but could not provide a clear explanation for the attack. According to the testimony, on 25 June di Marco was carrying some onions when one fell into the storeroom below, where he found Verzelli holding a harquebus. When di Marco inquired what Verzelli was doing with the firearm, Verzelli shot him without apparent provocation, as the two had never quarreled. While the motive remains unclear, it is plausible to conclude that the shot was accidental.

Despite the amputation, di Marco succumbed to a gangrenous infection on 5 July, which spread from his arm to his chest, neck, and shoulder blade. Verzelli denied firing the shot, claiming he was merely in the wrong place at the wrong time. He asserted that he had been asleep in the room, and was awakened by the

---

[242] ASR, Tribunale poi Governatore di Civitavecchia (1589–1913), b. 655, dossier 11, f. n.n.

sound of the gunshot. Testimony, however, from other sailors who had seen Verzelli holding the harquebus contradicted his account, leading to his conviction. As a result, he was sentenced to wear a double chain.[243]

4.2 Uncertain deaths: medical examinations and post-mortem judgments

Closely linked to assessing the severity of injuries was the task of examining bodies where the cause of death was uncertain and, where necessary, performing an autopsy. Identifying the cause of death was primarily a question of determining whether it was natural or violent. In the event of a natural death, it was necessary to find out whether the death resulted from some unknown disease, and, if so, whether it was contagious, so that appropriate precautions could be taken. Should the cause of death be violent, however, it was crucial to establish whether it occurred as a result of an accident or an act of aggression. In cases where death was the result of someone else's actions, it was essential to identify the perpetrator so that the most proportionate punishment could be applied.[244]

In the correspondence between Vittorio Vitolini, Scribe of the *Bagno*, and Giuseppe Prini, Minister of the *Bagno*, it is noted that, in 1709 a convict died in Livorno after complaining for three years of a pain beneath his right clavicle, at the level of his first rib. He had been hospitalized multiple times, but treatments proved ineffective, until he suddenly developed a high fever, difficulty breathing, and began coughing up blood. He died shortly thereafter. The attending medical staff were unable to determine the cause of death, as the patient had never previously exhibited any fever or signs of wasting. At Dr. Romanelli's request, the corpse was dissected, revealing an enlarged and inflamed pulmonary lobe, with a nodule that had grown so large that it occupied half the lung's surface. The nodule was fleshy and difficult to cut, with pulmonary gangrene observed underneath it. Although the exact diagnosis is not recorded, it seems reasonable to assume that the cause was a lung abscess, likely due to a tumor. Notably, there were no signs of tuberculosis, and no preventive measures were thought necessary.[245]

"Carbuncles" [*anthrax*] and tumors were particular concerns for medical staff. They were widely recognized as the most obvious symptom of plague after the emergence of buboes. These black-crusted tumors were often confused with or re-

---

[243] ASR, Tribunale poi Governatore di Civitavecchia (1589–1913), b. 656, dossier 20, f. n.n.
[244] De Ceglia, *Body of Evidence*; Park, Criminal and Saintly Body; Pastore, *Il medico in tribunale*, pp. 30–36, 81–87;
[245] ASF, MP, 2107, c.n.n.

sembled *petechiae*—small red or purple spots caused by subcutaneous bleeding. Whenever inmates died with these symptoms, a post mortem was conducted to ensure that the cause of death was not bubonic plague. For instance, in March 1731, the superintendent of the galley warehouse informed the Commissioner of the galley squadron that a slave had suddenly died in the *Bagno*'s infirmary due to a tumor in his throat. Whereupon, the hospital's two doctors and four surgeons were ordered to perform an autopsy, suspecting it might be a case of a "poisonous disease." Ultimately, the precaution proved unnecessary, as the medical team determined that the tumor was not indicative of anything malignant, but rather the consequence of *erysipelas*—a dermatological condition already recognized as non-contagious at the time.[246]

Post-mortem examinations were also crucial in determining the cause of violent deaths. On 9 March 1650, aboard the Tuscan galley *San Francesco*, a dead body was found in the *scandolaro*—a storeroom at the rear of the craft. The body was that of a man about 30 years old, with red hair, and dressed as a sailor. As noted by Giovanni Battista di Vincenzo Mugrone, the assistant surgeon of the galley, the body displayed multiple wounds: one above the right breast extending "six fingers" (approximately 11–12 cm) toward the right arm; another caused by a bullet to the right elbow; and a third on the left ribs. A final wound extended from above the left shoulder all the way down to the kidneys. The description of these wounds matched those on a mariner named Tommaso Napoletano, who had been treated five days earlier after a fight with another oarsman named Giuseppe di Giovanni. According to Mugrone, the wounds on Napoletano's arm and chest were so deep that they were likely fatal. The autopsy report confirmed that Napoletano had died from these two penetrating injuries. Della Bordigliera was subsequently found guilty of homicide and sentenced accordingly.[247]

On 16 November 1654, the *aguzzino* on the Papal galley *San Pietro* was awoken at around five a.m. by convicts shouting, warning him that one of their companions had thrown himself overboard. After searching all the benches, he discovered that Giuliano Careso was missing. Shortly thereafter, Careso's body was found, tossed by the waves against the stern of the galley *Padrona*. The body was immediately pulled aboard, but it was too late: Careso had already drowned. His corpse was laid out on the bow, with his clothes stripped off, no hat, no shoes, and his chains still in place. The galley surgeon was called to examine him. The *aguzzino* reported:

---

[246] ASF, MP, 2116, c.n.n.
[247] ASF, MP, 2132, part II, dossier 8, *Liburnen. Iurisdictionis, sommario* 4, n.n.

> There was a dead man, whose name I do not know, lying on the ground with his face to the sky, a blackened face with blood in his mouth and nose, all swollen, and even frothing. He had two shackles on his right foot, one of which had a chain with fourteen links, with the last link broken… I cannot say by what instrument the chain was broken, nor can I determine the cause of death. However, because he is swollen, I'd say that he had drowned.[248]

The report from Francesco Parentanio, the surgeon authorized to voice an official opinion, confirmed the *aguzzino*'s suspicions:

> I can clearly see this dead man, who, having neither bruises nor wounds, is entirely bloated, his face blackened. I judge that he drowned, because I observe it is unusual for a man to be so bloated […] I judge, in [good] conscience, as far as my knowledge allows.[249]

Once the cause of death had been determined, the next step was to understand how Careso had managed to escape from his bench while still wearing his shackles. To assist in this, two blacksmiths were summoned to examine the chain. Both agreed that it had neither been cut nor sawn through, and that it was in perfect condition. They concluded that someone must have loosened it, as Careso could not have escaped otherwise. Suspicion soon fell on Vincenzo, a *mozzo*—a trusted sailor or auxiliary—who had unfastened Careso a few days earlier. Vincenzo had been ordered the previous Monday to replace Careso's chain. Careso's fellow benchers were all questioned and each confirmed that they had seen Careso bribing Vincenzo to change his chain during the night. They reported that both men had been drinking together around half past midnight, on the night before Careso's death. Despite Vincenzo's denials that he had replaced Careso's chain with a lighter one or had any dealings with him, the testimony of the entire crew pointed to his guilt.[250] As a result, Vincenzo was found guilty not of causing Careso's death —which appeared to be accidental—but rather of bribery and of releasing a convict without authorization from his superiors. This was considered a serious

---

[248] BCR, 34B13, *Raccolta di notizie e scritture diverse sopra le galere pontificie, armamento di vascelli, fatto dal papa Alessandro VII per soccorso di veneziani contro il turco, fortezza e porto di Civitavecchia*, f. 320: "che v'era un huomo morto, che il nome io non lo so, disteso in terra con la faccia verso il cielo, quale era con la faccia negra e sangue nella bocca e nel naso tutto gonfio facendo anco la schiuma et li aveva due maniglie al piede diritto, una de quali maniglie haveva una catena di maglie 14 e ultima maglia rotta… non posso sapere con che strumento sia stata fatta detta rottura, e di che sia morto io non lo posso giudicare ma perché è gonfio dico che si sia affogato."
[249] Ibid., f. 323: "Io vedo benissimo questo morto che non havendo botte ne ferito e tutto gonfiato con la faccia negra giudico che sia affogato perché colgo esser difforme d'un huomo per essere tutto gonfio […] giudico in coscienza per quanto s'e rende il mio sapere."
[250] Ibid., ff. 320–329.

crime, punishable by corporal punishment, removal from office, and often incurred the same sentence as the convict: service in the galley.[251]

Finally, it should be noted that among the numerous sudden and violent deaths, several resulted from the rowers' dissipated lifestyles, which, even if unintentionally, brought about their own demise. In March 1709, the slave Macametto of Abdala from Constantinople and known as "Stambuli," was found dead inside the *Bagno*, lying on a table where he had been sleeping. According to the surgeon Franceschini, the cause of death was the slave's excessive alcohol intake and constant smoking.[252] In this instance, there was no one to punish, only a dead slave whose fate elicited pity.

---

[251] AAV, Misc. Arm. IV/V, b. 54, f. 85: "28. Nessuno di qualsivoglia stato, e conditione ardisca far'sferrare Forzati, ne altro huomo di catena senza ordine del Sig. Proveditore i quale non lo doverà fare se prima non li viene ordinato da Noi, ò dal nostro Luogotenente generale in nostra assenza, sotto gravi pene corporali ad arbitrio, e privatione dell'offitio [...] 45. Che niuno di qualsivoglia conditione ardisca far sferrare, e sferri, ò disponga de forzati, ne di altri huomini di catena senza licenza nostra sotto pena di pagare quel che fuggisse à nostro arbitrio, e se sarà con dolo, & a mal fine la pensa sarà della galera e a vita."

[252] ASF, MP, 2107, c.n.n.

Chapter 4
On the Sin Abhorred by Beasts: Sodomy aboard Early Modern Galleys

1 Criminals and immoral individuals: Cultural representations of early modern *galeotti*

The aim of assembling a crew of militarily disciplined and God-fearing oarsmen underpinned a whole series of measures designed to impose stricter controls over the *galeottis*' behavior and morals. However numerous and strict the edicts concerning the governance of the galleys and their crews may have been, the sources—both institutional and procedural—invariably depict convicts and galley slaves as individuals prone to a range of criminal and immoral actions—primarily escapes, fights, and murders, but also forgery, gambling, blasphemy, and sodomy. The image that emerges is one of a violent and undisciplined microcosm, whose members were incapable of self-correction and redemption. The severity of punishments for offenders, along with the abundance of regulations on this front, risks creating a distorted image of crime aboard the galleys, as it is not entirely clear whether these edicts simply reflect an increase in vigilance against behavior deemed dangerous, or whether they serve as a blatant symptom of how the rules were not effectively respected, while offenses continued to be numerous—if not actually increasing. It is necessary to interpret these sources from both perspectives.

From a methodological standpoint, Mario Sbriccoli's position against a purely quantitative reading of criminal sources stands out. Such an approach would rest on "the false assumption that judicial archives contain the history of crime, when in fact they only record that of criminal justice." This abundance of procedures concerning certain actions considered illegal at the time was, in fact, evidence not so much of the reality of that particular crime, but rather of a more thorough and vigilant social oversight. To quote an example given by Sbriccoli, if theft appears to increase during times of famine, this should be interpreted as an actual rise "not in the hunger of the thieves, but in the fear and therefore the vigilance of the owners."[1] We may not necessarily be experiencing a real increase in criminal activity, but rather a more intransigent attitude toward it. Sbriccoli, however, does not advocate abandoning the quantitative approach to studying crime, but instead

---

1 Sbriccoli, Fonti giudiziarie, p. 493.

recommends using it judiciously, bearing in mind that "trials deal with crime, but they reveal justice."[2]

Furthermore, it is clear that whenever the authorities attempted to impose strict discipline, they were confronted with a highly complex and multifaceted reality made up of individuals who could not be subjected to a single rule and pattern of behavior. This is all the more evident when one considers that galley crews generally numbered around 300 men, and to think that it would have been possible to control every one of them effectively is surely utopian. Despite the cramped conditions, even the most vigilant observer would not have been able to keep a close watch on the entire crew, especially in contexts where attempts to extort and bribe officials—who pretended not to notice what was going on in return for the promise of profit—appear to have been quite common.[3]

Furthermore, despite the regulation requiring each rower to carefully observe whether his fellow oarsmen on the bench—as well as those in front and behind—were obeying orders, and to notify the galley's officers if they were not, sources suggest that mutual observation and denunciation among rowers was surprisingly rare. Instead, rowers appeared to respond to each renewed attempt at control with alternative strategies in their bid to avoid the authorities' gaze and maintain as much freedom as possible. A kind of tacit, mutual agreement seemed to prevail between slaves and convicts: they did not hesitate to bear false witness on behalf of themselves and their comrades to avoid punishment. Rather than fostering discipline, the increasing promulgation of regulations aimed at controlling and correcting slaves and convicts only served to marginalize them further from society, thereby consolidating the galleys' reputation as a haven for criminal and blasphemous individuals, ironically undermining the very discipline they sought to enforce.

Sodomy is often cited as one of the most widespread offenses committed among rowers of the Tuscan and Papal galleys between the 16th and 18th centuries. While reports of such acts involving slaves and convicts date back to the early use of these laborers aboard, the issue came to be seen as increasingly severe over time. As a result, legal responses became progressively more stringent, with legislation designed not only to curb the practice but also to intensify its legal suppression. Despite earlier records of sodomy among inmates in the 16th century, it was not until the 18th century that authorities officially acknowledged the phenomenon, potentially giving rise to the mistaken belief that it was a relatively recent problem.

---

[2] Ibid., p. 494f.
[3] Santus, "Il turco," p. 126f.

This raises the question of whether the increasing frequency of edicts and the heightened legal action against sodomy indicate a genuine rise in the practice, or whether they signal a shift in societal attitudes—where the imperative to publicly denounce and suppress the offense became more pressing than its concealment. The latter interpretation is more convincing, since the archival records frequently feature complaints of this practice even in the earliest years of the navies' existence.

The examination of these sources also invites reflection on the apparent paradox surrounding the treatment of sodomy, a crime that was both socially and religiously condemned, yet did not seem to provoke significant alarm aboard the galleys. More specifically, it prompts an analysis of the ambivalence in the authorities' stance toward sodomy among rowers. While it was officially recognized as a grave offense that needed eradication, in practice, limited efforts were made to suppress it unless it was connected to issues that threatened to destabilize the internal dynamics and operations aboard the galleys. A striking example of this can be seen in cases where the *mozzi* and *aguzzini*, who were tasked with monitoring and disciplining rowers, themselves faced accusations of sodomy.

The study of the representation and suppression of sodomy aboard galleys offers valuable insights into the broader debate surrounding a supposed paradigm shift in attitudes toward the practice during the 18th century. While the intellectual currents of the Enlightenment did not acknowledge the sexual identity of those labeled as "sodomites," nor eliminate the social condemnation of sodomy as unnatural, it nonetheless marked a critical juncture in the discourse. Enlightenment thinkers called for the decriminalization of sodomy, particularly from an anti-Catholic perspective. However, the 18th century also saw the emergence of a new consciousness regarding identity and gender among those engaged in same-sex relations, leading to the formation of distinct communities of individuals who identified as "homosexuals" due to their shared attraction to persons of the same sex.[4]

Such a shift in self-identification and gender awareness provoked a reaction from political authorities, who began to devote greater resources to isolating and

---

[4] Grassi, *Sodoma*, p. 21; Davidson, *L'emergenza della sessualità*, pp. 48–51. In the early modern period, the term "sodomite" referred to an individual act—that of "sodomy"—rather than to a particular group, and it carried with it the notion of sin. In contrast, the category of "homosexual" only emerged in the late 19th century to define a member of a specific group who identified with same-sex love relationships, not necessarily with particular acts. According to this theory, the concept of homosexuality as a "perversion" arose from a new understanding of illness linked to the emergence of modern psychiatry. The sodomite was a moral and legal category, whereas the homosexual became a pervert and a patient.

controlling these individuals, while punishing them more discreetly than in the past.[5] Although the naval authorities only took an interest in the "act" of sodomy—and not in the convicts' alleged homosexual inclinations—one can nevertheless notice a greater concern and willingness to curb and to find a solution to the problem posed by sodomy, viewed both as a vice and a crime.

It is noteworthy that even in the earliest sources, the fact that the *galeotti* indulged in sodomy is not described with any degree of surprise. On the contrary, the notion that galley convicts and slaves practiced sodomy is presented as almost self-evident—not due to assumptions about their effeminacy, but rather because they were perceived as inherently criminal and amoral individuals, relegated to the margins of society—if not entirely outside it—by virtue of their actions, inclinations, or even their religious beliefs. These beliefs, particularly Islam—the faith to which the majority of slaves adhered—were often viewed as predisposing factors and frequently associated with sodomitic vice.

The idea of inherent "inclinations" toward sodomy can be traced in early modern thought—though not in terms of gender identity or sexual orientation as understood today. Instead, such inclinations were viewed as moral or behavioral tendencies—believed to stem from criminality or immorality—either seen as innate or acquired over time through exposure to corrupting influences. As Marina Baldassarri observed, in early modern Rome, sodomy was often regarded as a consequence of marginalization, violence, and deviant behavior. It was also widely believed that many individuals who engaged in sodomy had themselves been victimized in their youth, leading them to internalize the practice as a habitual part of their identity. This cycle of victimization and normalization contributed to the formation of a true *habitus*—an ingrained disposition toward the act.[6]

While the available sources on sodomite *galeotti* appear to confirm the "acts paradigm"—according to which the concept of sexual deviance prior to the modern era applied only to acts and not to persons or identities[7]—Umberto Grassi reminds us that the sources ultimately reveal only that "it was the judges who were obsessed with acts."[8] It is important to bear in mind that, in these circumstances, the naval authorities were dealing with individuals already deprived of their liberty. Trials were not intended to determine the moral or legal status of the acts themselves, but rather to decide whether punitive measures should be imposed,

---

5 Ibid., p. 161.
6 See, Baldassarri, *Bande Giovanili*. This was also an accepted theory on the medical level. Some critiques on this reductive position have been advanced by Scaramella, *Un doge infame*; Lagioia, Passione Amorosa.
7 Halperin, *History of Homosexuality*, p. 29.
8 Grassi, *L'offizio*, p. 26.

primarily to set an example for the rest of the crew. The authorities' overriding concern was not driven by a moral or ideological imperative but by the need to prevent and punish any form of promiscuity aboard so as to maintain the strict military discipline essential to the success of seafaring operations.

Throughout the early modern period, sodomy was condemned both as a crime and as a sin, with its denunciation deeply rooted in cultural, legal, and religious discourses and representations. This moral and legal framework created a theoretical gap in early modern medicine concerning the topic of sodomy, which was often deemed too vile and shameful to be addressed in medical discourse. However, contrary to traditional historiography, it is possible to argue that a form of "medicalization" of sodomy existed even before the 19th century. While the theoretical connection between medicine and sodomy appeared tenuous, in practice, the expertise of doctors was frequently sought during sodomy trials. In fact, sodomy between men[9] had to be "perfect" to be considered punishable—that is, involving anal penetration.[10] In many cases where the accused denied the act, it became necessary to conduct a meticulous physical examination to identify tangible signs of sexual intercourse, such as the presence of hernias, which were then thought to indicate penetration. Unsurprisingly, the only individuals equipped with the knowledge and skill-set to perform such examinations were doctors. That said, in many cities, doctors were legally required to report any signs of illicit sexual activity discovered while examining their patients[11]—a responsibility that extended to those aboard the galleys as well.

Medical examinations in sodomy cases reveal another layer of complexity in the doctor-patient relationship, marked by inherent ambiguity. Indeed, the sources often depict a dynamic in which the doctor and the accused engage in a kind of rivalry. Eager to avoid admitting guilt, the accused would frequently deny the findings of the medical examination, offering inventive—if often implausible—excuses. Beyond the dissimulation of sexual practices, language itself played a key role in obscuring the truth. In this context, medical expertise was crucial to uncovering the facts and ensuring that such deceptions did not go unchallenged.

---

9 It is important to note that in the early modern period the definition of sodomy also encompassed anal intercourse between a man and a woman, masturbation, and sexual acts with animals.
10 This was penetration performed in an illicit vessel (*penetrazione in vaso indebito*).
11 The first city which envisaged this type of measure was Venice in 1453. Canosa, *Grande paura*, pp. 107–115.

2 Sodomy and its condemnation in the early modern period

In spite of the numerous and profound attempts to impose discipline among rowers, early modern Italian galleys remained highly undisciplined spaces. The control exerted over convicts and galley slaves targeted not only their physical labor but also their behavior. Alongside discipline aimed at improving technical prowess, there were also efforts to regulate their moral conduct. This explains the critical role played by shipboard chaplains, who, alongside the naval authorities, were tasked with supervising both the slaves and convicts. Despite all the attention to the rowers' behavior, any infractions and disturbances that occurred were often described as regular occurrences. Among the most widespread disturbances, one in particular seems to have caused great concern for the galley chaplains: sodomy. While often concealed at an institutional level, only much later did it become a specific subject for regulation.

The frequency with which sodomy was practiced among galley convicts and slaves did not surprise the authorities. Throughout the early modern period, the "nefarious vice" was condemned not merely as a manifestation of effeminacy, but as an extreme expression of a violent and immoral nature—one that, in fact, led its perpetrators to reject every value on which society was based. Drawing on arguments from the Old Testament, as well as from St. Paul (4–64) and St. Thomas Aquinas (1225–1274), it was argued that, in God's natural order sexual desire should be solely directed toward reproducing the human species, and that, therefore, any such act not intended for procreation was a violation of God's plan.[12] Sex against nature became the symbol for rebellion against God's Divine Will, which, in turn, led to the attribution of "demonic" characteristics to sodomites.[13] Indeed, it was considered not only to be contrary to the natural order of creation [*peccatum*], but also contrary to the social and ethical order on which the civil community was based [*crimen*].[14]

2.1 Sodomy as a sin against nature

The term "sodomy," coined in the Middle Ages, carried deeply religious connotations due to its biblical origin. Sodom was, in fact, one of the five cities of the Pentapolis in Canaan, which—according to an account in the Book of Genesis—

---

[12] Lavenia, *Un'eresia indicibile*, pp. 8–11.
[13] Grassi/Marcocci, *Le trasgressioni della carne*, p. 12.
[14] See Grassi, *Sodoma*, pp. 53–55.

was destroyed by God, along with the city of Gomorrah, to punish its inhabitants for sexual transgressions. However, the homoerotic interpretation of the Genesis account seems to have emerged later, as the biblical episode never explicitly refers to homosexual rape.

Genesis 19 (1–25) does not explicitly mention the practice of homoerotic intercourse by the inhabitants of Sodom. Rather, it recounts the arrival of two angels at the city gates at nightfall, where they encountered Lot, who insisted on offering them shelter overnight in his dwelling. When it was time to retire for the night, Lot's house was surrounded by the people of the city. They demanded to see the strangers, as they wished to "abuse" them. Lot offered his daughters in exchange, but the Sodomites attacked him, prompting the angels to intervene, rescue Lot, and urge him to flee the city, which, having incurred the wrath of God, was destroyed by a rain of fire and brimstone.[15]

Up until the first centuries of the Christian era, the scourge of Sodom and Gomorrah was attributed to the violation of the obligations of hospitality and assistance to foreigners and travelers, and to the duty to protect the poor.[16] Later reinterpretations shifted the focus from social transgression to sexual ones. The first explicitly homosexual reading of this biblical passage came relatively late, and is thought to date back to the Jewish philosopher Philo of Alexandria (c. 25–50).[17] The idea that the sins of the inhabitants of Sodom were attributable to sexual activity, and in particular the idea that divine punishment by fire and brimstone was intended for the sins of the flesh [*scelera carnis*], was not officially accepted until Gregory the Great (540–604).[18]

Despite the absence of a direct condemnation of sodomy in Scripture, the anti-homosexual reinterpretation of the myth of Sodom and Gomorrah gained increasing traction, and was reinforced by the Church Fathers, who generally held a markedly negative view of sexuality. Libido—framed within the legacy of concupiscence—was conceived as a disordered and destructive force. Within this framework, it was not merely beyond the control of the will, but was indicative of a fallen nature, and thus came to be associated with what lay outside the bounds of rational self-mastery. Accordingly, its renunciation was interpreted as an expression of spir-

---

15 As stated in Boswell, *Christianity*, p. 94: "In only ten of its 943 occurrences in the Old Testament does it [sodomy] have the sense of carnal knowledge."
16 Boswell, *Christianity*, pp. 94–97.
17 Lv 18:22 "cum masculo non commisceberis coitu femineo, quia abomination est, cum omni pecore non coibis;" 20:13 "qui dormierit cum masculo coitu femineo uterque operati sunt nefas, morte moriantur."
18 D'Angelo, *Liber gomorrhianus*, p. 42.

itual strength and moral freedom.[19] Although Jesus does not explicitly condemn homosexuality in any passage of the New Testament, it is nevertheless there that we encounter the first reference to homoerotic relationships as acts "against nature," in the extended invective against the pagans contained in Paul of Tarsus' Epistle to the Romans.[20] In Romans 1:26–27,[21] one might read an explicit condemnation of sodomy in the censure of Roman sexual practices, which were linked with idolatry, and which St. Paul associated with pagan orgiastic rites in honor of false deities.[22] Similarly, two words from the Letters of St. Paul to the Corinthians (1 Corinthians: 6:9),[23] and from St. Paul to Timothy (1 Timothy 1:9–10),[24] have been interpreted since the twelfth century as clear indications of a condemnation which would result in the exclusion of "men who lie with males" [*masculorum concubitores*] from paradise.[25] For St. Paul, sex against nature was not a cause of divine punishment, but rather evidence of it: the visible consequences of a moral and spiritual disorder rooted in idolatry. Modern scholarship has revisited these interpretations to challenge long-standing assumptions. Moreover, as John Boswell argued in his seminal work *Christianity, Social Tolerance and Homosexuality* (1980),[26] St. Paul did not appear to condemn homoeroticism *per se*; rather, he sought to protect young males from the pederastic practices typical of Greco-Roman society.[27]

Boswell identified and categorized the core accusations against homosexuality made by the Church Fathers. First, the accusation of "bestiality," which finds its clearest expression in the Epistles of Barnabas (70–132). Although now considered apocryphal, these epistles—included in the *Codex Sinaiticus*, one of the most renowned manuscript copies of the Bible—were accepted as Scripture

---

19 Grassi, *Sodoma*, p. 29.
20 Ibid., p. 25.
21 "propterea tradidit illos Deus in passiones ignominiæ nam feminæ eorum inmutaverunt naturalem usum in eum usum qui est contra naturam "similiter autem et masculi relicto naturali usu feminæ exarserunt in desideriis suis in invicem masculi in masculos turpitudinem operantes et mercedem quam oportuit erroris sui in semet ipsis recipientes."
22 "nam feminae eorum immutaverunt naturalem usum in eum usum, qui est contra naturam. Similiter autem et masculi relicto naturali usu feminae exarserunt in desideriis suis invicem, masculi in masculos turpitudinem operantes et mercedem, quam opportuit, erroris sui in semetipsos recipientes."
23 "iniqui regnum Dei non possidebunt [...] neque molles neque masculorum concubitores."
24 "sciens hoc quia iusto lex non est posita sed iniustis [...] sceleratis et contaminates paricidis et matricidis homicidis fornicariis masculorum concubitoribus."
25 Boswell, *Christianity*, p. 106 f.
26 Although outdated, John Boswell's thesis still represents a starting point for the study and reflection of sodomy during antiquity and the early modern period. See Kefler, *The Boswell Thesis*.
27 Benigno/Lavenia, *Peccato o crimine*, pp. 100–103.

during the early centuries of Christianity and influenced the writings of many prominent Church Fathers for generations. This symbolic linking of animal traits with sexually illicit conduct found echoes in classical natural history." Referring to notions widely accepted in the ancient world and found in works such as Pliny's *Naturalis Historia* (ca. 77–78), the author justified Moses' ban on eating the flesh of certain animals by linking them to various sexual sins.[28] The hyena, for example, was believed to be able to change sex and, like the hare or the weasel, was obsessed with sexual intercourse. Over time, these animals became metaphors for homosexuality, and Moses' prohibition against consuming them came to be interpreted as a covert condemnation of homoerotic relationships. The influence of the Epistles of Barnabas is evident in Clement of Alexandria's *Paedagogus* (c. 150), a manual for Christian parents, in which the author advocated for the "Alexandrian rule" according to which, in order to be moral, sexual relations had to be directed solely and exclusively toward procreation.[29]

Similarly, in the medieval *Physiologus* (2nd–3rd centuries A.D.) a text of anonymous authorship—commonly known as the "bestiary"—comparable condemnations can be found; likewise, Bernard of Cluny in the 12th century denounced men engaging in homosexual relations by likening them to hyenas.[30] Notably, the belief that the hyena was capable of switching gender was also accepted by pre-modern medicine, due to the particularly developed clitoris in females of the species which often gave rise to confusion in distinguishing between the sexes. In fact, "the enlarged clitoris, through which she urinates, copulates, and gives birth, resembles a penis due to pseudo-phallic erections; the labia are shaped much like testicles."[31] As early as the Middle Ages, the hyena had become a symbol of hermaphroditism and of sexual mores devoid of fixed gender roles, and was therefore perceived as a sign of deviation from the natural order of creation.[32]

Alongside accusations of bestiality, homosexuality was also associated with other reprehensible behaviors—such as the tendency to molest children, and even more gravely, paganism.[33] Homoerotic relations thus became a symbol and a distinguishing mark of barbarians and heretics who, by rejecting the sacred Word of God, also defied it in their conduct, thereby violating the nature of the di-

---

[28] Boswell, *Christianity*, pp. 137–139.
[29] Ibid., p. 140.
[30] Ibid., pp. 141–145.
[31] In this regard, please refer to Zuccolin, Hyena, p. 672.
[32] Ibidem.
[33] Boswell, *Christianity*, pp. 143–145.

vine plan.³⁴ The theological weight of these condemnations was further solidified in the works of influential early Church thinkers. In this regard, St. Augustine's (354–436) reflection on the "Nature of Good" is central. As Boswell noted, the expression "against nature" can be misleading, since the idea of "nature" as a universal law or metaphysical truth was largely absent at the time. The term appears, rather, to refer to the character—the personal nature—of the individuals in question—to the innate disposition of human beings in the sense of what was "natural" to them.³⁵ Consequently, whenever St. Augustine employed the term "nature," he was not referring to an abstract ideal or metaphysical category, but rather to the intrinsic and "normal" characteristics of the human species. His condemnation of sodomy was seen not as a crime against an abstract law of nature, but as a distortion of human nature itself, which God had created for the purpose of procreation—not for homoerotic intercourse.³⁶

St. Augustine, who had accepted the hypothesis linking male homosexual relations with the sin of the sodomites in *De Civitate Dei* (413–426), explicitly condemned sodomy as being against nature in his *Confessiones* (398), arguing that "when the nature created by God was violated by such perversity, it was the very union between God and man that was compromised." Given that the telos of human nature is procreation, St. Augustine also condemned as shameful any kind of sexual act not directed toward that end, and in such cases advocated abstinence—even between spouses—as the preferable option. Lust, according to St. Augustine, was the first consequence of Original Sin and thus became symbol of human decadence. Had Adam and Eve not succumbed to the serpent's temptation, they would have enjoyed sexual pleasure, aimed purely at procreation, and the penis would have been subject to human will like any other limb. Instead, Original Sin irretrievably corrupted sex, which, now totally driven by desire, became the symbol of everything contrary to reason. Accordingly, all sexual intercourse not directed at procreation was to be condemned, and sodomy, which was not only stemming solely from sexual pleasure but also involved a reversal of gender roles, with men assuming the role of women—deserved the harshest condemnation.³⁷

Despite this early condemnation by the Church Fathers, up until the eleventh century, controlling sodomy did not seem to be a particularly pressing issue on the political agendas of ecclesiastical and secular institutions. The social and political

---

34 The association between sodomy and heresy—and, in particular, between sodomy and the Muslim faith—will be discussed in more detail in the following pages.
35 Boswell, *Christianity*, p. 110 f.
36 Ibid., pp. 150–152.
37 Grassi, *Sodoma*, p. 29–31.

changes over the following centuries, however, radically altered this situation. A key manifestation of the shift in the Roman Church's attitude toward sodomy occurred at the Council of London of 1102, when it was officially defined as a sin for the first time, equated with fornication, and pederasty.[38] At the Third Lateran Council in 1179, Canon XI introduced legal provisions to defrock any cleric found guilty of sodomy, while also prescribing excommunication and banishment for laymen. These provisions were later included in the *Decretales* of Pope Gregory IX (r. 1227–1241), which, in turn, became part of the *Corpus Iuris Canonici*.[39]

The scholastic period would take these theological foundations and codify them into a coherent moral framework. The Catholic Church's condemnation of sodomy as an act "against nature" was enshrined and systematized in the work of St. Thomas Aquinas (1225–1274). In *Summa Theologica*, a veritable systematization of Christian doctrine, Aquinas asserted that God had directly communicated—and made discernible in the human mind—the law of nature, which, put simply, was "the participation of rational creatures in the eternal divine law." According to the Thomistic conception, sin constituted an unnatural deviation that distanced mankind from God and reduced the human condition to a level closer to that of beasts. Sins against nature are analyzed in the *Summa* in the section on lust, specifically in Articles XI and XII of *quaestio* 154 of the *Secunda secundae*. These vices were denounced as especially grave—not only contrary to reason and nature but also contrary to God—and were considered more serious than sacrilege, since the order of nature even precedes Christian revelation.[40]

In redefining sexual behavior from a theological standpoint, Aquinas distinguished *fornicatio*, a less grievous sin, from *abominatio*, which he identified as a true act against nature and the most serious form of lust. Aquinas listed several sexual behaviors that fell under the category of "against nature," presented in increasing order of gravity: *immunditia* or *mollities*, associated with masturbation; *bestialitas*, sexual intercourse with non-human creatures; *sodomia*, same-sex acts, male and female; and more broadly all forms of sexual intercourse considered unnatural, especially anal intercourse [*in vaso indebito*].[41] As several historians have noted, it is significant that, within the Thomistic framework, sodomy and "sin against nature" were not synonymous; the former was merely a subcategory of the latter.[42] In any case, Aquinas classified such acts as sins against the Sixth Commandment—"Thou shalt not commit adultery"—though with partic-

---

38 Baldassarri, *Bande giovanili*, p. 109.
39 Ibid.
40 Lavenia, *Eresia indicibile*, p. 14.
41 Canosa, *Grande paura*, p. 18.
42 Grassi, *Sodoma*, p. 51 f.

ular gravity, due to what Aquinas referred to as *specialis ratio deformitatis*, that is, a specific reason for the act's deformity which rendered the venereal act especially indecent. Moreover, sodomy was also seen to contravene the First Commandment—"I, the Lord, am your God. Thou shalt not have other gods besides me"—insofar as it not only offended God but also implied a fundamental "error" in judgment.[43]

2.2 Sodomy as a crime against nature

From the twelfth century onward—and even more markedly with the development of inquisitorial justice in the thirteenth century—the assimilation of religious and social deviance, including sexual deviance, became increasingly pronounced. As Pope Lucius III (r. 1181–1185) declared in his decretal *Ad abolendam*, issued in 1184, the Church's task was to define heresy, while it was the secular authorities' duty to eradicate it. This principle was later confirmed by a decretal of the Fourth Lateran Council (1215), which obliged secular authorities to expel heretics from their territories.[44] Sodomy, already condemned as a sin, soon came to be prosecuted as a political crime, assuming the characteristics of *lèse-majesté*, and was thus considered a threat to the security of the state.[45] As a result, political authorities were compelled to react, seeking ways to eliminate the sodomitic vice from their cities, both to preserve social order and to avoid incurring divine wrath.

This intertwining of the moral and criminal spheres was fully justified by the legal system's increasing tendency—already evident in the Middle Ages but reaching its apogee in the 16th century—to politicize even religious or moral transgressions. Behaviors traditionally defined as sins became political offenses, ceasing to be matters of private conscience and instead assuming the status of public concern. The result was a blurring of boundaries between crime and sin, and the emergence of the belief that criminality and immorality were necessarily perceived as two sides of the same coin.[46] With the rise of the Inquisition, the distinction between sin and crime became increasingly fluid, and any transgression could thus potentially be reclassified as a crime—an expression of disobedience to constituted authority.[47] The classification of sodomy and blasphemy as political

---

43 Lavenia, Sessualità, islamofobia, p. 105 f.
44 Grassi, *Sodoma*, p. 53 f.
45 Grassi, *L'offizio*, p. 181 f.
46 Sbriccoli, *Crimen laesae maiestatis*.
47 Prodi, *Storia giustizia*, pp. 93–97.

crimes led to both sins being treated as conscious, subversive acts of defiance against the moral values underpinning social coexistence. As such, they came to be regarded as acts of grave insubordination, undermining the moral values and sovereign authority upon which social order depended.[48] Consequently, justice could not tolerate either blasphemy or the outright denial of God, and was compelled to respond in an intransigent manner.[49] As Paolo Prodi observed, traditional sins were increasingly publicly punished by sanctions, as "civil religion" came to be a key factor in fostering social cohesion throughout these centuries.[50]

Recognizing the social order as a reflection of the natural order—which, in turn, mirrors a higher divine order—the imposition of what Sbriccoli termed a "liturgy of obedience" became essential for maintaining the prevailing hierarchy.[51] This order was not merely political or social, but moral, natural, and therefore cosmic. Anyone who contravened it committed a mortal sin against the very structure of creation. Non-conformity thereby took on a political dimension, and warranted both earthly and heavenly punishment—social and moral sanctions alike.[52] The gravity of such deviation was understood as a disruption of the prevailing social order, upon which all rules, roles, and hierarchies rested.[53] Transgressions related to the sexual sphere were considered particularly dangerous, as they undermined not only the principles of Christian morality but also the very pillars of societal stability. For this reason, no leniency could be granted: suppression and punishment had to be swift and rigorous. Condemning these crimes became a civic moral duty, as it was in the public interest that any offense against the social order be met with unequivocal retribution.[54]

The criminalization, suppression, and punishment of sodomy, even on a political and legal level—though it reached its peak in the early modern period—had very ancient origins. As early as *Leviticus,* the prescribed punishment for sodomites was death.[55] Reflections on the criminal nature of sodomy became increasingly profound and precise over the centuries. A case in point: in the latter half of the 12th century, the French theologian Peter Cantor equated sodomy with murder in a chapter of his work *Verbum Abbreviatum* [The Abridged Word], dedicated to the sodomitic vice [*vitium sodomiticum*]. Since God had created man and woman

---

48 Sbriccoli, *Crimen laesae majestatis.*
49 Grassi, *L'offizio,* pp. 181–184.
50 Prodi, *Storia giustizia,* p. 170.
51 Sbriccoli, *Crimen laesae majestatis,* pp. 117–127.
52 Ibid., p. 102f.
53 Prodi, *Storia giustizia,* p. 292.
54 Ibid., pp. 133–135.
55 Boswell, *Christianity,* pp. 101–103.

for the purpose of procreation, both the murderer and the sodomite were perceived as guilty of destroying humankind—the former by killing, and the latter by failing to procreate.[56]

Similarly, from the late 13th century onward, sodomy was classified as an "atrocious crime" [*enorme delictum*] on a par with murder, repeated theft, and counterfeiting, and was therefore subject to corporal punishment[57] as evidenced in the Code of Alfonso X, issued in 1255, though only enforced in the 14th century. Criminalization also occurred in France: the *Livres de Jostice et Plet* (also known as the *ancienne coutume d'Orléans*, circa 1260), and the *coutumes de Beauvaisis*, compiled by the jurist Philippe de Beaumanoir in 1285, both prescribed corporal punishment for sodomites. The *Livres de Jostice et Plet* imposed castration for the first offense, amputation of the penis for the second, and burning at the stake for the third.[58] By the late 13th century, almost all European legislative codes contained specific laws and provisions targeting sodomy,[59] reflecting a broader consensus around its legal and moral condemnation.

The perceived gravity and impiety of sodomy is also reflected in the etymology of another term commonly used to refer to the *vizio nefando* [nefarious vice]. Alongside its classification as a sin or a crime, sodomy was regarded as so abhorrent and detestable that it could not—indeed, should not—even be named. The term "nefarious" derives from the Latin *nefandus*, composed of the prefix *ne* and the suffix *fandus*, meaning "that which must not be spoken."[60] It denoted an act so vile that it defied description.[61]

By the 14th century, the Latin term had been adapted into Italian and used to describe not only sexual relations between men but, more broadly, to denote all acts considered "against nature." This encompassed "any kind of libidinous coitus either with an improper person or in an improper orifice."[62] Despite this generality, the term typically referred to male same-sex relationships. Indeed, sodomy

---

56 Canosa, *Grande paura*, p. 15 f.
57 Rocke, *Forbidden Friendships*, p. 20.
58 Canosa, *Grande paura*, p. 19.
59 Grassi, *Sodoma*, p. 54.
60 *Fandus* is the gerundive of *fari*, meaning "to speak with solemnity." In Latin grammar, the gerundive is a verbal adjective expressing obligation or necessity, typically with a passive force and found in both active and deponent transitive verbs. Often called a "participle of necessity," it conveys an action that must be done—an imperative here explicitly negated by the prefix *ne*, thus rendering *nefandus* as "that which must not be spoken."
61 Alfieri, Il discorso su tribadi e sodomiti, p. 21.
62 Cattaneo, "Vitio nefando," p. 57: "coitus libidinosus vel cum persona indebita vel in vaso indebito."

in medieval and early modern discourse encompassed a wide spectrum of non-reproductive sexual acts, including anal intercourse between men, between men and women, masturbation, pederasty, and even bestiality—all of which were viewed as transgressions against the natural order and thus subject to moral and legal condemnation.

It is notable, as many historians since the late 1970s have observed, that the severity of anti-sodomy legislation was not always matched by comparable levels of rigor in its enforcement, which was often relatively moderate. This leniency might have taken the form of severe but infrequent penalties, or conversely, of more frequent condemnations tempered by relatively mild sentences.[63] Several explanations have been advanced for this apparent discrepancy. The most widely accepted among historians is that the authorities sought to keep the extent of sodomy's spread in a given area hidden from public view, primarily for reasons of decorum and to preserve a city's good reputation, as appears to have been the case in the Papal States. In other contexts, such as Florence and its territories, this more tempered approach may have functioned as a pragmatic strategy aimed not at the outright suppression of the nefarious vice, but at its containment.

3 Moral order and sodomy: sexual transgressions among early modern *galeotti*

It is, therefore, not surprising that early modern *galeotti*, who were forced to row either because of past crimes or because they were slaves belonging to a heretic religion, would engage in illicit sexual relations. The immoral acts they committed were seen as a further expression and confirmation of their inherently perverse and evil nature. Nor should we forget the persuasiveness of the, albeit reductive, "plumbing thesis," which—if the sources are to be believed—enjoyed considerable traction at that time. According to that view, the marked incidence of sodomy in highly rationalized and supervised all-male segregated settings, such as prisons, galleys, and the army, inhibited heterosexual impulses and redirected them toward homosexuality.[64] Forced into all-male environments, the *galeotti*, and more broadly those working aboard ships, practiced sodomy regularly, and according to precise, codified patterns.[65]

---

[63] To cite some works: Courouve, *La répression*; Lever, *Les bûchers*. Similar conclusions can be derived from the data collected in Rocke, *Forbidden Friendships*; Grassi, *L'Offizio*, Baldassarri, *Bande giovanili*.
[64] Rowson, Omoerotismo, p. 50.
[65] Burg, *Pirati e sodomia*, p. 214; Calcagno, "Brutale libidine," p. 176.

Whatever the reasons the *galeotti* practiced sodomy, what concerns us here is the fact that they did so and that they usually managed to evade the strict control systems in place at sea, aboard the galleys, and ashore—in hospitals or within the *Bagno* compound. In response to continued concerns, from the 18th century onward, the naval authorities began to envisage stricter oversight and more intransigent measures of repression. The 18th century thus represents a juncture in terms of the control and suppression of irregular sexual mores among rowers. The mounting severity of the punishments meted out and, more generally, the proliferation of sources on the subject—especially procedural ones—should not lead us to believe that sodomy was an absolute novelty; quite the contrary. The extant sources on sodomy aboard the galleys must be analyzed as evidence of a heightened vigilance toward this practice, and simultaneously as a symptom of a persistent disregard for the rules already in force, and perhaps, of an ever-increasing number of violations.

3.1 Vice and vigilance: sodomy and social control in the Tuscan maritime state

Florence's reputation as a "sodomite city" was already well established throughout Europe in the 14th century—so much so that the popular German term for "sodomite" was *Florenzer*, and the verb *florenzen* meant to "sodomize."[66] This stereotype was confirmed by an innumerable series of testimonies, not only from foreign authors but also—and more tellingly—from Italians themselves. For example, the Bolognese chronicler Matteo de' Griffoni attributed the cause of the Arno's flooding in October 1333 to the divine wrath provoked by the spread of the vice in Florence.[67] Similarly, in 1376, Pope Gregory XI, in his denunciation of usury and sodomy—named them as two of the characteristic sins of Florentines.[68]

As already anticipated in Chapter 2, aboard the galleys of the Medicean fleet, sodomites accounted for only 1.1% of those condemned to life at the oars, despite the fact that this offense was the first crime ever punished with a galley sentence in Tuscany. This can be explained both by the restriction of such criminal sentences to repeat offenders and by the relatively limited number of surviving trial records—due largely to the difficulty of prosecuting such cases.[69]

---

66 Canosa, *Grande paura*, p. 24.
67 Ibid.
68 Rocke, *Forbidden Friendships*, p. 3.
69 Lo Basso, *Uomini da remo*, table p. 341.

As with the Papal galleys during the same period, the directives governing the Medici fleet included no explicit provisions to prevent illicit acts of this nature—with only a few rare exceptions. This omission should not be read as evidence of the crime's absence, but rather a deliberate effort to keep it hidden. To acknowledge the existence of such practices would have constituted a serious scandal—not only for the perpetrators themselves but also for the crew as a whole and, above all, for the authorities under whose jurisdiction the vessels operated. Furthermore, to publicize the fact that such acts were widespread would have only exacerbated the stereotype of Tuscany as a land of depravity. Multiple documents attest that the authorities explicitly sought to punish sexual misconduct "without publicity and solemnity," shielding proceedings from the indiscreet gaze of the populace.[70] The decision not to prosecute sodomy may also have been motivated by a pragmatic concern: to avoid the loss of oarsmen vital to the fleet's operation. After all, though it was deemed an abominable practice, if practiced discreetly, it caused minimal disruption to navigation; it is therefore plausible that the authorities chose to turn a blind eye to such conduct aboard.

The earliest reference I have found to such behavior aboard Tuscan galleys appears in the instructions issued in 1590 by Francesco da Montauto, Knight of St Stephen and General of the Galleys. These directives prescribed life sentences at the oar for convicts and two years' imprisonment for any knights of the Order caught engaging in illicit sexual acts.[71] While it is impossible to determine why de Montauto took this extra precaution, certain assumptions can be ventured. Reading the instruction's *incipit* reveals that the fleet was about to set sail for Sicily and that de Montauto had agreed to allow Don Diego Enríquez de Guzmán, Count of Alvadeliste—who had previously served as Viceroy of Sicily—[72] to travel aboard one of his galleys.[73] Could it be that the presence of this illustrious guest compelled the General to exert stricter control over the crew's behavior, with the dual aim of not inconveniencing his distinguished passenger and, at the same time, of preventing his own name from being tarnished by reports of a failure to maintain discipline? It strikes me as a rather plausible hypothesis. As has already been argued, there was an awareness that they were dealing with slaves

---

70 See for example ASF, MP, 2109, cc.n.n.: "purchè ciò segue di notte e senza pubblicità."
71 ASF, MP, 2131, dossier 3, c. 31r.
72 Alba De Liste (Alvadeliste), Diego Enríquez de Guzmán conte di, in *Dizionario biografico Treccani*, consulted in May 2023.
73 ASF, MP, 2131, dossier 3, c. 30r. The practice of embarking passengers on galleys was frequently documented. The captain made sure that guest passengers felt at ease and that crew members behaved appropriately.

and criminals "who are morally inclined, either through licentiousness or hardened habit, to live badly."[74]

Evidence of the vice within the *Bagno de' forzati* is comparatively more abundant. According to the aforementioned Father Filippo of Florence, the practice was encouraged not only by the general promiscuity prevailing within the compound—where slaves and convicts slept with each other without restriction—[75] but also by the guards who accepted bribes in exchange for their silence. While the *Bagno* operated under a regime of militaristic discipline, disturbances and violence remained commonplace. Significantly, even in this context, no specific measures were introduced in the early years to curtail such conduct. According to the instructions issued by the Captain of the *Bagno* in July 1609, the only preventative rules were that young prisoners should not leave their assigned quarters at night, and that neither prostitutes nor "suspicious" young men should be allowed to enter the building freely—since their very presence, it was feared, would arouse the inmates' libido. Deprived of women on whom to satisfy their carnal desires, the inmates would inevitably turn their attention to these young men.[76]

That these rules were poorly enforced—and that the *Bagno* authorities often permitted this state of affairs—is demonstrated by a 1648 investigation ordered by the Grand Duke. This inquiry followed accusations of corruption lodged by eleven slaves and three *buonavoglia* against the Captain of the *Bagno*, Marco Fabbroni, who, in turn, deflected the charges toward the Vice-Commissioner of the galleys and Scribe General, Francesco Pepi. To resolve the matter, Rector Domenico Puccini was dispatched to Livorno. Upon his arrival on 2 July, he issued a proclamation calling on all oarsmen and officers to testify regarding the conduct of the two ministers. The investigation concluded with the conviction of Captain Marco Fabroni, who was found to have orchestrated an elaborate system of bribery and violence, in which a significant portion of the income was derived from payments extorted from convicts and slaves in exchange for sexual access to young inmates.

It is notable that the prisoners who provided this testimony were all slaves: the first to speak was Romadà of Seit from Tunis, who carried out his reporting

---

74 Ibid., dossier 6, *Ordini da osservarsi nelle Galere del ser.mo G.duca di Toscana*, c. n.n.: "i quali sono moralmente inclinati ò dalla Licentia, o dall'habito confermati à mal vivere."

75 Bernardi, *Relazione*, p. 17.

76 ASF, MP, 2132, dossier 7, *Costituzione, et ordinazione dell'Offitio del Capitano del Bagno*, cc.n.n.: "capo XXII. Et sopra tutto proveda che li Giovani stiano nelle stanze assegnateli, se parimente di giorno, e più particolarmente la notte [...] capo XXIIII. Non lasci entrare nel Bagno donne di partito, ne praticarci Giovani di sospetto, accio si rimedi all'inconvenienti, che ne succede."

work with the utmost scrupulousness, denouncing all those involved without regard to their religion. This shows that—although systems of tacit agreement and forms of complicity certainly existed among the *galeotti* to evade official oversight—ultimately each man had to act in his own best interest, especially in a reality where brute pragmatism governed all conduct.[77]

This state of affairs changed with the arrival of the Capuchins at the head of the *Bagno* and, in particular, following the establishment of the *vigilanti di Maria*. To eliminate the vice, the Capuchins sought to resolve the problem of promiscuity among the *galeotti* by introducing a rigorous system of denunciation and by preventing direct contact during sleep: wooden boxes—open at the front—were installed in the dormitories, within which each convict was required to sleep alone. Any slave or convict caught *in flagrante delicto* was publicly and severely punished: he would be pilloried and subjected to 50 or 100 lashes, depending on the gravity of his sexual misconduct.[78] To "arouse terror in the others," and thus dissuade them from future transgressions, sodomites were also punished by having a "ring" affixed to their foot with the "iron sock."[79]

Tellingly, the sources documenting such acts within the *Bagno*—albeit initially limited—increased once the Capuchins took control. On one hand, this reflects the heightened vigilance regarding the *galeotti*'s behavior; on the other, it reveals that the system of intimidation introduced by the Capuchins did not fully eradicate the vice among inmates. Furthermore, it is reasonable to assume that numerous cases of sodomy were not voluntarily recorded or prosecuted, due to the authorities' desire to keep such disorders as discreet as possible. However, just as riots and violence persisted, so, too, did sodomy.

As early as 1680, the *Bagno*'s *aguzzino* was instructed to segregate young rowers from the rest of the crew and to isolate them in a special section called the *Bagno de' Giovani* [youth prison], in an effort to limit sodomy, which was believed to be incited by the mere presence of young men—reflecting the early modern association of this offense with pederastic paradigms. These precautions were not entirely effective, as sodomy could just as easily occur between two adult men.[80] Though less frequently documented, sodomy between young men also

---

77 ASF, MP, 2168, part 1. The episode has been analyzed in Santus, *"Il turco,"* pp. 127–130.
78 Bernardi, *Relazione*, pp. 33–37.
79 See, for example, ASF, MP, 2112, 2114, 2117 cc.n.n.
80 ASF, MP, 2099, c. 107r–v; ASF, MP, 2130, cc.n.n. Presumably, this was the section of the *Bagno* called Sant'Antonio. For the early modern period, criminal sources on sodomy show a clear predominance of male pederasty. This distortion is likely due to the violent nature of such acts, which were therefore prosecuted. Consensual same-sex relations are rarely recorded, and when

took place—as in 1681, when two young convicts caught having sexual intercourse were punished with 25 lashes every morning for a month.[81] Nevertheless, in 1694, Father Ginepro complained that the *Bagno de' Giovani* was "a den of vice and turpitude." To address this persistent problem, the internal layout of the *Bagno* was reconfigured: prisoners were to sleep alone, each confined to a wooden chest, open at the front.[82]

There is evidence, however, of sodomy among galley rowers as late as the 18th century, and its frequency would appear to have increased. As in earlier periods, there are accounts of officers—and in particular, the leader of the *vigilanti di Maria*—who actively encouraged such behavior by promising to feign ignorance in exchange for a bribe.[83] In January 1715, for example, the *galeotti* denounced the guardian of the *Bagno* of Sant'Antonio, Bartolomeo Piovani, who was accused of offering the older rowers money to eat in other canteens, thereby allowing him to *"fare raddotto di gioventù"*—to concentrate all the young rowers in one place, presumably exposing them to potential abuse by older inmates.[84]

Similarly, in 1719, the convict Giovan Francesco Sartori was removed from his quarters after being caught touching the "shameful parts" of another convict, Francesco Mancarelli.[85] Again in 1735, two slaves were beaten for being caught sodomizing each other.[86] There were also numerous cases of such behavior which were recorded but went unpunished. One example is that of a Jewish convict imprisoned in the *Bagno* for kidnapping, marrying, and impregnating a young Flemish girl under false pretenses: he posed as a Christian and then abandoned her. Although he had sodomized a young man in his tavern in 1719, he was not punished, as he was on friendly terms with the guards. It was for this reason that the Scribe bypassed regular channels and went directly to the Commissioner of the Galleys, insisting that the Jew be beaten and punished, as would have been the case for any other convict.[87]

Again, in 1736, an appeal was made to the Commissioner of the Galleys to punish the slave Alì of Tripoli, who had been caught sodomizing a child. Although he

---

they are, it is often in a state of flagrancy. See, Benigno/Lavenia, *Peccato o crimine*; Scaramella, Storia omosessualità, p. 8 f.
81 ASF, MP, 2099, c. 293r.
82 ASF, MP, 2130, *Proposizioni del Pre Ginepro per la riordinazione, et aggiustamento del Bagno*, cc.n.n.
83 ASF, MP, 2106, cc.n.n.
84 ASF, MP, 2108, cc.n.n.
85 Ibid., cc.n.n.
86 Ibid., cc.n.n.
87 ASF, MP, 2111, cc.n.n.

had previously worked in the Commissioner's household, and had been found guilty of lechery on several occasions, his transgressions had repeatedly gone unpunished. Officially, this was attributed to the poor reputation of his accusers; unofficially, it was widely understood to be due to the Commissioner's protection. The renewed demand for justice in this instance was likely due to the identity of Alì's most recent victim: a boy not yet seven years old, and the nephew of Corporal Leonardo Nardi.[88]

Sometimes, the authorities resorted to medical expertise to obtain conclusive findings about the likelihood of sodomy. On 26 October 1697, a trial was held in which two slaves were accused of having sodomized a 23-year-old Flemish youth in exchange for money. The man was examined by a surgeon, who found that not only was his anus dilated and soft, but also that his shirt was also soiled with a substance resembling semen. Given that the same two slaves were accused of attempting to sodomize another 22-year-old boy two days later, the governor was asked to banish them.[89]

The surgeon's opinion was again sought in 1699, when the slave Isuff Abdouman of Tripoli was accused of having sodomized a 17-year-old German boy. According to the plaintiff, the two had been caught in the bedroom of Abdouman's tavern, just as the slave was about to insert his penis into the naked boy's anus. However, as they had not been caught in the act, and only the suspicion of sodomy remained, the governor ordered that the boy be examined by a surgeon. According to the medical report, there was no physical signs from which it could be concluded that the penetration had occurred. As the slave's guilt could not be established, he was not punished. However, to prevent such cases from arising in the future, the governor issued an edict forbidding slaves from employing young Christians under the age of 25 as servants in their shops and taverns.[90]

Another report dates from 1705 and was written by the surgeon Carlantonio Franceschini, who had been instructed to examine the slave Alì Buzzi Buzzi from Tripoli, accused of sodomy and sentenced to the chain. Extraordinarily, the report concerns the body of the active party rather than the passive one, and was requested by the Scribe of the Galleys, Vitolino Vitolini, with the aim of securing the suspect's acquittal—not his condemnation. The surgeon's statement is unambiguous: Buzzi could not have committed sodomy because his penis was almost entirely consumed by ulcers, thus rendering him incapable of copulation with either sex.[91] While the specific disease afflicting the slave is not

---

[88] ASF, MP, 2117, cc.n.n.
[89] ASLi, Capitano poi Governatore poi Auditore Vicario, 1550–1808, b. 3086, trial n. 64, f. n.n.
[90] ASLi, Capitano poi Governatore poi Auditore Vicario, 1550–1808, b. 3088, trial n. 172, f. n.n.
[91] ASF, MP, 2106, c.n.n.

identified in the source, it is plausible to assume it was syphilis. Corrosion of the penis and the nasal cartilage were hallmark symptoms of this venereal disease. Curiously, syphilis does not seem to have been widespread in the *Bagno*. Very few documented instances emerged from the sources I examined, and none involved convicts.[92] As noted in 1724, the disease was more prevalent among the *buonavoglia*, owing to their greater freedom of movement.[93]

Caused by the bacterium *Treponema pallidum*, syphilis is a sexually transmitted infection that, left untreated, can detrimentally affect the entire body. The pathogen responsible for the disease, however, was unknown at the time, as was the fact that the genital organs of either sex could serve as vectors of infection. Syphilis was widely regarded as a disease specific to female genitalia, and particularly associated with women of ill repute—prostitutes.[94] With fewer opportunities to freely exit the *Bagno* and meet prostitutes, convicts were likely less exposed to the disease.

It is notable that no reference can be found to slaves with syphilis, despite procedural sources—such as trials held before the Governor's Court—demonstrating how frequent sexual intercourse was between Muslim slaves and Christian prostitutes. It is important to underscore that such encounters between a slave and a prostitute were not punished because they involved a relationship between an unfree man and a free woman, but rather because they constituted a relationship between a heretic man and a Christian woman.[95]

While we now understand that syphilis can be transmitted through sexual contact, the explanation provided by the authorities at the time was internally coherent within their worldview. It was not so much the absence of sexual relations with prostitutes that limited infection among forced rowers, but rather the near-total lack of sexual activity in general—strictly forbidden both aboard the penal fleet and within the *Bagno*.

Let us now return to the role of medical expertise. The following example reflects a pattern also observed in Civitavecchia. It concerns a sentence pronounced in October 1710 against the *buonavoglia* Tommaso di Gio Rossi, who, upon admission to the hospital, underwent a thorough examination by the surgeons. During this examination, they noted that he complained of pain in his posterior parts. Confronted with the findings of the medical evaluation, the oarsman confessed

---

[92] The first documented case dates back to 1690, when two *buonavoglia* were hospitalized after being infected with syphilis, see, ASF, MP, 2101, c.n.n.
[93] ASF, MP, 2113, cc.n.n.
[94] Tognotti, *L'altra faccia di Venere*, pp. 30–49.
[95] See ASLi, Capitano poi Governatore poi Auditore Vicario, 1550–1808, bb. 3086, trials n. 475, n. 637, n. 167; b. 3088, trial n. 232. See also Brogini, *L'esclavage à Malte*, p. 143.

to having been sodomized. As a result, he was imprisoned and dismissed from his post to prevent future occurrences.[96]

Of particular interest is a case that appears to contribute in part to the hypothesis that, well before the 19th century, sodomy could be conceived not merely as a sexual act but also as an expression of genuine emotional and erotic attraction between members of the same sex. The case dates from July 1705, and involves two forced rowers: Francesco Mosti—convicted in 1695 for prostituting his wife—and a second man whose name is unknown, but who was nicknamed "Braccino." Despite repeated threats from the guards, the pair had reportedly been "making love" for over two years. Suspicion that they were engaged in sexual relations was eventually confirmed by one of the *vigilanti di Maria*, who had been explicitly instructed to monitor them closely. One night, the vigilante hid and caught the two "lovers" [*innamorati*] in the act. Although no other witnesses were present, the *vigilante*'s reputation was unimpeachable, and both men were subsequently punished with one hundred lashes each.[97]

What stands out most in this case is the use of explicitly amorous language: the authorities spoke of "love" and "lovers," rather than portraying the men as criminals engaging in illicit acts. Yet it is equally clear that the authorities' attention remained squarely focused on the act of sodomy itself—an offense strictly forbidden and subject to punishment. The possibility that terms such as "lovers" and "frolicking"[98] were used ironically—or perhaps mockingly—cannot be excluded, though this would be surprising given the gravity of the accusation. Comparison with comparable sources suggests that no such irony was intended. Rather, the language reflects a genuine recognition of the emotional bond between the two men involved—an affection that, while acknowledged, in no way mitigated the perceived gravity of the offense in the eyes of the authorities.[99]

3.2 Repressing sodomy in the Papal States

Rome represented a different reality from Florence, as the seat of the Holy See never had a magistracy specifically dedicated to suppressing sodomy. In the Holy City, sodomy's ambivalent character as a *crimen mixti fori*[100] [crime of dou-

---

96 ASF, MP, 2107, cc.n.n.
97 ASF, MP, 2106, c.n.n.
98 In the Italian source: *"innamorati," "amoreggiare."*
99 On the possibility of writing a history of emotions using archival sources, see, Lagioia, Passione amorosa; Scaramella, Amori nascosti.
100 Prosperi, *Tribunali della coscienza*, pp. 476–484.

ble jurisdiction] attained its fullest expression during the early modern period, as this immoral practice fell under the jurisdiction of three separate courts: the city governor—responsible for civil and criminal justice; the Inquisition—tasked with stamping out heresy; and the Cardinal Vicar, who presided over cases involving sin and crime. The governor's court handled the *forum externum*—treating this immoral act as a criminal offense; the Inquisition dealt with the *forum internum*, addressing it as a sin, while the Cardinal Vicar's tribunal was the first-instance court for *mixti fori* crimes.[101]

This overlap among three distinct magistracies should not be surprising, as each contributed, in a complementary way, to the suppression of sodomy on the criminal, theological-religious, and moral levels. Cooperation between Church and State in this domain was already clearly visible in the 16th century, when Pope Pius V issued two constitutions—*Cum Primum* in 1566 and *Horrendum illud scelus* in 1568—stipulating that individuals convicted of this unnatural vice by the Church were to be handed over to the secular authorities for execution, namely, by burning at the stake.[102] Despite grave concerns about the vice—deemed an extremely serious offense—Roman authorities appeared to make little effort to prevent and suppress it. In most cases, the Governor's Court only acted after someone had been formally denounced. This likely reflects the prevailing tendency toward leniency and discretion in handling such cases, motivated by the desire not to disturb public order and to preserve the Holy City's respectable image.[103]

According to trial records preserved in the Roman State Archives, the "nefarious vice" was thought to have been particularly widespread among the lower classes and in the city's more popular quarters. Typically, these cases involved adult men as the active party and children or adolescents as the passive one, following the classical model of pederasty. For this reason, most denunciations originated with the parents of the children involved. Sometimes, the parents informed the Governor's Court after the children confided in them; at other times, they did so after the child complained of pain in the buttocks, which led to an examination by a doctor who confirmed the hypothesis of anal intercourse. In many other instances, however, proceedings were initiated *ex officio*, either when the perpetrators were caught in the act by police officers [*birri*], or upon receipt of a medical

---

101 Fosi, *La giustizia*, pp. 21–32. Prodi, *storia giustizia*, p. 332. In the early modern period, sodomy was condemned both as a sin and as a crime "against nature," thus emerging as a *crimen mixti fori*.
102 Cattaneo, "Vitio nefando," pp. 56–59.
103 Ibid., p. 76.

report.[104] In Rome, doctors were legally required to notify the Governor whenever they examined a wounded or injured person.[105]

At the Inquisition tribunal, it was often the accused who came forward, hoping for leniency—frequently naming their sexual partners in the process. The Roman Inquisition, however, was less concerned with the crime of sodomy *per se* than with the potential presence of any heretical positions associated with it. When no such heretical element was found, the Inquisition typically referred the matter to the court of the Cardinal Vicar, who, if the accused was found guilty, would then pass the case to the Governor's Court for execution of the sentence.[106]

The desire to keep this reality hidden from public view has been rightly identified as a key reason for the relative scarcity of sodomy trials in the Roman archives.[107] Once again, we should avoid interpreting the sources from a purely quantitative perspective. Instead, we should consider this marked absence as an indirect indicator of the authorities' overall stance toward such illicit acts. Indeed, the judicial sources preserved in the fonds of the Governor's court suggest not only that cases of sodomy were relatively rare, but that they were confined to the city's lower classes. Members of the higher social strata are entirely absent from the records. As Marina Baldassarri has argued, this should not be taken to mean that they did not engage in such practices, but rather that they resolved such matters privately, away from public scrutiny.[108] The Roman authorities thus adopted a strategy of limited publicity but exemplary punishment in response to this unnatural vice: throughout the early modern period they sought to limit scandal while using harsh penalties as a deterrent.

The widespread problem of sodomy—alongside blasphemy—aboard the Papal galleys was denounced very early on as a matter of grave concern. The author of the aforementioned report on the disorders aboard Papal galleys, commissioned by Pope Pius V in 1571, described as commonplace the practice of youths being brought aboard specifically for the purpose of having them sodomized by members of the crew.[109] Aristocrats—captains and high-ranking officers[110]—would embark with teenagers, referred to as "galley nobles," under the pretense

---

[104] Baldassarri, *Bande giovanili*, pp. 120–123.
[105] Ibid., p. 131 f.
[106] Cattaneo, "Vitio nefando," p. 68.
[107] Ibid., p. 76.
[108] Baldassarri, *Bande giovanili*, p. 161 f.
[109] AAV, Misc. Arm. II, b. 110, *Secondo Avvertimenti sopra I disordini delle galere di S. Santità occorsi nell'anno passato 1571 dati da certe religiose persone et da bene con I rimedij necessari et opportune per emendargli*, ff. 394–396.
[110] Crescentio, *Nautica mediterranea*, p. 94.

that they were their relatives or pupils. Young apprentices were recruited from among the crew to be turned into prostitutes, while in the rowing chamber, younger convicts were stationed near the hatches below deck so that they could easily disembark whenever clients approached.[111] The practice of sodomy, "abhorred even by beasts," was fiercely condemned by the chaplains responsible for the spiritual care of the *galeotti*—not only because of the sheer unnaturalness of the act, but also due to its heretical nature, as such immoral conduct was believed to risk provoking divine wrath. The report consequently recommended that only adults be permitted aboard to avoid "the great dishonor to Christianity caused by doing as the Turks do, who take young men (on board) with them."[112]

This situation appears not to have improved over the years, as in virtually every record of official visits to the Papal galleys by representatives of the Vicar of Rome—including that of 1668—this illicit practice continued to be identified as one of the most serious and widespread disorders on board.[113]

Significantly, however, it was not just sodomy that was forbidden aboard galleys, but sexual relations in general, as it was believed that fornication weakened discipline and rendered soldiers effeminate. After all, galleys were primarily military spaces, and the crews manning such spaces were made up of forced rowers and soldiers who had to fight in armed confrontations. As stated in one of the memories on the good governance of the fleet, "war is not the time for fornicating, but for fighting vigorously without falling in love."[114] The presence of prostitutes

---

[111] AAV, Misc. Arm. II, b. 110, *Secondo Avvertimenti sopra I disordini delle galere di S. Santità occorsi nell'anno passato 1571 dati da certe religiose persone et da bene con I rimedij necessari et opportune per emendargli*, ff. 394–396; BCR, Ms, 34D18, *Memorie e scritture diverse appartenenti alle galere Pontificie e condannati nelle medesime Principalmente in tempo del tesorierato di mons. Lorenzo Corsini poi cardinale, e papa col nome di Clemente XII*, f. 35.

[112] Ibid., f.396: "ne senza tanto disonore al Christianesimo che si faccia come fanno i turchi che menano li giovani." See also Malcolm, Forbidden Love, p. 18.

[113] Please note that the spiritual care of the *galeotti* on the Papal galleys was the responsibility and fell under the jurisdiction of the Vicar of Rome, as explained in Chapter Three. ASVR, Atti della segreteria, b. 74, *Giurisditione dell'e.mo vicario sopra le galere pontificie ed, in Civitavecchia, sopra l'ospedale di S. Barbara e su di alcune chiese di Civitavecchia (1722–1773), Acta visitationis triremus (1670)*, f. 62r: "Di vitij pubblici e scandalosi fra questi galeotti io ho osservato predominare e frequentarsi quella della bestemmia [...] L'altro vitio è la sodomia quale pure si castiga quando si sa ma che e materia occulta molti ne scappano impuniti la causa però di questa furfanteria procede principalmente dal mandarsi persone giovane in galera, e questi vagabondi ragazzacci, e questo è quanto posso riferire per verità e per la notitia che ne ho per essermi passati per le mani e per essere pubblico notorio."

[114] AAV, Misc. Arm. II, b. 116, *Ricordo per il buongoverno delli soldati nell'armata*, f.189: "Secondo perché il tempo della guerra non è per fornicare, se non per battagliare animosamente senza innamoramenti."

was strictly forbidden, as not only would they distract the crew from their key duties, but also because they were thought to indirectly foster more illicit sexual conduct. Since it was impossible to provide enough prostitutes to gratify the entire crew, the risk of too many men resorting to illicit practices to satisfy their sexual needs was considered extremely high.[115]

At first glance, this decision by the papal navy might seem surprising, given that, since the 14th century, prostitutes had been commonplace in land armies, often accepted, even welcomed, as a means of reducing sexual violence and deterring men from preying on civilians or each other, thereby helping to avoid incidents of war rape.[116] As the anonymous author of the report observes, however, prostitution—long tolerated in Italian cities as a "necessary evil" to discourage men from committing such illicit acts—would have only created further problems aboard galleys. It was believed to encourage vice and weakness among the crew and, in the confined quarters of a ship, would have provoked moral scandal given the lack of private space for fornication.[117]

Alongside the presence of young men and prostitutes, sodomy aboard the galleys was further facilitated by excessive promiscuity and the freedom of their living conditions. Indeed, not only did their cramped spaces force oarsmen to sleep on top of each other, but it also appears that, contrary to orders, the convicts were often free to roam the galleys at night rather than being chained to their benches —enabling them to engage in forbidden acts.[118]

---

115 Ibid.: "Primo perché le meretrici sono poche, et non potendo tutti haverle, et vedendole li da occasione di cercare, et fare altri peccati nefandi;" BCR, Ms, 34D18, *Memorie e scritture diverse appartenenti alle galere Pontificie e condannati nelle medesime Principalmente in tempo del tesorierato di mons. Lorenzo Corsini poi cardinale, e papa col nome di Clemente XII*, f. 35: "Non permetter che entrino Donne nella darsena, e molto meno nelle galere. Ne meno ragazzi, o sbarbati scapoli sotto qualsivoglia pretesto, anche di cercar la limosina et in nessun conto permettere che vi dormino dentro di notte."
116 Barbagli, *Comprare piacere*, pp. 118–130.
117 Ibid., pp. 182–184. AAV, Misc. Arm. II, b. 116, *Ricordo per il buongoverno delli soldati nell'armata*, f.189: "Quanto al terzo articolo delle meretrici, dico, che si permettono bene per la fragilità, et necessità nelle città, et terre grosse: ma nell'armata christiana sarebbe cosa perniciosa et scandalosa, et causa di più, et maggiori peccati.[...] Terzo perché li soldati, et altri vanno per mare, et per terra à pericolo della vita; quarto perché il luogo delle galere è stretto, et publico, et sarebbe grande scandalo attendere à fornicare; quinto perché il luogo è pericoloso del mare, et tempesta, et borrasche, et in terra fanno gl'huomini effemminati à combattere."
118 ASVR, Atti della segreteria, b. 74, *Giurisditione dell'e.mo vicario sopra le galere pontificie ed, in Civitavecchia, sopra l'ospedale di S. Barbara e su di alcune chiese di Civitavecchia (1722–1773)*, *Acta visitationis triremus (1670)* f. 62v; BCR, Ms, 34D18, *Memorie e scritture diverse appartenenti alle galere Pontificie e condannati nelle medesime Principalmente in tempo del tesorierato di mons. Lorenzo Corsini poi cardinale, e papa col nome di Clemente XII*, f. 35; ASVR, Atti della segreteria,

Strikingly, the accounts do not express any degree of astonishment that the *galeotti* were engaged in sodomy. On the contrary, it was simply portrayed as a sinful practice to be eliminated, and there is no sense of dismay in their descriptions—almost as though such behavior was taken for granted. Over time, however, the tone of these reports switched from detached to increasingly desperate. Based on the sources, the prevalence of sodomy among galley rowers was attributed to criminals' amoral disposition. It was believed to reflect either their innate moral corruption[119] or their heretical religious beliefs—particularly in the case of Muslims—which were thought to predispose them to such behavior.

Indeed, as a practice that violated the divine order of creation and disrupted natural law, sodomy was regarded as a rejection of God. It thus came to be treated as not only blasphemy, but as a heresy in itself. Accusing someone of sodomy thus became a way to purge the enemies of Christianity—not only for committing the act, but also for promoting the idea that it was not particularly sinful. The first public condemnation of sodomy as a form of heresy was recorded in Siena in 1262. The issue had nevertheless already been addressed thirty years prior: a 1232 Papal act signed by Pope Gregory IX authorized Dominican inquisitors to prosecute the "nefarious vice," alongside witchcraft, on Austrian soil.[120] In the Sienese Statutes of 1262, the persecution of sodomites was explicitly tied to persecuting heretics. Using the analogy of the political body, both heresy and sodomy were likened to an infectious disease: getting rid of them was deemed necessary to prevent the rest of the social body from being corrupted.[121] The association between sodomy and heresy is further evident from a simple analysis of the vernacular terms used across Europe to identify deviant sexual behavior. Thus, for example, in the Germanic world, the noun *Ketzerie*—which has its root in the term denoting the heretical group of Cathars—denoted both sodomy and heresy, while the verb form *ketzern* described unconventional sexual relations. Similarly, in England, France, Spain, and later in Italy, terms such as *bugger* (English), *bougre* or *bouggeron* (French), *bujrron* (Spanish), and *buggeratore* (Italian) were used to designate sodomites—words all derived from the Latin *bulgarus* which

---

b.74, *Giurisditione dell'e.mo vicario sopra le galere pontificie ed, in Civitavecchia, sopra l'ospedale di S. Barbara e su di alcune chiese (1722–1773)*, dossier n.n., 2 August 1781.

119 See, for example, ASVR, Atti della segreteria, b. 74, *Giurisditione dell'e.mo vicario sopra le galere pontificie ed, in Civitavecchia, sopra l'ospedale di S. Barbara e su di alcune chiese di Civitavecchia (1722–1773), Acta visitationis triremus (1670)* c. 62v: "Lo scandalo grave e commune in queste galere e il vitio nefando quale procede dalla mala inclinatione di questa gente condannata."

120 Lavenia, *Eresia indicibile*, p. 14 f.

121 Pastore, *Le regole dei corpi*, pp. 17–35.

referred to the Bulgarian heresy of Bogomilism.[122] Likewise, the term *erite* could be used indiscriminately to label both heretics and sodomites.[123]

The Protestant Reformation of the 16th century revived Catholic anxieties surrounding heresy and sodomy as practices against God. The accusation of sodomy, with all its negative connotations, became a powerful weapon of religious propaganda. While Catholics hurled such accusations against heretics—a category that included Protestants, Jews, and Muslims—Protestant Reformers returned fire, often accusing Roman clergy of moral depravity and sodomitical conduct.[124] Martin Luther, for instance, was portrayed as a deviant reformer, intent on undermining moral standards, depicted either as a sodomite or a libertine intent on "defiling the nuns."[125]

Blasphemy, simony, and sodomy eventually fell under the jurisdiction of the Roman Inquisition, further reinforcing their theological link.[126] It was Pope Paul IV (r. 1555–1559) who included this vice among the offenses that the Inquisition of 1557 was authorized to pursue.[127] Interestingly, over time, the Inquisition's concern narrowed, limiting its attention to cases that involved heresy. This is made explicit on 12 October 1600, when the Roman Inquisition, in the presence of Pope Clement VIII (r. 1592–1605), declared that *"super crimine nefando non proceditur in Sancto Officio"*—that is, the Holy Office would not proceed in cases of nefarious crimes unless heretical elements were present.[128]

The association between sodomy and heresy found its most potent expression in the enduring stereotype of the "sodomite Turk."[129] Throughout the early modern period, the term Turk did not specifically refer to someone from Turkey, but rather was used broadly to designate any Muslim, cast as Christianity's quintessential enemy. Anti-Muslim accusations cast Muslims as embodying a range of demonic traits, chief among them sex against nature—a transgression viewed as the ultimate rebellion against God and the natural order.[130] This propaganda was particularly prevalent in Iberia, shaped by centuries-long *Reconquista* (ca. 718–1492), yet the sodomite Muslim trope eventually gained traction through-

---

122 Grassi *L'offizio*, p. 131.
123 Grassi, *Sodoma*, p. 55.
124 Grassi *L'offizio*, p. 133.
125 Ibid., p. 141.
126 Baldassari, *Bande giovanili*, p. 113.
127 Cattaneo, "Vitio nefando," p. 59.
128 Lavenia, *Eresia indicibile*, p. 48 f.
129 See also Babayan/Najmabadi, *Islamic Sexualities*, Malcolm; *Forbidden Desires*; Massad, *Desiring Arabs*.
130 Grassi/Marcocci, L'intreccio dei desideri, p. 12.

out early modern Christendom, reaching its symbolic peak during the Cyprus War —a confrontation emblematic of the struggle between the Christian West and the Muslim East.[131]

The conflation of Islam with illicit sexual activity was initially articulated during the Crusades. One of the earliest depictions of Muslims as sodomites appears in a 10th-century hagiographic legend, the *Passio Sancti Pelagii*, according to which, around 925–926, the Emir 'Abd al-Rahman III was struck by the beauty of Pelagius of Cordóba, a 13-year-old boy who had offered himself as a hostage in exchange for the release of his uncle, Bishop Hermogius. The boy was martyred, courageously refusing to renounce his faith or submit to the Emir's carnal advances.[132] Over the centuries, a genuine cult emerged around the figure of Pelagius, who was later canonized. Iberian anti-Muslim propaganda spread throughout Europe and the story of Pelagius eventually reached northern Europe, as demonstrated by the near-contemporary account of the subject written in Saxony by Hroswitha of Gandersheim.[133]

Over time, anti-Muslim sentiment was further fueled by alleged reports of rape and violations of natural law by Muslims; these accounts had been circulating as early as the First Crusade, including a likely spurious appeal to Christian princes by the Byzantine emperor Alexius I Comnenus (r. 1048–1118), which accused Muslims of perpetrating atrocities, including homosexual rape. In the 12th century, the Abbot of Clairvaux, Henri de Marcy (1136–1189), invoked similar fears when he preached the necessity of a Third Crusade to prevent the nefarious vice from spreading westwards. Similarly, in his *Tractatus contra haereticos*, the theologian Alain de Lille (1125–1202) portrayed Muslims as obsessed with carnal pleasures.[134]

Although these commentaries were clearly shaped by negative stereotypes harnessed for propaganda purposes, they often distorted a much more complex reality. While the Qur'an adopts negative positions toward homosexuality, Islamic societies had, for centuries, tolerated certain forms of homoerotic practice with varying degrees of discretion. This tolerance was recast by Western propaganda as open endorsement; Muslims were accused of indulging in the most unbridled sexual freedom and of disregarding the laws of nature—particularly in their perceived failure to uphold gender distinctions.

This fusion of moral panic and ethnographic distortion continued into later centuries. A widely cited example is the treatise written around 1318 by the French

---

131 Capponi, *Lepanto*.
132 Lavenia, Sessualità, islamofobia, p. 108.
133 Grassi, *Sodoma*, p. 63.
134 Lavenia, Sessualità, islamofobia, p. 109.

Dominican Guillaume Adam, in which the author denounced Muslims for their alleged indulgence in transvestitism and for buying young slaves to be turned into prostitutes. Here again, these claims reflect distortion: the *mukhannaths* [transvestites], held a recognized, though socially marginalized, place in many Muslim societies, while the recruitment practices of the Mamluk regimes in Egypt and Syria (1250–1517) involved the purchase of young non-Muslims as slaves, who were then converted to Islam, trained as soldiers, and incorporated into the ruling elites, and sometimes became court favorites.[135]

However, the stereotype of the sodomite Muslim did not really take hold in the Iberian Peninsula until the 15th century, and Turks were never portrayed as effeminate. On the contrary, they were seen as a righteous scourge sent by God to punish the sins of Christians.[136] In time, sodomy came to be seen not as a sign of softness, but as yet another attribute of violent, evil masculinity. Notably, this trope was not exclusively propagated by Christian sources: various Arab authors and chroniclers also described the spread of sodomy in Muslim lands.

An especially illustrative example can be found in *Della descrittione dell'Africa et delle cose notabili che ivi sono* (1550) by Johannes Leo Africanus (c. 1485–1554). Born l-Ḥasan ibn Muḥammad al-Wazzan al-Fāṣī, he was a Berber ambassador and geographer who, after being captured by Christian corsairs in 1518, was brought to Rome as a slave and entrusted to Pope Leo X (r. 1513–1521). Baptized and welcomed into his court, he was commissioned to write what was one of the most authoritative early modern text on Africa.[137] In it, several pages describe homoerotic practices as widespread, especially in large urban centers such as Fez, and notably in settings marked by alcohol consumption such as banquets.

What is most striking is not merely Africanus's description of these practices but his ambivalent attitude: he refrains from openly condemning them.[138] While Islamic jurisprudence prohibited a wide range of sexual behaviors, authorities in the Ottoman Empire were rarely as aggressive in prosecuting sodomy as their Catholic counterparts in Southern Europe.[139] This difference was partly due to greater leniency toward consensual homoerotic relations, and partly to Islamic legal norms, which—unlike those in Europe—forbade anonymous denunciations and disallowed confessions extracted under torture.[140]

---

135 Rowson, Omoerotismo ed élite mamelucca, p. 24. On the *mukhannats*, see Rowson, The Effeminates, pp. 671–693; Instituzionalized Transvestitism, pp. 45–72.
136 Grassi/Marcocci, L'intreccio dei desideri, p. 14.
137 Zemon Davis, *Leone l'Africano*, (translation) p. 3 f.
138 Ibid., pp. 218–221.
139 Malcolm, Forbidden Love.
140 Marotta, Muslim Friend, p. 232 f.

The proponents of the anti-Muslim rhetoric soon felt the need to learn more about Islamic culture in order to attack it more effectively from within. After the Abbot of Cluny, Peter the Venerable (1092–1156), travelled to Iberia in 1142, his notary Peter of Poitiers, assisted by a group of translators and commentators, was tasked with compiling a collection of translated texts, known as the *Corpus Toletanum*. Among its central components was the Latin translation of the Qur'an by Pedro of Toledo, Herman of Carinthia, and Robert of Ketton.[141] Once the compilation was completed, Peter of Poitiers proposed producing another text, aimed at exposing the errors of Islam, in which the sixth paragraph of the second book was to be devoted to refuting what Muhammad allegedly taught about "the foulest act of sodomy," in the Qur'an, quoting: "Men, plough your women on whichever side you like." Ketton's translation lent itself to a distorted interpretation of a passage in *Sura* 2 (*Al-Baqara*, v. 223: "Your brides, for you, are like a field. Come to your field as you will") which appeared to sanction unrestricted sexuality freedom—even sodomy—between spouses.

Peter of Poitiers's work was ultimately not included in the text that would be titled *Liber contra sectam sive haeresim Saracenorum* [Treatise Against the Sect or Heresy of the Saracens], but circulated widely in manuscript form and proved relatively influential. Its echoes can be found in subsequent works, such as Jacques de Vitry's *Historia Orientalis* (1219–1226).[142] Peter of Poitiers' interpretation of the *Sura* 2 was later revived and popularized by the friar Alfonso de Espina in his *Fortalitium fidei* [Fortress of Faith, 1458], whose sixth book, *De bello saracenorum* [On the War Against the Saracens], was entirely devoted to the crusade against Judaism and Islam.[143] In this work, Espina claimed not only that the Qur'anic verse authorized sodomy—including homosexual relations—but that this permission applied to all Muslims, who therefore were inherently prone to vice and, for that reason, deserved to be burned at the stake,[144] in accordance with the legal prescriptions found in Alfonso X of Castile's royal charter, *Fuero Real* (1255), which prescribed the death penalty for those convicted of sodomy, as a model to follow.[145]

Despite the chaplains' and the galley officials' concerns—who both had to deal with inmates on a daily basis—no practical solutions were adopted at an institutional level, and official decrees concerning the efficient management of the Papal galleys made no reference to sodomy. This state of affairs changed, however,

---

[141] Lavenia, *Eresia indicibile*, p. 17 f; Grassi, *Sodoma*, p. 64.
[142] Lavenia, Sessualità, islamofobia, p. 110 f.
[143] Ibid., pp. 115 f.
[144] Ibid.
[145] Ibid., p. 112.

3 Moral order and sodomy: sexual transgressions among early modern *galeotti* — 223

with the edict issued on 15 January 1709 by the General Commissioner of the Navy, Carlo de Marini, which specifically targeted the problem of sodomy and blasphemy among the *galeotti*.[146] While earlier proclamations had clearly condemned blasphemy, they had only implicitly censured sodomy. In contrast, more than half of the 1709 proclamation is devoted to the high incidence of illicit sexual activity among convicts and slaves. It remains unclear whether this frequency was due to an actual rise of such practices, or whether the edict signals a shift in institutional awareness—namely, a decision to confront the matter openly rather than suppress it to avoid public scandal. The latter explanation appears more convincing. The text of the edict offers the clearest expression of this shift.

Declaring sodomy an abominable vice, "by which one offends both the Creator and nature, and which, more than any other sin, provokes the wrath of God," the edict called for decisive action. Those found guilty, whether active or passive participants, were to be condemned to the gallows. The death penalty also applied to accomplices, while those who failed to notify the authorities were to be fined and sentenced to five years in the galleys. This directive targeted, in particular, the bench-mates and the oarsmen on the fore and aft benches, who were instructed to denounce and report any transgressions to the *aguzzino*. Similarly, surgeons aboard were required to notify the authorities if any patient displayed signs even remotely suggesting involvement in sodomy.

Aware that this crime was committed in private and with the utmost discretion—rendering it difficult to detect—the naval authorities sought to establish a rigid and far-reaching system of mutual vigilance among crew members, combining severe punishments for offenders with generous rewards for informants.[147] In

---

146 AAV, Misc. Arm. IV/V, b. 54, f. 105.
147 Ibid.: "l'estirpazione degl'altri vizij, anche con il mezzo de' più severi castighi, & essendo tra essi il più abominevole quello della sodomia, con il quale offendendosi il Creatore, e la Natura insieme, più di ogni altro peccato provoca l'ira di S.D.M., perciò affinché ognuno si astenga dal commetterlo, siano Schiavi, Forzati, Vagabondi, Bonavoglia, Marinari, Soldati, & ogn'altro sottoposti alla nostra giurisdizione, con il presente pubblico Editto di ordine di Nostro Signore, come sopra; si notifica. Tutti, e singoli, come in occasione di tal delitto si procederà irremissibilmente contro di loro alla pena della Forca, agenti, ò pazienti, che siano, purché maggiori di anni dieci otto.

Nella quale sopradetta pena di morte incorrerà ogn'altro, che farà mezzano, ò causa, ò in qualsivoglia modo darà aiuto, commodità, consiglio, ò in altra qualunque maniera coopererà al commettersi detto vizio.

[...] E per venire più facilmente in notitia de' Delinquenti; si ordina à Comiti, Sottocomiti, Aguzzini, e Mozzi, & altri rispettivamente, che sopra i questo particolarmente invigilino, e succedono alcuno di detti delitti, ed in qualunque modo avendone da loro sentore, debbano subito

addition to harsh denunciations and exemplary punishments, further preventative measures included a ban on young men boarding the galleys, a prohibition against sleeping together under the same blanket and—most notably—the nightly release of convicts from their chains.[148] These strategies undoubtedly increased the levels of vigilance aboard the ships, yet they do not appear to have led to any meaningful reduction in sodomitical activity, which continued to be reported and condemned, particularly by the religious authorities charged with the *galeotti*'s spiritual welfare.

4 Criminal prosecution of sodomy in 18th-century Civitavecchia

Despite the impossibility of fully eradicating sodomy aboard the Papal galleys, notable developments occurred in Civitavecchia during the 18th century. In the wake of the 1709 edict, and with the declared aim of punishing sodomy more effectively, such cases began being prosecuted through formal criminal trials held before the Court of the Governor of Civitavecchia. Among the surviving sources are 14 sodo-

---

denunciarlo nel Tribunale di Monsignor Governatore suddetto, sotto pena di scudi cinquanta, di anni cinque di Galera, & altre ad arbitrio di Sua Signoria illustrissima.
   Nelle quali pene parimente incorreranno li Barbieri, Barbierotti, Chirurghi, & altri che cureranno, ò saranno richiesti di curare mali, che in qualunque modo possino essere stati causati da questo vizio ogni qual volta non daranno pontualmente la relazione in scritti nel Tribunale suddetto di chi li haverà richiesti per essere curati, ò di chi averanno medicato con la distinzione del nome, cognome, e luogo di permanenza, acciò meglio possa venirne in cognizione."
148 See, for example, ASVR, Atti della segreteria, b. 74, *Giurisditione dell'e.mo vicario sopra le galere pontificie ed, in Civitavecchia, sopra l'ospedale di S. Barbara e su di alcune chiese (1722—1773)*, c. 62v: "Lo scandalo grave e commune in queste galere e il vitio nefando quale procede dalla mala inclinatione di questa gente condannata, e dallo stare uno sopra l'altro a dormire, ma particolarmente perché si mandino ragazzi in galera;" BCR, Ms, 34D18, *Memorie e scritture diverse appartenenti alle galere Pontificie e condannati nelle medesime Principalmente in tempo del tesorierato di mons. Lorenzo Corsini poi cardinale, e papa col nome di Clemente XII*, ff. 35—36, 43—44: "Troppa libertà hanno i forzati, e così con pagare un grosso, o un giulio si sferrano e vanno dove gli piace per la calata della darsena, con soverchia facilità gli si permette che vadino anche nella città, se bene accompagnati. Così è in piacer loro di fare mille furfanterie in materia di senso, et altre anche più esecrande [...] di notte tempo stanno anche molti sferrati, si de forzati, come delli Buonavoglia, e schiavi, massime di quelli forzati, che hanno qualche officio, o che servono gli officiali, o che stanno a qualche servitio particolare, come da dispensa e così vanno a trovar chi vogliono e dormire con chi gli pare [...] XII. Non lasciarli dormire più sotto una comune coperta, ma darne una per uno. Mutargli spesso da banchi, massime li sospetti. Non permettere che li scapoli entrino nelle stanze delle galere, e che stiano sferrati alla notte."

my trials and surgeons' reports initiated against convicts and galley slaves between 1738 and 1781.[149]

In cases of suspected sodomy, it was the prerogative of the court-appointed physician to examine the defendant's anus. Venice was the first city in Italy where the central role played by doctors, surgeons, and barbers in attempts to eradicate this vice from the city was explicitly codified in law. On 8 August 1453, following news that a young boy had been sodomized and displayed lesions on his anus—and that the doctors treating him failed to report their findings to the judiciary—it was proposed that, henceforth, all physicians be required to notify the *Consiglio dei Dieci* if, upon inspection, they observed children with lesions suggestive of sodomy. Failure to notify such findings would result in a penalty of one year's imprisonment and permanent banishment from Venice. This proposal was not adopted, however.[150] Matters changed in 1468 following a proclamation issued on 7 January, which required the city's doctors and surgeons to report suspected cases to the *Consiglio dei Dieci* "just as they report gunshot wounds to the *Signori di Notte*, so they must denounce anyone who is injured in these parts [...] due to a laceration in the posterior parts caused by a male sexual organ."[151] As Priori later reported, even in late 17th-century Venice "the Surgeon or Midwife who sees a boy or girl broken in those shameful parts [sign of violent rape] must notify the authorities under oath."[152]

While Venice was unique in explicitly regulating the involvement of medical experts in such cases, recourse to a doctor's expert opinion in the prosecution of sodomy is also attested in other regions. Given that judges could not rely solely on the defendant's statements—as these were often categorical denials—medical examination became the only viable method of establishing the truth in cases of suspected sodomy.

Like other topics on which forensic medicine focuses, sodomy is also addressed in Paolo Zacchia's *Quaestiones medico-legales*. Although this alleged crime frequently required medical expertise—as attested by both contemporary archival sources and legal treatises, such as those by Farinacci and Savelli—no mention is made in the *Methodus Testificandi* of how the consummation of sodo-

---

[149] The trials are conserved in ASR, Tribunale poi Governatore di Civitavecchia (1589–1913).
[150] Canosa, *Grande paura*, p. 107.
[151] Ibid., p. 115: "allo stesso modo in cui denunciano i colpi d'arma da loro riscontrati ai Signori di Notte, così essi debbono denunciare chi è rotto in quelle parti [...] per una lacerazione nelle parti posteriori causata da un organo sessuale maschile."
[152] Priori, *Prattica criminale*, p. 11. "nel stupro nel putto, o nella putta per forza violata le parti vergognose rotte, le quali siano vedute per la Comare, o per il Chirurgo, riferendo il loro credere per giuramento."

mitic intercourse could be verified; even the space Zacchia devotes to male sodomy is limited to no more than a single page. He addresses the topic in Book IV, Title II, which is dedicated to virginity and rape [*De Virginitate, & Stuprum*], in *Quaestio* V focused on signs of rape in boys [*Constuprati Pueri signa*].[153]

Significantly, sodomy is consistently referred to as *stuprum*—a term generally understood to mean penetration. In its original Latin usage, *stuprum* signified "dishonor," and only over time did it acquire the broader meaning of penetration, and in particular "sodomitical anal penetration."[154] According to Zacchia, a medical examination—conducted on the body of the individual who had been penetrated—could reveal a range of valuable information, including whether anal intercourse had taken place, whether it occurred recently or in the past, and whether it happened once or on multiple occasions.[155] The more frequently and the more recently the act had taken place, the easier it was to detect anatomical indications, often in the form of anal lacerations called *raghadiae*. These lesions typically resulted from violent penetration by an external object—especially when the boy was young and the *stuprator* [penetrator] possessed a thick penis.[156]

Wherever possible, Zacchia recommended careful inspection of the victim's anus and the perpetrator's penis. The presence of bruising around the anus, dilatation, or fleshy protuberances called *carunculae* (or vulgarly, *crestas*) might also serve as indicators of anal penetration. Although such *crestas* could result from hemorrhoids, Zacchia argues that when other aforementioned symptoms were present, these bumps constituted compelling evidence that illicit sexual acts had taken place.[157] In support of his conclusions, he cites the authority of the Sephardic physician Amatus Lusitanus (1511–1568), who, in his *Curationum medicinalium centuriae septem* (1551), was likely the earliest—and for a long

---

153 Zacchia, *Quaestiones*, pp. 251–260.
154 Rousseau, Policing the Anus, p. 77.
155 Zacchia, *Quaestiones*, p. 260: "de recenti commissum, quaedam vero à multo tempore, & cursus quae stuprum unica vice patratum, aut de raro, quaedam quae fraequentatum ostendunt, oedem quoque quoque pacto de hujus stupri signum distinguendum est."
156 Ibid.: "stuprum in puero, quod de recent patatrum fuerit, ut ante mensem, praecipue si fraequentatum sit, indicare manifeste possunt, Ani scissurae Rhagadiae dictae, qua facile ob illatam vim apparent, praecipue si puer tenoriis aetatis sit, & stuprator crassa mentula dotatus."
157 Ibid.: "in lividum colorem [...] & partis dilatation, quae etiam multo post tempore perdurare potest, & indicare insimul stuprum frequentatum, etiam à multo tempore commissum, quod multò magis significant quaedam carunculae, seu carneae excrescentiae, quas vulguo cristas vocant, quae maxime ex frequenti Sodomia originem habent [...] quamplures Medicos pro ipsis haemorrhoidibus accipiantur."

time the only—European author to investigate the potential medical consequences of anal intercourse between men.[158]

In most instances, the available sources consist solely of medical reports or summary trials, in which both the verbal testimony of the accused and the criminal sentences are largely absent. Nevertheless, for the purposes of this research, these sources yield a significant amount of valuable information. First, they attest to a clear shift in the authorities' stance aboard the Papal galleys. While most proceedings were not brought to completion, the very fact that the judicial apparatus was activated demonstrates greater institutional commitment than in the past. It must be emphasized, however, that in the absence of final judgments in most cases, it remains unclear whether or how the offenders were punished, or whether they were sentenced to death, as prescribed in the 1709 edict.

These sources also provide insight into aspects of daily life aboard the galleys, revealing a reality in which structured violence coexisted with codified patterns of interaction, at times giving rise to genuine acts of solidarity among the rowers. The trial records also underscore the crucial role of medical expertise in both detecting and suppressing this nefarious crime—one that could not, and should not, have been permitted to occur within the confines of the Papal fleet. The following analysis will begin by focusing on the role of the medical professionals who were summoned to offer expert opinions in such cases.

The earliest available source—a report by Francesco Faraone, a surgeon at the Ospedale di Santa Barbara—dates from 1728 and concerns a Muslim slave from the *Patrona* galley, Amettino from Biserta. While the reason he was admitted to the hospital remains unclear, the surgeon's diagnosis is unambiguous: Amettino exhibited multiple excrescences around his anus, which, in his medical opinion, could only have been caused by the nefarious vice.[159]

Faraone's diagnosis raised further questions. Were they truly the result of anal penetration? Could they have been mistaken for hemorrhoidal swellings? Alternatively, were they genital warts, a common symptom of venereal diseases such as gonorrhea and syphilis—both of which were already widespread throughout the Mediterranean region at the time, though not yet definitively linked to

---

158 Rousseau, Policing the Anus, p. 86.
159 ASR, Tribunale poi Governatore di Civitavecchia (1589–1913), b. 642, dossier 10, f. n.n.: "Io sottoscritto ho visitato nell'Ospedale delle galere di Nostro Signore Ametto di Biserta detto il tignoso schiavo della Galera Patrona et avendo viste e ben considerate le di lui parti posteriori e pudende, et avendo ritrovato in torno alla circonferenza del ano varie escrescenze carnose dette Porrifichi o creste le quali secondo la mia peritia dico essere stati prodotti da reiterati atti del vitio nefando."

same-sex intercourse?[160] Since no record of the trial has surfaced, it remains uncertain whether one ever took place.

Exceptionally, Amettino's story has not been completely lost, as two other documents concerning him have been found within the same archival collection. The first, dating to 1724, predates the surgeon's report. It, too, was penned by Faraone and details a separate incident: Amettino had been hospitalized after being kicked in the stomach by the galley's subordinate boatswain—the *sottocomito*—an injury that caused him to spit blood and collapse. According to Amettino, the officer's aggression stemmed from frustration at his inexperience, as he had just recently joined the crew.[161] While this earlier report sheds no light on the allegations of sodomy, it offers a revealing glimpse into the brutal and often violent conditions endured by slaves, helping to contextualize Amettino's life within the broader reality of maritime servitude.

The second source, dating from 1738, is more striking: it was written a full decade after the initial report that had confirmed Amettino's alleged sodomitic behavior. This time, however, the offense took place on dry land rather than at sea. Over the intervening years, Amettino completed his service aboard the galley and later spent five years working in the governor's palace in Civitavecchia. He maintained a shack in the city's dockyard and owned a cheese shop within the city. He was accused of repeatedly sodomizing the 15-year-old Antoniuccio Bengardi of Lucerne—an unemployed ex-servant who frequented Amettino's tavern, where he drank coffee and alcohol. The trial began after the city's *birri* raided Amettino's bedroom, where they discovered him wearing only a shirt, and in the company of the half-undressed Bengardi. The raid was justified by Amettino's bad reputation; he had been accused of "fornicating" and luring young boys to "satisfy his brutal libido."[162]

That Amettino had engaged in sexual relations with other boys is corroborated by prior investigations, during which he had earned the nickname "Beauty of Tunis" [*Bella di Tunisi*]. What is especially striking, however, is how in this case his bad reputation was implicitly reinforced by his Muslim identity—thus perpetuating the enduring stereotype that associated Islam with sodomy. This is further reflected in the testimony of Nesan of Amor from Bizerte, Amettino's companion in the governor's palace. When asked why Amettino had been arrested, he initially feigned ignorance, but later admitted he had been present during the raid and

---

160 See, for example, Berco, *Syphilis and the Silencing of Sodomy*; Siena, The Strange Medical Silence.
161 ASR, Tribunale poi Governatore di Civitavecchia (1589–1913), b. 642, dossier 10, f. n.n.
162 ASR, Tribunale poi Governatore di Civitavecchia (1589–1913), b. 643, dossier 8, f. n.n.: "per sfogare la sua brutale libidine."

knew that Amettino had been found in bed with Bengardi. Unprompted, Nesan insisted that this "nefarious practice" was condemned not only by Christianity but also by Islam, adding that other Muslim slaves viewed Amettino with hostility because they disapproved of his conduct. As Achille Marotta suggested, Amor likely engaged in a distancing strategy—disassociating Amettino from the broader Muslim community in Civitavecchia, and thereby implicitly acknowledging how easily the accusation of sodomy could be extended to others within that group.[163]

Further evidence of this underlying anxiety lies in the fact that Bengardi had previously been warned by local beer makers of the risks of associating with slaves—a social group widely regarded as both dangerous and immoral. In this case, the presence of eyewitnesses, Amettino's bad reputation—as confirmed by multiple testimonies—and most significantly, Bengardi's own confession were sufficient to establish the charge. Expert medical opinion was deemed unnecessary, as it would have merely corroborated what was already accepted as fact.

The trial concluded with Bengardi—already incarcerated—sentenced to exile from Livorno, under threat of corporal punishment should he fail to comply.[164] Surprisingly, Amettino was neither convicted nor formally acquitted, even though a harsher sentence might have reasonably been expected given his sexually active role. Nevertheless, this episode reinforces the prevailing stereotype of galley slaves as inherently immoral and fundamentally incorrigible individuals.

As the Amettino episode demonstrates, not every expert report or sodomy trial involving forced rowers preserved in the Civitavecchia governor's fund refers to incidents that occurred aboard the fleet's vessels. While convicts and oarsmen were theoretically isolated from broader society, in practice, opportunities for interactions with civilians were far from rare. These men were often allowed to work in the port of Civitavecchia whenever ships were docked for extended periods. Though regulations required them to remain in chains and under the watchful eye of their overseers—the *aguzzini*—the archival sources reveal a reality in which such constraints were routinely disregarded, and inmates frequently moved about the city with relative freedom.

In the 1763 criminal trial involving the rape of eleven-year-old Angelo, attempts were made to trace the identity of the accused convict. Following what could be described as a "traditional" procedure, the trial was instigated when the boy's mother, Bartolomea Grassino, became suspicious after her son was unable to defecate properly and had been complaining of pain in his genital area for a month. She then took him to see the city surgeon, Bartolomeo Ridolfi, whose re-

---

163 This hypothesis was first advanced by Marotta, The Muslim Friend, pp. 230–252.
164 ASR, Tribunale poi Governatore di Civitavecchia (1589–1913), b. 643, dossier 8, f. n.n.

port confirmed the suspicion that the pain was the consequence of repeated rectal penetration. He instructed the mother to notify the governor. When questioned, Angelo confessed that he had been approached on several occasions by a convict, who had lured him with the promise of money and taken him behind the soldiers' quarters. He described in detail to the magistrates how the conscript had pulled down his trousers, forcing him to straddle on the ground, and then penetrated him anally, ejecting semen in the process—a matter of the utmost gravity, as rectal intercourse involving ejaculation was the gravest form of sodomy.[165] According to the boy's description, the man was around 40 years old, "fair" in complexion, stocky at the waist, with black hair and eyes. No fewer than seven inmates matched this profile. One particular telling detail emerged during the investigation: Angelo identified the man as a forced rower not by his chains—which he did not wear—but by his clothing. This small yet revealing observation starkly illustrates the extent of the freedom some galley rowers evidently enjoyed. He also noted that he had not seen the man in roughly two months, and had not been assaulted recently.[166] As with many other cases of this kind, the final judgment is absent from the archival records, leaving it uncertain whether the suspects were ever questioned or whether the matter was quietly dropped without resolution.

Returning to the central role played by medical expertise in suppressing sodomy, one particularly revealing episode concerns the six denunciations recorded in 1769 for the "supposed nefarious vice."[167] Each case originated with a report by the galley surgeon, who, during routine visits to galley slaves recovering in the dock hospital, had been explicitly instructed by the superintendent of the galleys to conduct meticulous inspections of the "posterior parts"—especially in those who complained of a "pain in the buttocks." These examinations formed part of a broader directive to "pursue every possible means to eradicate the nefarious vice, which is unfortunately deeply embedded" within the galleys.[168] In certain

---

[165] ASR, Tribunale poi Governatore di Civitavecchia (1589–1913), b. 656, dossier 2, f. n.n.: "e poi mi fece vedere una manciata di denari che disse volermi dare, e che gl'avessi fatto fare quello lui voleva, cosiché mi calò li calzoni, e poi mi fece mettere colle mani per terra a pecoroni, e lui calatosi il suo membro, che era duro me lo appuntò nel culo, e facendo forza me lo mise dentro sentendo del dolore, e lo menava avanti, e dietro, e poi mi sentii dentro del culo della materia calda, e del brugiore."

[166] Ibid., f. n.n.

[167] ASR, Tribunale poi Governatore di Civitavecchia (1589–1913), b. 658, dossiers 4–9.

[168] Ibid., dossier 4, f. n.n.: "siccome il signor sopraintendente alle galere ripieno di zelo per l'onor di Dio, tenta ogni strada di estirpare il vizio nefando purtroppo radicato."

cases, and in an effort to obtain greater certainty or to corroborate the initial assessment, the opinions of additional surgeons were also sought.

The complexity of the doctor-patient relationship in a setting such as the galley—situated somewhere between the will to rehabilitate and the will to control—is noteworthy here. I have discussed at length the importance of mutual trust in a patient's path to recovery. Yet in this context, such trust was all but nonexistent, or, at best, fragile. Physicians were acutely aware that they were treating criminals or heretics; thus, while they were obligated to care for them, they often did so reluctantly, and with a constant suspicion of deceit. Above all, physicians were expected to set aside their professional vows of confidentiality and denounce their patients should they detect any signs of violent injuries—and, in cases like this, signs of rectal trauma. One can imagine that medical professionals rarely hesitated to report such patients as individuals of dubious morality. The very precautions and attention meant to safeguard the health of the rowers were, in practice, turned against them—transformed into instruments of control, surveillance, and condemnation.

It is also striking that in most archival sources no testimony or depositions by the convicts can be found—except, in some cases, to answer the perfunctory question of whether they knew why they were in court. Evidently, such inquiries and the ensuing investigations were only initiated once sodomy was already suspected. If the offenders had been caught *in flagrante* and eyewitnesses were available, they were often condemned directly without further questioning. This was the case, for instance, in the trial held on 26 February 1762 against the life convict Andrea Pighi and his neighbor, the vagabond Luca Bianchi, who—thinking they could take advantage of the cover of darkness—were discovered by a companion "committing the sin of sodomy" on Christmas Eve and promptly denounced to their *sotto aguzzino*.[169]

When the veracity of the act had to be ascertained, the governor's court proceeded in two stages: first, a medical report was commissioned from the galley surgeon who was instructed to examine the convict's anus for any signs of hernias, ulcers, ruptures, and other trauma indicative of penetration. Second, the accused and the witnesses were interrogated. From the sources examined, it is clear that the primary purpose for interrogating the accused was to extract a confession. If the convict's version of events deviated from the expected narrative, his testimony was considered worthless. Such silence—or compliance—is hardly surprising, given that the accused were not free citizens but individuals already

---

[169] ASR, Tribunale poi Governatore di Civitavecchia (1589–1913), b. 655, dossier 24, f. n.n.

stripped of their liberty, and, by extension, of their civil rights—whether temporarily or permanently.

It is nonetheless notable that in every trial of this kind, the rowers consistently denied having been sodomized, often resorting to a remarkably similar justification: they claimed to have been wearing borrowed or ill-fitting shirts or trousers that caused irritation and itchiness around the anus, resulting in injuries from constant scratching.[170] This defense was repeated almost verbatim by the convict Giovanni Battista Petrini in 1769. Unsurprisingly, such excuses were rarely believed. As the surgeon's report explicitly stated in Petrini's case, "some excrescences and callosities around the orifice of the anus, produced by the *vitium nefandum*, were to be observed." So confident was the surgeon in his diagnosis that he even suggested resorting to "the operation of fire," a cauterization procedure believed to cure anal lesions caused by rectal penetration.[171] What is perhaps most revealing is the standardized nature of the *galeotti*'s excuses, thus raising the question: why was this particular defense—centered around borrowed clothing—so widely adopted? Was it rumored to be credible? And, if so, on what basis? Indeed, given the rowers' notoriously poor hygiene and wretched living conditions, it is not entirely implausible to believe that shared garments could have transmitted infections or fungus-related skin conditions. Nevertheless, in the eyes of the authorities, such claims were insufficient to override the surgeons' report.

As with many of these cases, the archival sources do not indicate whether the suspected *galeotti* were ultimately convicted or acquitted. Still, it is plausible to infer that most were found guilty, as the surgeon's report would have been taken as definitive. That many trials were left unresolved may also reflect pragmatic considerations: had the accused been convicted, they would have likely faced the death penalty—an outcome that would have been counterproductive for the operational capacity of the fleet, given the constant need for a steady source of manpower.

Yet the systematic involvement of medical professionals in these criminal proceedings underscores just how crucial their role had become in determining the truth. A case in point is that of convict Vincenzo Buganti, who admitted during his trial that he had once been sodomized—but only in the past, and not while he was

---

[170] See, for example, ASR, Tribunale poi Governatore di Civitavecchia (1589–1913), b. 658, dossier 5, f. n.n.: "io sono tre giorni che sto ammalato in questo letto essendoci stato portato dalla galera Capitana dove sono forzato a tempo, per motivo che 20 maggio addietro essendomi un paio di calzoni che non so dirgli di chi fossero mi attaccarono al culo il male che è stato visitato."
[171] Ibid., dossier 6, f. n.n.: "ho riconosciute alcune escrescenze, e callosità intorno all'orifizio dell'ano, prodotte dal vitio nefando, per cui gli si deve fare l'operazione del fuoco."

serving aboard the galley. The surgeon's report, however, contradicted this account, attributing his injuries to "repeated sodomy," thus implying that the abuse had been recent.[172] Similarly, the record is silent on the verdict, but the case vividly illustrates how medical testimony could decisively override the voices of those accused.

One of the rare instances in which significant space was afforded to the convicts' testimony dates back to 1773 in a trial that followed a convict's voluntary denunciation. Alessio Giovanni Celebrini presented himself at the Ospedale di Santa Barbara in February 1773, claiming he had been sexually attacked on board by another convict, Domenico Mortelli, who worked as a *mozzo*. The hospital surgeon's report confirmed that Celebrini's injuries were consistent with rectal intercourse. Once recovered, he was sent back to the ship to await trial, which took place two months later. The very fact that a formal trial was instituted raises an important question: why, unlike in so many other sodomy cases, was this particular instance investigated and the accused potentially punished? While a definitive answer remains elusive, one plausible explanation is that voluntary denunciation—particularly one supported by the testimony of trusted medical professionals—was more difficult for the authorities to ignore or suppress. Furthermore, this case was far from routine; it had the potential to disrupt the fragile equilibrium within the galley. Mortelli was not a common oarsman, but a *mozzo*, tasked with assisting the *aguzzino* in overseeing discipline among the crew. His role lent the case particular gravity: leniency was simply not an option. After all, how could the naval authorities expect order to be upheld if those entrusted with enforcing it were themselves the first to violate it?

Despite his initial voluntary denunciation, Celebrini altered his version of events during the trial. Contrary to the surgeon's findings, he now claimed that the irritation to his anus resulted from an overly aggressive cleansing and from a pair of poor-quality trousers he had purchased months earlier from a Bolognese merchant in the port—recycling a well-known and frequently invoked excuse. He formally withdrew his accusations against Mortelli, who, in turn, denied ever having any kind of relationship with Celebrini. Both men supported their revised accounts by claiming that the rest of the crew could vouch for them. The questioning of their shipmates, combined with Celebrini's retraction, strongly suggests that once back aboard the vessel, he was likely subjected to threats by his companions—indirectly revealing the existence of a tacit pact among crew members to

---

[172] ASR, Tribunale poi Governatore di Civitavecchia (1589–1913), b. 658, dossier 8, f. n.n.: "per causa di reiterata azione sodomitica e non per altro motivo."

keep such practices hidden from the eyes of the authorities.[173] This is further supported by the fact that few sodomy trials appear to have been initiated by formal complaints from fellow crew members—at least not in the surviving records.

Ultimately, although the surgeon continued to insist that this was a clear case of sodomy, the trial culminated with the acquittal of both men and Mortelli's transfer to another galley, on the grounds that there was no evidence of any friendship between the two, and that Celebrini's now-treated injury could no longer be medically examined.[174]

As before, the situation was not resolved through the use of force. On the contrary, it seems that none of the trials examined ended with the application of corporal punishment. Several plausible explanations can be proposed. At first glance, it might seem that aboard the Papal fleet the political and judicial authorities were indeed committed to suppressing sodomy—though not through violent means. What seemed to matter more—and perhaps proved more effective—was not punishing those who had already committed the offense but rather deterring others from doing so. It may have been enough to simply open an investigation or initiate legal proceedings—both to instill fear among the crew and to signal that a functioning apparatus of control was actually in place.

One could argue that there was, in fact, a reluctance to prosecute illicit sexual practices in earnest, and that the initiation of partial or unresolved trials was ultimately a performative gesture—meant more to deflect accusations of negligence than to genuinely root out the practice. Indeed, reading the archival sources, one is struck by the sense that the authorities were not seriously invested in eradicating sodomy among convicts. The archival evidence suggests that even when criminal proceedings were initiated, they were rarely pursued with any real vigor or with a clear desire to uncover the truth. A related possibility is that it was simply a period of transition during which the naval authorities hesitated about whether to prosecute sodomy rigorously, torn between the duty to administer justice and the desire to avoid scandal or risk the loss of skilled manpower aboard seafaring vessels.

It should not be forgotten that the galleys were a context in which the accused were forced laborers at the oars. Perhaps the authorities were reluctant to impose the death penalty, since doing so would simply have meant depriving the galleys of much-needed free labor. Moreover, the Governor's Court's lack of interest in prosecuting suspects may have stemmed from the social status of the accused—men already condemned to society's margins, such as convicts and slaves. In prac-

---

[173] A similar hypothesis is suggested by Calcagno, "Brutale libidine," p. 180.
[174] ASR, Tribunale poi Governatore di Civitavecchia (1589–1913), b. 660, dossier 8, f. n.n.

tice, the governor had little or no direct contact with the rowing crews or knowledge of their crimes. If the alleged act occurred aboard a vessel rather than in the city streets, it may not have elicited particular concern, despite falling under his jurisdiction. Nevertheless, the mere fact that procedural records related to sodomy committed by the rowing crews have survived—and that investigations were in fact carried out—points to a significant shift at the institutional and judicial level, and underscores the indispensable role that medical professionals played in prosecuting such crimes.

5 Theoretical gaps in early modern medicine: the case of sodomy

Trials against sodomitic galley rowers open up another crucial topic alongside the doctor-patient relationship: the relationship between medicine and sodomy. It is widely accepted that in the pre-modern era, Western European medicine offered no formal theorization of sodomy. The only setting in which it directly addressed the topic was legal—when physicians were required to physically examine suspected sodomites in search of indications of sexual intercourse. Recent historiography has revealed that a science of homosexuality—though both terms are anachronistic—already existed in the early modern period. They remain useful, provided homosexuality is understood as same-sex relations and science as a broad form of knowledge. Indeed, well before the 17th century, one finds medical and forensic discussions of anal intercourse, including anatomical, mental or physiological causes; classification of homoerotic relationships into sub-genders, proposed therapies, and more.[175]

The persistent idea that same-sex relations were not subject to scientific investigation until the modern period stems from a historiographical tendency to privilege—perhaps to overemphasize—law and theology as the sole discursive arenas for addressing sodomy in the early modern period. This view underpins the *acts paradigm*, which holds that—prior to the 19th century—sexual deviance was understood solely in terms of individual acts, not as reflective of personal identity. From the modern period onward, however, the figure of the sexual pervert emerges. In the first case, one speaks of "sodomy;" in the second of "homosexuality." This marks a stark dichotomy between the early modern "sodomite"—a legal and moral category—and the modern-day "homosexual"—

---

[175] Borris, The Prehistory of Homosexuality, pp. 4–6.

conceived as an individual defined by an innate perversion of mind or body and therefore a subject of scientific scrutiny.[176]

By the 19th century, sexual deviance was no longer framed in terms of isolated acts, but as the expression of membership in a particular social group—one defined by experiencing emotional and sexual attraction toward members of the same sex. Beginning in the 18th century, the study of same-sex relationships shifted from focusing solely on illicit anal intercourse to examining its perceived pathological effects on the body and the personality. Scientific interest in sexuality expanded, particularly with the rise of psychiatry, which sought to explain why and how one became "homosexual"—a term coined in 1869 by the Hungarian journalist Karl-Maria Kertbeny—and whether, and how, it could be "cured." Throughout the 19th century, homosexuality increasingly came to be viewed not as a temporary deviation from the norm, but rather as a chronic, pathological condition.[177]

And yet, a natural, identity-based conception of sodomy can be traced as far back as antiquity. Theories of innate same-sex affinities circulated in ancient Greek and Roman sciences, especially in medicine and astrology, and were later revived during the Middle Ages. Since antiquity, various authors had argued that the sin "against nature" had, if fact, natural causes. Given the sensitivity of the subject, however, such works were few in number, written mostly in Latin, and often censored. The moral, legal, religious, and social condemnation of sodomy was so severe that it discouraged scientific inquiry and prevented these texts from being translated into the vernacular. The risk of exclusion from the scholarly community was too high—even in the name of the love of knowledge.[178]

Why, then, was it problematic to recognize sodomy as a subject worthy of medical attention? Didn't the humoral theory, after all, allow for a natural explanation of such practices. As discussed in Chapter 1, an individual's temperament was thought to be predicated upon a unique balance of the four humors. There was no single, universal temperament [*complexio*]; rather, it varied with sex, age, climate, and geographical region. As Borris has noted, "complexional physiology was widely applied to explain the range of human deviations from perceived norms of biological sex, gender, and sex roles."[179]

Given that health was not defined by strict humoral equilibrium, and that each individual was believed to possess a distinct constitution, it followed that dif-

---

176 Ibid.
177 Watson, *Forensic Medicine*, p. 117f.
178 Borris, The Prehistory of Homosexuality, p. 9f.
179 Ibid., p. 17f.

ferent sexual orientations were not only possible, but not inevitably classified as pathological. Individuals whose inherent temperament led them to prefer same-sex relationships were not—within this framework—considered ill.[180] Even the Hippocratic-Galenic theory of sexual physiology allowed for intermediate sexual forms that challenged the male-female binary. According to this theory, there was only one sex—male—and female genitalia were simply internalized male organs, the result of insufficient heat during intrauterine development.[181] Other disciplines, such as physiognomy and astrology, also contributed to normalizing individual idiosyncrasies. They reinforced the idea that certain tendencies, though contrary to the norm, were both natural and innate—attributed to malign celestial influences and visibly inscribed on the body.[182]

What emerged from these diverse frameworks was a conception of nature not as a static system governed by immutable laws, but as dynamic and variable. Deviations from the norm were not exceptions to nature but expressions of it—which, by its very nature, never acted randomly, nor aimlessly.[183] This was the paradox of so-called monstrous creature: their "unnatural natures" were not "in keeping with nature" [*secundum naturam*], yet neither were they "contrary to nature" [*contra naturam*].[184]

The answer is simple: recognizing sodomy as a medical condition required situating it within the realm of nature rather than that of morality. Yet to seek natural explanations for this crime *against* nature—to "naturalize the unnatural"—would have not only implied a teleological justification for such practices, but also risk placing them "beyond the boundaries of vice."[185] As Aristotle argued in Book VII of the *Nicomachean Ethics*, moral virtue and vice are not innate but acquired through habit and training. If sodomy were deemed a natural condition—however defective, like blindness or disease—it could no longer be condemned as a moral failing, and thus not a vice in the true sense of the term.[186]

Whether sodomy was attributed to physiology [*natura*], to a corrupted humoral balance [*egrotative*], or to habitual behaviors [*consuetudine*], it was consistently framed in medicalized terms, as though akin to illness. Fourteenth-century scholars such as Nicole Oresme and Heinrich von Friemar regarded same-sex acts as products of bad habits, no more morally significant than biting one's nails or

---

180 Cadden, *Nothing Natural*, p. 152.
181 Laqueur, *Making Sex;* Montecón Movellán, Oltre la repressione, p. 149.
182 Borris, The Prehistory of Homosexuality, pp. 20–28.
183 Cadden, *Nothing Natural*, p. 35 f.
184 Ibid., p. 62 f.
185 Ibid., pp. 140–175.
186 Ibid., pp. 144–146.

fiddling with one's hair.[187] Some argued that medicine should avoid these illicit practices due to the *turpitudinem*—the shame associated with the topic—others insisted that it was both the right and the duty of natural philosophers, including physicians, to investigate all phenomena in nature. As Walter Burley argued: "Nothing natural is shameful, all things in the world are pure." For a philosopher, even the most repulsive and vile phenomena held intrinsic value in the pursuit of truth.[188]

Despite such views, the stigma surrounding sodomy was so deeply entrenched across early modern European society that medical scholars largely avoided further investigating the subject. While there was no formal censorship, the implicit threat of undermining its moral and legal condemnation—and with it, the system of values underpinning social order—served as a powerful deterrent. Nevertheless, throughout the 16th and 17th centuries, some clerics accepted the possibility that sodomy might have physiological causes. For example, Manuel do Vale de Moura, the Lusitanian representative on the Inquisition, argued that it could stem from early exposure to vice, malign celestial influences or an inherent temperament—especially one found among "barbarians." Moura claimed that some regions and individuals with humoral imbalances were predisposed to such acts. This naturalization of vice, however, did not mitigate moral judgment; rather, it served as a tool to reinforce religious condemnation and fuel hatred.[189]

One of the earliest, and perhaps most noteworthy, attempts to understand the causes of sodomy in terms of natural processes can be found in the ancient text *Problemata* (Book IV) attributed to Pseudo-Aristotle. In Question 26,[190] he specifically addresses the issue of homosexual lust, beginning with the phrase: "Some individuals, compelled by nature, commit the sodomitical sin [*sodomiticum peccatum*]."[191] This collection of questions [*problemata*], dealing primarily with natural phenomena, was divided into 38 books [*particulae*], each covering a specific subject. Book IV included 32 questions related to "venereal matters," among which Pseudo-Aristotle seeks to answer: "Why do some men enjoy the passive role in sex, while others enjoy both active and passive roles?"[192] He offers two explanations: some men are born with anatomical defects that divert semen from its proper path; others develop this inclination as a habit formed through past experiences. Thus, sodomy between men was included among those phenomena

---

187 Ibid., pp. 159–161.
188 Ibid., p. 182.
189 Lavenia, *Eresia indicibile*, pp. 34–38.
190 See Blair, Autorship, pp. 189–227.
191 Cadden, *Nothing Natural*, p. 35.
192 Ibid., p. 3.

which, though not inherently virtuous, were nonetheless part of nature—much like the sensation of dizziness when drunk or the tendency to fall asleep while reading.[193]

Despite the renown of the author traditionally attributed to the *Problemata*, especially Book IV, it garnered scarcely any scholarly attention, likely due to both the subject matter and the work's excessively dense and complicated arguments. The first Latin translation of the text, by Bartolomeo da Messina in the mid-13th century, appeared much later than other works in the Aristotelian corpus.[194] Nevertheless, during the 14th century, a small number of eminent scholars began to engage with Book IV, notably Pietro d'Abano (c. 1250–1315), Jean de Jandun (c. 1285–1338), Walter Burley (c. 1275–1345), and Evrart de Conty (c. 1330–1405). These scholars—all natural philosophers closely linked to universities where medicine and natural philosophy were taught principally through the study of Aristotle's works—also spent considerable time in Paris.

The most prominent—and earliest—commentator on Book IV was Pietro d'Abano, a natural philosopher and physician, whose career spanned the universities of Padua and Paris. His works reflected and were shaped by the rationalist tendencies of 13th-century Padua, which aimed to provide causal explanations for all natural phenomena.[195] D'Abano's commentary on *Problemata* was likely completed around 1310, roughly contemporaneous with his influential *Conciliator* which sought to reconcile differences between medicine and natural philosophy.[196] His commentary begins with an inquiry into sexual pleasure, broadening from specific men's desires into a universal account, thereby laying the groundwork for understanding sodomy as a variation of a natural process. Sexual pleasure, in his view, results from the expulsion of superfluous matter—*semen*—through ejaculation, which helps maintain humoral balance. In this sense, ejaculation is akin to other expulsions from the body—such as defecation or the shedding of tears—all of which serve to ensure bodily harmony.[197]

Under normal circumstances, semen is expelled through the penis. D'Abano, however, theorizes that some men's "pores" (which refer to channels in the body) may be "not constituted according to nature either because those pores in the penis are blocked [...] or alternatively the humidity flows into the anus and exits that way." In such cases, the passages for sperm—the "pores"—may be ar-

---

193 Ibid., p. 8f.
194 Ibid., p. 14f.
195 See Piaia, *Pietro D'Abano*.
196 Siraisi, *Medieval Renaissance Medicine*, p. 60; Ronzoni/Piaia, *Pietro D'Abano*.
197 Cadden, *Nothing Natural*, pp. 39–42.

ranged "not according to nature."[198] He proposes two possibilities. The first is the vessels at the base of the penis may be blocked—a condition similar to that found in eunuchs or effeminates. The second is that moisture may accumulate and be discharged through the anus instead of the penis, resulting in the evacuation of semen through an anatomically incorrect orifice. He likens this anomaly to blindness, which he suggests may arise from a thick humor obstructing the optic nerve.[199]

Thus, D'Abano identifies two types of anatomical deviations in the male sexual organs: one where the natural expulsion of excess bodily humors is obstructed and another where semen is misdirected to the anus, where, contrary to nature, the contraction during sexual intercourse does not take place in the buttocks but in the upper part of the penis. As a result, no pores open in the penis; instead the fluid is expelled through openings around the anus, flowing in the wrong direction.[200]

The desire to engage in sodomy could thus arise not only from anatomical defects but also from repeated practice, regarded as a form of "second nature." According to D'Abano, individuals referred to as sodomites might become such after becoming accustomed to the act from an early age.[201] It is notable that the idea that one could become a sodomite through habituation in adolescence is also widely attested in archival sources. A deeply ingrained early modern belief held that one of the principal dangers of indulging in such practices—following the model of classical pederasty—was that it rendered young people effeminate and inured them to the practice.[202] D'Abano's conclusion is clear: while sodomy is an abominable practice, it is a thoroughly natural phenomenon and "beyond the boundaries of vice," while those who practice it cannot be considered immoral from an ethical standpoint.[203] Though he acknowledges sodomy's intrinsic criminal nature, his language remains notably ambiguous: he never explicitly states his opinion on whether these men and their inclinations should be condemned.[204] Later scholars, on the contrary, argued in favor of condemning such acts and advocated censoring Book IV of the *Problemata* as too unseemly a subject matter for natural philosophy. Sex could—and should—only be discussed in the context of

---

198 Ibid., p. 46 f.
199 Ibid.
200 Ibid., p. 49 f.
201 Ibid., p. 74.
202 See Baldassarri, *Bande giovanili*.
203 Cadden, *Nothing Natural*, p. 159 f.
204 Ibid., p. 190.

promoting procreation and preserving good health, with all its other aspects relegated to morality.²⁰⁵

Over the course of antiquity, and, later, in the early modern period, two disciplines closely related to medicine sought to explain sodomy on a scientific and natural basis: astrology and physiognomy. As outlined in Chapter 1, medical astrology formed one of the key branches of ancient medicine. It was believed that the position of the planets could influence a person's state of health, and therefore, studying the stars could provide the answer in the search for the most appropriate remedy for physiological or psychological disturbances.²⁰⁶ In addition, horoscopic astrology was believed to be an essential tool not only in medical practice, but also in theories of generation. Since the Middle Ages, the notion that the stars—with their heat, light, and occult effects—influenced the creation of the embryo had become firmly rooted. An individual's mental and physical constitution was thought to be determined by their horoscope not only at the time of conception but also at birth.²⁰⁷ One's sexual orientation, too, was included among those innate traits subject to astral influences. As early as Ptolemy, in aphorism VIII of the *Centiloquium* and in book IV of the *Quadripartium*, harmful sexual inclinations were attributed to an inauspicious conjunction of Venus or Jupiter, unmediated by the presence of Saturn.²⁰⁸

Physiognomy—the science of interpreting bodily signs as indicators of personal character traits, state of health, and even life and death—was also condemned on similar grounds. Early modern physiognomy, based in part on ancient astrology and complexional theory, posited that internal conditions manifested themselves through physical signs that could be read on the body. At the same time, such signs could also indicate future changes. Consequently, one's sexual inclination—which was also a natural and internal characteristic—could be traced through external signs, which in turn could be identified and studied.²⁰⁹ Moreover, physiognomy was an even more deterministic science than astrology, as it assumed that the types and applications of the characteristics studied were fixed and universally applicable across all historical periods.²¹⁰

---

205 Ibid., pp. 198–202.
206 Siraisi, *Medieval Renaissance Medicine*, p. 111.
207 Ibid.
208 Cadden, *Nothing Natural*, pp. 97–102.
209 Ibid., p. 21.
210 Borris, *Sodomizing Science*, p. 137.

5.1 Medicine and sodomy in the courtroom: some reflections

In conclusion, when discussing the complex relationship between medicine and same-sex acts throughout the early modern period, we cannot avoid addressing the field of forensic medicine. Perhaps the only context in which these topics were unanimously seen as interlinked was in criminal trials involving suspicion of such offenses. The impossibility of readily establishing whether the illicit activity actually occurred meant that judges often had to rely on the opinions of doctors summoned to testify as medical experts. Although medical reports did not explore the motives and causes—whether natural or otherwise—underlying the act, they nonetheless reveal something about the early modern conception of the subject. From a purely legal standpoint—focused on its condemnation—same-sex intercourse was not an abstract or physiological condition but a specific act: rectal penetration, especially between men.[211] Paradoxically, at a time this transgression was not widely regarded as a subject worthy of medical interest, medicine was the only discipline thought capable of providing certainty as to whether the act had been committed, thus placing this offense both within and beyond the competence of medical practitioners.

The analysis of same-sex offenses among early modern *galeotti* thus confirms the role of medical practitioners in the courtroom, demonstrating how the doctor's expertise was central in identifying and addressing this vice. First, doctors helped maintain strict discipline among rowers by notifying the naval authorities of any physical sign indicative of illicit sexual behavior—ulcers, lesions and similar indicators. A thorough examination of the suspected sodomite's anus could provide certainty about the alleged illicit activity, compensating for the lack of control by galley officials over the crew, and the ineffectiveness of punitive threats. The obligation for doctors to cure the patient, examine him carefully, and report their findings to the authorities was central in cases involving such infractions, demonstrating the intricate, and at times contradictory role medical practitioners played in maintaining social and moral order aboard early modern galleys. This intersection of medical practice, legal authority, and social discipline sheds light on early modern attitudes toward non-normative sexuality and underscores the deeper complexity of the doctor-patient relationship—both aboard galleys and within society at large—where medical expertise functioned as a tool of moral governance and a mechanism of control.

---

211 Rousseau, Policing the Anus, p. 78.

Conclusion

As the purposes for which vigilance is exercised change over time or in response to evolving circumstances, the very meaning of vigilance also transforms. It emerges as a highly multifaceted and nuanced concept, with meanings that may differ—sometimes simultaneously—depending on one's perspective.[1] In the medical sphere, this multivalent nature of vigilance became especially apparent. Unlike traditional forms of surveillance, vigilance studies lack both the normative rigidity of control and reference to an external, superior observer. Nonetheless, the physician remains a figure of authority, acting not only as a "private citizen" but also as an agent bound by institutional frameworks. Recourse to the method of vigilance studies has made it possible to better grasp these dynamics. Observation, control, and inspection were not merely top-down impositions on doctors; they also emerged as natural response to the need for security aboard galleys—security that could only be ensured if all parties fulfilled their duties and cooperated in maintaining order.[2] Thus, "medical vigilance"—defined as attentiveness to even the slightest sign manifested or reported by the patient—could take on both positive and negative valences: positive when directed toward healing, negative when used to produce evidence for prosecution.

This shift in the doctor's gaze—from the illness itself to the patient *as* illness—mirrored broader political discourses of the time. The metaphor of the political body, which gained enormous traction in medical and political treatises from the fifteenth and sixteenth centuries onward, framed the health of the body in anatomical terms. To preserve the integrity of the collective body, it was deemed necessary to physically separate "dangerous" individuals from other citizens, thus eliminating—as a surgeon would—the corrupt parts from the mystical body of society.[3] The analogy between the human body and the political body is also articulated by de Castro in Chapter IV of Book IV of the *Medicus Politicus: Corpus Humanum mirificam cum Republica bene ordinata similitudine repraesentare*.[4] In this allegorical model, the State is imagined as a physical body, with the *princeps* [the ruler] as the head and the subjects as the limbs, which must obey and submit to his will. The *princeps*, however, was not only likened to the head, but also to the physician, for both were entrusted with defending and maintaining the body's

---

[1] Brendecke/Molino, *Cultures*, p. 11 f.
[2] Brendecke, Warum Vigilanzkulturen?, pp. 10–17. On the limits of the concept of "disciplining", see Schiera, Prodi, "Disciplinamento", pp. 349–351.
[3] Pastore, *Le regole dei corpi*, pp. 17–35.
[4] De Castro, *Medicus Politicus*, pp. 238–242.

equilibrium—whether political or physical. Just as rulers were tasked with equipping the city with gates and ramparts to defend its citizens from external attacks and render it impregnable, so too were physicians responsible for maintaining the balance of the humors by carefully regulating what the body ingested and what it expelled.[5] And just as the proverbial wise doctor was called upon to prescribe effective—albeit often unpleasant—remedies, so too was the skilled ruler obliged to apply severe penalties to heal the mystical body of the city.[6]

The analogy of the wounded body in need of healing became an instrument through which the State could legitimize more active and repressive measures. This metaphor also helped reinforce the role of the doctor as an undeniably central political and social figure, extending far beyond his homespun and traditional therapeutic duties. By assisting in the task of prosecuting criminals, the medical establishment aligned itself with political power, demonstrating just how indispensable its role was in safeguarding social order. The ultimate result seems to have been a dual process of legitimization: both of the social and political functions of medicine throughout society, and, in turn, the transformation of medicine itself into an instrument for legitimizing the repressive and persecutory policies advocated by political authorities. This is not, of course, a dismissal of medicine's primary function as therapeutic, but rather an exploration of how pivotal the political recognition of doctors as "experts of the body" was for their professional affirmation. This reflection may strike some as echoing the Foucauldian thesis on the disciplinary nature of medicine. My aim is to emphasize that a doctor held both a political and therapeutic role, thereby underscoring the complexity of early modern medicine.

The multifaceted nature of medical vigilance was particularly evident in settings where doctors had to deal with dangerous or suspect individuals. In this regard, the galleys provide an exceptional context for analysis, given the marginal status of their rowers—who were primarily slaves and convicts. Doctors had to balance the need to maintain their health with a reluctance to treat them given that they were criminals and heretics, while also being obligated to monitor their behavior and report any illegal activities. Medicine, thus positioned between curing and controlling the body, played an essential role not only in safeguarding the crew's health but also in maintaining order and discipline among the rowers.

As comparative studies of the Tuscan and the Papal contexts have demonstrated, the galleys were effectively configured as spaces for vigilance—both vertical and horizontal—with the primary objective of ensuring strict military and

---

5 Ibid., p. 238 f.
6 Pastore, *Le regole dei corpi*, p. 32.

behavioral discipline aimed at successful seafaring. The galleys were hybrid spaces: part military institution, part penal institution, and part religious mission. The need to impose military discipline went hand in hand with the necessity of having rowers adhere to Christian doctrine. This led to an intense program of catechization, aimed not only at slaves, who were heretics, but also at Christian convicts sentenced to serve aboard the galleys due to criminal actions or blasphemous behavior. At the same time, the rowers' moral re-education was complemented by the convicts' penal correction. Although a reintegrative concept of punishment was lacking, the notion of a transformative punishment for the individual had existed at least since the 16th century. Serving as an oarsman in a galley was regarded as the most degrading condition a man could endure in his lifetime. It epitomized a form of social death, reducing the individual to a mere instrument of labor. While stripped of their freedom and legal status, galley oarsmen were nonetheless recognized as individuals, especially from a moral and religious perspective. Far from being merely a source of free labor, galley rowers were individuals to be cared for, re-educated, and reprimanded.

The ultimate goal was not so much their social reintegration as their transformation into obedient and God-fearing rowers serving purely technical and military purposes. In this sense, they required constant monitoring, with punishment administered in a cautionary manner whenever an offense was committed. For this system of control to be fully effective, it had to rest upon a reliable apparatus of surveillance and denunciation, one that, in turn, hinged upon a system of mutual observation. Indeed, keeping an eye on 300 men at once was no easy task, even in such extremely confined spaces as the galley. Precisely for this reason, various strategies were devised to ensure that every crew member could become the constant object of attention by those with whom they shared the galley. In addition to being monitored by their officers, rowers were required to keep a watchful eye on their bench mates, as well as on those seated in front of and behind them, and to report any improper behavior to the authorities. Similarly, convicts and slaves were required to check that their officers behaved properly and, if they failed to do so, to notify the authorities of any form of abuse or harassment. The galley and its associated shore facilities thus became spaces in which everyone—officers and crew alike—was held to a strict code of behavior, constantly under the watchful eye of other crew members. Medical vigilance was therefore just one of many forms of surveillance enacted upon early modern galley slaves and convicts.

The analysis of the practices and strategies of vigilance aboard galleys also contributes to the ongoing debate on the role and meaning of deprivation of liberty in the early modern period, as well as on the supposedly neat distinction between "free" and "unfree" rowers.[7] Indeed, the decision to study slaves and convicts as a unique category of "patients"[8]—to whom medical attention was directed—led me to question their social status and ultimately convinced me of their commonality. This suggests that there was no substantial difference between the status of convict and slave aboard early modern Italian galleys. Indeed, the same sources often refer to convicts as *servants or slaves for punishment* [*servi/ schiavi di pena*]. As Giovanna Fiume noted, the sentence to the galley—albeit with due differences—rightly fell within the category of penal servitude [*servitus poenae*] as theorized by Roman Law.[9] Deprivation of liberty, in fact, could take three forms: slavery and servitude, under the power of a master; isolation and imprisonment, under the power of justice; and captivity, under the power of a political and/or religious enemy.[10] Deprivation of liberty while in the custody of justice should thus be considered, in itself, a true condition of slavery. In particular, I am convinced that the characteristics of slavery theorized by Claude Meillassoux in 1986—depersonalization (the transformation of human being into an object of property), desocialization (the forced insertion of a foreigner into an alien society), and decivilization (the reduction of social relations to a sole bond of subjugation to a patron)[11]—were equally applicable to slaves and convicts for the duration of their sentence. As with slaves, galley convicts were stripped of all civil rights, relegated to society's margins, and considered beasts of burden.[12]

Regarding galley doctors, the available sources unfortunately only allow for a partial reconstruction of some of their biographies. Nevertheless, it is reasonable

---

7 De Vito/Schiel/van Rossum, From Bondage to Precariousness?, pp. 644–662. See the research of the COST Action CA18205 "Worlds of Related Coercions in Work" (WORCK), emerging from the working group "Free and Unfree Labor" of the European Labor History Network, and the research by the Bonn Center for Dependency and Slavery Studies (BCDSS).
8 Physicians were required to treat rowers even when they were healthy, such as when examining the body to assess its condition. It should be rembered, however, that their primary function was, of course, therapeutic.
9 On the concept of penal servitude, see Beggio, "*Servitus poenae*"; McClintock, Dal Servus Poenae, pp. 1072–1085.
10 Fiume, Schiavitù mediterranee, p. Xf. Recently, the research group at the Bonn Center for Dependency and Slavery Studies proposed a model that transcends the dichotomy of free/not free, freedom/slavery, using the analytical concept of "strong asymmetrical dependency," of which slavery is only one facet. The intrinsic features of these systems of dependency include the use of violence and the possibility for the dominated to exercise "(inter) agency", understood "not merely in terms of (violent) opposition or resistance but rather as the opportunity to act within relations of asymmetrical dependency." See Winnebeck, Sutter, Hermann, Antweiler, Connermann, Asymmetrical Dependency, p. 21f.
11 Meillassoux, *Anthropologie de l'esclavage*.
12 Chizzolini, Navigating Ambiguities.

to infer that the importance of their role was reflected in the social and professional prestige attached to the post, which, in addition to guaranteeing a fixed salary, likely represented a form of public recognition of the holder's skills. It has been noted, however, that the office was not only an honor but also demanded significant sacrifices. Despite the harshness of the task, the prospect of a steady income and professional and social advancement must have weighed more heavily, as the position was evidently highly coveted—confirmed by the various petitions preserved in the archives. Confronted with an intensely competitive medical marketplace, a public and permanent position was perhaps the best prospect to which an ordinary doctor could aspire.

From my analysis, it is feasible to argue that the functions of galley doctors were threefold. First, and most obviously, was their therapeutic role, aimed at keeping the crew healthy and fit for service. In the aftermath of the Battle of Lepanto, the authorities were forced to confront just how poor the levels of medical care aboard the galleys had been. Their inability to efficiently treat and rehabilitate the crew not only contributed to high mortality among soldiers and oarsmen—and thus to a significant loss of cheap labor—but also fueled a broader sense of disorder, which resulted in numerous escapes as well as a pervasive state of despair. It therefore became necessary to improve hygienic and sanitary conditions aboard the galleys, as well as on land, with the opening of hospitals. These facilities provided an ideal setting for developing original hygienic and medical-organizational measures. Ultimately, public slaves and convicts received better healthcare than private slaves and the majority of the city's residents, as their health was considered paramount for technical and military purposes.

Their second function was to observe and examine the rowers' bodies for technical, military, and economic ends. In these contexts, physicians did not treat rowers as medical patients, but rather as objects to be valued for their economic utility and operational efficiency. Finally, a doctor could be consulted *a posteriori* to determine the causes of death or the severity of injuries in the case of murders or violence. In the eyes of galley doctors, rowers at times thus appeared as patients, at others as subjects of technical evaluation, and still at other times simply as bodies—dead or alive—on which to pronounce a judgment with procedural weight.

The doctor's role was inherently ambivalent: it required vigilance over galley slaves and convicts, who had to be both cared for and controlled. Oarsmen were to be treated with the utmost zeal and charity—after all, seafaring was untenable without a healthy and able crew. Yet the presence of forced rowers meant that doctors had to be constantly on alert, approaching them with extreme caution —if not outright suspicion. These were not ordinary patients; on the contrary, their status as slaves and convicts often took precedence over their medical condition.

From a medical standpoint, there was no distinction between convicts and slaves. As the Hippocratic Oath states, physicians should not discriminate between patients, regardless of their status: "Into whatsoever houses I enter, I will enter to help the sick, and I will abstain from all intentional wrong-doing and harm, especially from abusing the bodies of man or woman, bond or free."[13] The only real difference lay in where they were treated—a difference rooted in religious, not medical, considerations. In practice, it is likely that physicians were more hostile or ill-disposed toward slaves than toward the rest of the crew. Nevertheless, when we examine the specific duties and attitudes expected of doctors in relation to enslaved and condemned rowers—as well as the archival sources at our disposal—this difference seems minimal. This apparent uniformity in treatment prompts a reconsideration of the supposedly neat divide between free and unfree labor aboard early modern galleys, revealing just how blurred and nuanced those boundaries truly were. Nowhere was this ambivalent character of medical vigilance more evident than the galley hospital—a structure ambiguous by its very nature, where medical personnel operated simultaneously in collaboration and in competition with religious staff. Such hospitals functioned partly as places of therapy, partly as sites of religious instruction and moral rehabilitation, and partly as instruments of confinement and disciplinary control. The rowers hospitalized there were, at one and the same time, patients to be cured—physically and spiritually—heretics to be converted or re-educated in Christian morality, and slaves and convicts to be monitored and controlled.

Among the various criminal trials conducted against slaves and convicts aboard the galleys, those concerning sodomy undoubtedly warrant special analysis. The examination of legislative, procedural, and diplomatic sources presents an image of early modern *galeotti* as individuals deviant both in habits and behavior. They are depicted as prone to a range of crimes and vices, in particular blasphemy and sodomy—thus casting the *galeotto* as an inherently immoral figure, and reinforcing, inter alia, the traditional association between sodomy and heresy.[14] While such representations can be traced as far back as the 16th century, it was only from the 18th century onward that this image was also formally institutionalized, leading to the mistaken assumption that it represented a novel development—an interpretation which, as has been demonstrated, is unfounded. An analysis of the representation and suppression of sodomy aboard Italian galleys, particularly

---

13 MacKinney, Medical Ethics, p. 31.
14 Grassi, *Sodoma*, p. 55.

those of the Papal States, confirms that, during the 18th century, the authorities developed a renewed and multifaceted sensitivity toward the issue.

As historians of sexuality and religion have noted, the Enlightenment marked a transitional period in the punishment of sodomy, characterized by a gradual reduction in the number of trials and the severity of punishments—likely reflecting a growing desire to avoid publicizing such matters.[15] And yet, in the context of the galleys, the opposite seems to have occurred. We observe a broader willingness to prosecute sodomy more intransigently and publicly—a crime which, although long described as widespread, particularly aboard the Papal galleys, had previously been deliberately kept out of public view. Not only did the majority of sodomy cases go unnoticed and unpunished, but many of those that did reach the courtroom went unresolved, suggesting a lack of real interest in punishing sodomites. One is left with the impression that, while sodomy was deemed an abominable crime to be eradicated from the galleys, political and judicial authorities did not regard it as particularly serious—at least not when committed by individuals already marginalized, if not entirely outside society. This attitude seems especially true when those accused represented a forced or free labor force whose punishment would have entailed the loss of manpower, and thus, a disadvantage to the State. Paradoxically, marginality and infamy granted these individuals a kind of limited impunity. Already persecuted and relegated to the margins of society, *galeotti* were punished with less severity than the law prescribed.

Furthermore, sodomy among the *galeotti* represented yet another sphere of conflict between the officer class and the chaplains tasked with the fleet's spiritual governance. For the officers—primarily interested in technical discipline for navigational purposes—sodomy was undoubtedly an issue, but not an urgent one. For the galleys' chaplains, however, more invested in disciplining souls, the technical and moral spheres could not—and should not—be separated. The sources underscore the importance of medical expertise whenever a case of sodomy was suspected, as uncovering the truth was notoriously difficult. Because sodomy happened in private and typically met with vehement denials, it became necessary to rely on instruments that could produce certainty. Since the accused's testimony could not be trusted, recourse was often made to expert medical opinion. Medical practitioners thus became invaluable in prosecuting sodomy, both through their expert assessments and their duty to report illegal acts—though their role, of course, was far from infallible.

---

15 See Alfieri, Il discorso su tribadi e sodomiti; Alfieri/Lagioia, *Infami macchie*; Casanova, Meglio non dire, pp. 32–43.

Early modern medical vigilance can be understood as a complex and multifaceted concept with far-reaching practical implications, reflecting the intricate nature of human interactions, power dynamics, and the governance of bodies within early modern society. The analysis of the medical vigilance exercised by galley doctors over slaves and convicts should not be confined to the context of the galleys alone; rather, it should serve as a foundation for a broader reflection on early modern society as a whole. These penal ships, as a case study, reveal a context that, through its specificities, magnified practices and ideologies that were otherwise prevalent—though often less visible—in everyday social life.

By exploring the role of medical vigilance within this maritime penal system, we not only uncover the diverse duties of doctors but also gain insight into the intersection of medical authority, social control, and the governance of bodies in early modern Italy. Ultimately, the practices aboard the galleys provide a unique lens through which to explore the broader dynamics of power and discipline in early modern society.

Bibliography

Primary Sources

Manuscripts

Archivio di Stato di Firenze (ASF): Mediceo del Principato, 627, 638, 1802–1803, 1824, 1836–1837, 2077–2078, 2082–2084, 2099–2118, 2130–2132, 2168, 2426, 5760; Gabinetto 111.

Archivio di Stato di Livorno (ASL): Capitano poi Governatore poi Auditore Vicario, 1550—1808, bb. 3082–3083, 3086–3088, 3100, 3110, 3122, 3166, 3233–3234; Magistrato poi Dipartimento di Sanità (1696–1860), bb. 52, 69–72, 78–79, 85, 109, 644–645.

Archivio Apostolico Vaticano (AAV): Misc.Arm. II, bb. 86, 101, 110, 116; Misc. Arm. IV/V, bb. 47, 54, 61, 66, 68, 70; Misc. Arm. IX, bb. 13,50; Segr. Soldati, bb. 1, 22, 26, 31; Segr. Stato, Particoalri, b. 34; Pio, b. 112.

Archivum Romanum Societatis Iesu (ARSI): Fondo gesuitico, manoscritti, n.10; Instit.234; Instit.50; Ital. 139.II; Rom. 132.I; Rom. 138.

Archivio di Stato di Roma (ASR): Camerale II – Sanità, bb. 9, 10, 24; Camerale II – Commissariato delle soldatesche e galere, bb. 646–654, 682; Camerale III – Comuni, Civitavecchia, bb. 807–810, 827, 846; Tribunale Criminale del Governatore, Congregazione della visita alle carceri, bb. 29–34; Governo poi tribunale di Civitavecchia, bb. 639–666.

Archivio Storico del Vicariato di Roma (ASVR): Atti della segreteria, bb. 74, 81.

Biblioteca Apostolica Vaticana (BAV): R.G. Storia.2.IV.3764; Stamp.Cappon.V.681 (int.9, int.16bis); Stamp.Cappon.V.682 (int.44); Stamp.Cappon.V.683 (int.30,38,93); Stamp.Chig.II.108 (int.81); Stamp.Chig.S.108(3), fig.6; Stamp.Ferr.III.2046 (int.3); Stamp.Ferr.IV.8532 (int.4).

Biblioteca Corsiniana Roma (BCR): 34B13; 34D11; 34D18.

Printed Sources

Anazarabeo, Dioscoride: *Della materia medicinale*. Libro Secondo. Tradotto da Marcantonio Montignano da S. Gimignano medico. Florence 1547.

Barbaro, Antonio: *Pratica criminale*. Venice 1739.

Benedetti, Alessandro: *De medici et aegri officio libellus aphorismus, in De re medica opus insigne & apprime medicinae canditatis [sic] omnibus utile … hoc ordine digestum*. Basel 1549.

Bernardi, Filippo da Firenze: *Relazione di quando i cappuccini furono deputati alla cura spirituale del Bagno e delle galere di Livorno, 1706* (APCF, ms. non catalogato). Edited by Salvadorini, Vittorio. Canterrano 2020.

Bindi, Ioannis Baptista: *Loemographiae Centumcellensis, siue De Historia pestis contagiosae, quae anno intercalari MDCLVI in ecclesiastica ditione primùm Ciuitatem veterem inuasit, & inde in Pontificiarum Triremium Ducem fuit illata, libri quinque. In quibus omnium pene Casuum, qui evenerunt, Observationes: primaria Origo: admiranda Propagatio: Essentia: horrenda Symptomata: Causae: Curatio: Praeservatio, & Praesidia medica, quae profuerunt, & obfuerunt, laconice, at perfunctorie describuntur*. Rome 1658.

Bonomo, Giovanni Cosimo: *Osservazioni intorno a' pellicelli del corpo umano fatto dal dottor Gio. Cosimo Bonomo, E da lui con altre osservazioni scritte in una Lettera all'illustrissimo Sig. Francesco Redi*. Florence 1687.
Botallo, Leonardi: *Commentarioli duo, alter de medici, alter de aegroti munere*. Lugduni 1565. Edited by Carerj, Leonardo/Bogetti Fassone, Anita. Turin 1981.
Cantini, Lorenzo: *Legislazione Toscana raccolta e illustrata dal dottor Lorenzo Cantini socio di varie accademie–Tomo I*. Florence 1800–1808.
Castelli, Petrii: *De optimo medico*. Rome 1637.
Castiglione, Baldesar: *Il libro del Cortegiano*. Edited by Barberis, Walter. [Venice 1528] Turin 1998.
Celsus, Aulo Cornelio: *De Medicina*. With an English translation by Spencer, W. G., in three volumes. Vol. III Cambridge 1979.
Codronchius, Giovan Baptista: *De Christiana, ac tuta medendi ratione*. Ferraria 1591.
Codronchius, Giovan Baptista: *Methodus Testificandi*. Imola 1597.
Corradi, Alfonso: *Annali delle epidemie occorse in Italia dalla prime memorie fino al 1850. Compilati con varie note e dichiarazioni [da] Alfonso Corradi*. Bologna 1863.
Cospi, Antonio Maria: *Il giudice Criminalista, dove con dottrina teologica, canonica, civile, filosofica, medica, storica e poetica si discorre di tutte quelle cose, che al giudice delle cause criminali possono avvenire*. Venice 1681.
Crescentio, Bartolomeo: *Nautica mediterranea di Bartolomeo Crescentio Romano ... Nelle quale si mostra la fabrica delle galee', galeazze, & galeoni ... S'insegna l'arte del nauigar nell'uno, e l'altro mare ... Et un Portolano di tutti i porti da stantiar vascelli co i loghi pericolosi di tutto il mare Mediterraneo*. Rome 1601.
Cuggiò, Antonio: *Della giurisdittione e prerogative del Vicario di Roma. Opera del canonico Nicolò Antonio Cuggiò segretario del tribunale di Sua Eminenza*, ms. 1678. Edited by Rocciolo, Daniele. Rome 2004.
Dalla Croce, Gio. Andrea: *Cirugia Universale e Perfetta. Di tutte le parti pertinenti all'ottimo Chriurgo. Di Gio. Andrea Dalla Croce Vinitiano. Nella quale si contiene la thoerica, & prattica di ciò che può essere nella Criugia necessario: come più ampiamente nel Sommario si dichiara*. Venice 1583.
Davies, William: *True relation of the travailes and most miserable captivitie of William Davies, barber surgion of London, under the duke of Florence*. London 1614. Translated and edited by Neri, Angelina. Pisa 2000.
De Castro, Rodericus: *Medicus Politicus, sive de officiis medico—politicis tractatus, quatuor distinctus libri*. Hamburg 1614.
Deciani, Tiberii: *Tractatus Criminalis*. Augustae Taurinorum 1593.
Fei, Giovan Pietro: *Liburnensis Iurisdictionis*. s.l., s.d. (ASF, Mediceo del Principato, f. 2132).
Farinacci, Prosperi: *Praxis et Theoricae Criminalis*. Lugduni 1635.
Fidelis, Fortunati: *De Relationibus Medicorum libri quatuor*. Leipzig 1602.
Fracastorii, Hieronymii: *De sympathia et antipati rerum–Liber I, De contagionibus et contagiosis morbi set eorum curatione–Libri III*. Ventiis 1546. Translated and annotated by Busacchi, Vincenzo. Florence 1950.
Fragoso, Juan: *Cirugia universal, aora nuevamente emendada, y añadida en esta sexta impression ... Y mas otros quatro tratados. El primero es una suma de proposiciones contra ciertos avisos de cirugia. El segundo, de las declarationes acerca de diversas heridas, y muertes. El tercero, de los Aphorismos de Hippocrates, tocantes a cirugia. El quarto, de la naturaleza, y calidades de los medicamentos simples*. En Alcala de Henares 1621.

Gastaldi, Nieronimi: *Tractatus de avertenda et profliganda peste politico— legalis eo lucubratus tempore, quo ipse Loemocomiorum primo, mox Sanitatis Commissarius Generalis fuit, Peste Urbem invadente.* Bologna 1684.

Guglielmotti, Alberto: *Storia della Marina Pontificia dal secolo ottavo al decimonono.* Rome 1856.

Guglielmotti, Alberto: La Guerra dei Pirati. In: Guglielmotti, Alberto: *Storia della Marina Pontificia dal secolo ottavo al decimonono.* Rome 1856.

Guglielmotti, Alberto: La Squadra Permanente. In: Guglielmotti, Alberto: *Storia della Marina Pontificia dal secolo ottavo al decimonono.* Rome 1856.

Ingrassia, Johanne Philippo: *Liber, quo multa adversus barbaros medicos disputantur, collegiique modus ostenditur ac multae quaestiones tam physicae quam chirurgicae discutiuntur.* Venetiis 1547.

Ingrassia, Johanne Philippo: *Methodus dandi relations pro mutilatis, torquendis, aut a tortura excusandis: pro deformibus, [...].* 1578. Edited and printed by Curcio, Giovanni. Catania 1938.

Marteilhe, Jean: *Memoires d'un galerien du Roi-Soleil.* Edited, annotated and prefaced by Zysberg, André. Paris 1989.

Martini, Angelo: *Manuale di metrologia, ossia misure, pesi e monete in uso attualmente e anticamente presso tutti i popoli.* Turin 1883.

Moroni, Gaetano: *Dizionario di erudizione storico ecclesiastica da S. Pietro sino ai giorni nostri.* Venice 1861.

Pallavicino, Pietro Sforza: *Descrizione del contagio che da Napoli si propagò a Roma nell'anno 1656; e de' saggi provvedimenti ordinati allora da Alessandro VII.* Rome 1837.

Pantera, Pantero: *L'armata Navale del Capitan Pantero Pantera Gentil'uomo Comasco & Cavalliero dell'hanbito di Cristo. Divisa in doi libri. Ne i quali si ragiona del modo, che si ha à tenere per formare, ordinare, & conservare un'armata maritima.* Rome 1614.

Pera, Francesco: *Curiosità Livornesi. Inedite o rare.* Livorno 1888.

Perrando, *Prefazione* preface of Ingrassia, Johanne Philippo: *Methodus dandi relations pro mutilatis, torquendis, aut a tortura excusandis: pro deformibus, [...].* 1578. Edited and printed by Curcio, Giovanni. Catania 1938

Petrarca, Francesco: *Invective contra medicum. Testo latino e volgarizzamento di Ser Domenico Silvestri.* Edited by Ricci, Pier Giorgio. Rome 1978.

Priori, Lorenzo: *Prattica criminale secondo il ritto delle leggi della Serenissima Repubblica di Venetia.* Venice 1663.

Ramazzini, Bernardino: *De morbis artificum diatriba.* Mutina 1700.

Redi, Francesco: *Lettere di Francesco Redi Patrizio aretino.* Vol. 1. Florence ^{2}1779.

Redi, Francesco: *Opere di Francesco Redi gentiluomo aretino e accademico della Crusca.* Vo. 4. Naples 1741.

Savelli, Marcantonio: *Pratica Universale.* Venice 1697.

Scetti, Aurelio: *The Journal of Aurelio Scetti. A Florentine Galley Slave at Lepanto (1565-1577).* Translated and edited by Monga, Luigi. [Pisa[1] 1999] Tempe 2004.

Selvatico, Giambattista: *De ijs qui morbum simulant deprehendendis liber.* Milano 1595.

Siccus, Ioannis Antonij: *De optimo medico: caput primum libri primi De antiqua medicina ejusdem auctoris: quæ in margine capita notantur, in Galeni Codicibus, ex nouissima Iuntarum editione requirenda sunt.* Venice 1551.

Villanova, Arnaldo: De Cautelis Medicorum. Translated by Sigerist, Henry E. and annotations by McVaugh, Michael. In: Grant, Edward (ed.): *A source book in medieval science.* Cambridge (Mass.) 1974.

Zacchia, Paolo: *Quaestiones medico-legales, Opus Iurisperitis appriem necessarium,medicis perutile, cateris non injucundum*. Avenione[4] 1655.

Zerbi, Gabrielis: *Opus Perutile De Cautelis Medicorum*. Padua 1495.

Secondary sources

Ago, Renata/Parmeggiani, Antonio: La peste del 1656–57 nel Lazio. In: *Società italiana di demografia storica. Popolazione, società e ambiente. Temi di demografia storica italiana (secc. XVII–XIX)* (1990), pp. 593–612.

Agrimi, Jole/Crisciani, Chiara: *Edocere Medicos. Medicina scolastica nei secoli XIII–XI*. Naples 1988.

Alessi, Giorgia: *Il processo penale. Profilo storico*. Rome/Bari 2004.

Alessi, Giorgia: Pene e remieri a Napoli tra Cinque e Seicento. Un aspetto dell'illegalismo d'ancien régime. In: *Archivio Storico per le provincie napoletane* XV (1977), pp. 231–251.

Alfani, Guido/Melegaro, Alessia: *Pandemie d'Italia: dalla peste nera all'influenza suina l'impatto sulla società*. Milano 2010.

Alfieri, Fernanda: L'età della disciplina cristiana. Confronti e comparazioni. In: Lavenia, Vincenzo (ed.): *Storia del Cristianesimo III. L'età moderna (secoli XVI–XVIII)*. Rome 2015, pp. 321–349.

Alfieri, Fernanda: Il discorso su tribadi e sodomiti in età moderna. Tra volontà di punire e difficoltà di dire. In: Grassi, Umberto/Lagioia, Vincenzo/Romagnani, Gian Paolo (eds.): *Tribadi, sodomiti, invertite e invertiti, pederasti, femminelle, ermafroditi.... per una storia dell'omosessualità, della bisessualità e delle trasgressioni di genere in Italia*. Pisa 2017, pp. 19–28.

Alfieri, Fernanda/Lagioia, Vincenzo (eds.): *Infami macchie. Sessualità maschili e indisciplina in età moderna*. Rome 2018.

Alfieri, Fernanda: Forme e usi della memoria nel corpus gesuitico del XVII secolo. Fra istruzione, spiritualità, missione. In: *TIEMPOS MODERNOS* 13(47) (2023), pp. 338–356.

Altieri Biagi, Maria Luisa/Basile, Bruno (eds.): *Scienziati del Seicento*. Milano/Naples 1980.

Andretta, Elisa: *Roma medica. Anatomie d'un système médical au XVIe siècle*. Rome 2011.

Angeletti, Luciana Rita/Cavarra, Berenice/Gazzaniga, Valentina: *Il De urinis di Teofilo Protospatario, centralità di un segno clinico*. Rome 2009.

Angiolini, Franco: *I Cavalieri e il Principe. L'Ordine di Santo Stefano e la società toscana in età moderna*. Florence 1996.

Angiolini, Franco: Il Granducato di Toscana, l'Ordine di Santo Stefano e il Mediterraneo (secc. XVI–XVIII). In: *Ordens militares. Guerra, religião, poder e cultura. Actas do III encontro sobre Ordens Militares*. Vol. 1 (1999), pp. 39–61.

Angiolini, Franco: Slaves and Slavery in the Early Modern Tuscany (1500–1700). In: *Italian History and Culture* 3 (1997), pp. 67–82.

Angiolini, Franco: La pena della galera nella Toscana moderna (1542–1750). In: Antonelli, Livio (ed.): *La polizia in Italia e in Europa. Punto sugli studi e prospettive di ricerca*. Soveria Mannelli 2006, pp. 79–115.

Antonelli, Livio (ed.): *La polizia in Italia e in Europa. Punto sugli studi e prospettive di ricerca*. Soveria Mannelli 2006.

Arrizabalaga, Jon: Medical ideals in the Sephardic diaspora. Rodrigo de Castro's Portrait of the Perfect Physician in early Seventeenth-Century Hamburg. In: *Medical History* 53 (S29) (2009), pp. 107–124.

Ascheri, Mario: Consilium sapientis, perizia medica e res judicata. Diritto dei dottori e istituzioni comunali. In: Kuttner, Stephan/Pennington, Kenneth (eds.), *Proceedings of the fifth International Congress of Medieval Canon Law: Salamanca, 21-25 September 1976*. Città del Vaticano 1980, pp. 533-579.

Audouin-Rouzeau, Frédérique: *Les chemins de la peste. Le rat, la puce et l'homme*. Rennes 2003.

Aymard, Maurice: Chiourmes et galéres dans la seconde moitié du XVIe siècle. In: Benzni, Gino (ed.): *Il Mediterraneo nella seconda metà del '500 alla luce di Lepanto. Atti del convegno di studi*. Florence 1974, pp. 71-95.

Aymard, Maurice: Epidémies et médecins en Sicile à l'époque moderne. In: *Annales Cisalpines d'Histoire Sociale* 4 (1973), pp. 9-36.

Babayan, Kathryn/Najmabadi, Asfaneh: *Islamicate Sexualities. Translations across Temporal Geographies of Desire*. Harvard 2008.

Baker, Robert B./McCullough Lawrence: *The Cambridge World History of Medical Ethics*. Cambridge 2008.

Baldanzi, Francesco: Nell'Ospedale di "Santa Maria Nuova di Firenze a imparare il cerusico": origini e primo consolidamento della Scuola Medica e Chirurgica (XVI-XVIII secolo). In: *Archivio Storico Italiano* CLXXVII/2 (2019), pp. 273-304.

Baldassarri, Marina: *Bande giovanili e "vitio nefando". Violenza e sessualità nella Roma barocca*. Rome 2005.

Bamji, Alexandra: Health Passes, Print and Public Health in Early Modern Europe. In: *Social History of Medicine* 32/3 (2019), pp. 441-464.

Barbagli, Marzio: *Congedarsi dal mondo. Il suicidio in Occidente e in Oriente*. Bologna 2009.

Barbagli, Marzio: *Comprare piacere. Sessualità e amore venale dal Medioevo a oggi*. Bologna 2020.

Barker, Hannah: *That Most Precious Merchandise. The Mediterranean Trade in Black Sea Slaves*, 1260-1500. Philadelphia 2019.

Beggio, Tommaso: *Contributi allo studio della "Servitus poenae"*. Bari 2020.

Belardelli, Alessandra: Il governo della peste. L'esperienza romana del 1656. In: *Sanità, scienza e storia* 1 (1987), pp. 51-79.

Benedetti, Marina: *Medioevo inquisitoriale. manoscritti, protagonisti, paradossi*. Rome 2021.

Benedetti, Roberto: Dalla galera all'Ergastolo. Storia del carcere per ecclesiastici criminali. In: *Ricerche di storia sociale e religiosa* 81 (2012), pp. 15-69.

Benedetti, Roberto: Dell'Ergastolo o Pia Casa di penitenza e correzione in Corneto. Storia di un carcere dimenticato (1627-1874). In: *Giornale di Storia, Dialoghi sul carcere* 38 (2021), pp. 1-64.

Benedetti, Roberto: Il Gran teatro della giustizia penale. I luoghi della pubblicità della pena nella Roma del XVIII secolo. In: Boiteux, Martine/Caffiero, Marina/Marin, Brigitte (eds.): *I luoghi della città. Roma moderna e contemporanea*. Rome 2010, pp. 153-197.

Benedetti, Roberto: Servi introvabili e schiavi visibili. Un'analisi delle fonti giuridiche dello Stato della Chiesa (secoli XVI-XVIII). In: Di Nepi, Serena (ed.): *Schiavi nelle terre del papa. Norme, rappresentazioni, problemi a Roma e nello Stato della Chiesa in età moderna. Dimensioni e problemi della ricerca storica. Rivista del Dipartimento di studi storici dal Medioevo all'età contemporanea dell'Università La Sapienza di Roma* 2 (2013), pp. 53-80.

Benedetti, Roberto: Tribunali e Giustizia a Roma nel Settecento attraverso la fonte delle liste di traduzione alla galera (1749-1759). In: *Roma moderna e contemporanea* 3 (2004), pp. 507-538.

Benigno, Francesco/Lavenia, Vincenzo: *Peccato o crimine. La chiesa di fronte alla pedofilia*. Bari/Rome 2021.

Benvenuto, Grazia: *La peste nell'Italia della prima età moderna. Contagio, rimedi, profilassi*. Bologna 1996.

Benzoni, Gino (ed.): *Il Mediterraneo nella seconda metà del '500 alla luce di Lepanto*. Florence 1974.

Berco, Cristian: Syphilis and the silencing of sodomy in Juan Calvo's Tratado del morbo gálico. In: Borris, Kenneth/Rousseau, George (eds.): *The Sciences of Homosexuality in Early Modern Europe*. London/New York 2008.

Bergdolt, Klaus: The Discourses of Practitioners in Medieval and Renaissance Europe. In: Baker, Robert B./McCullough, Lawrence (eds.): *The Cambridge World History of Medical Ethics*. Cambridge 2008, pp. 370–378.

Bernardi, Walter/Guerrini, Luca (eds.): *Francesco Redi. Un protagonista della scienza moderna. Documenti, esperimenti, immagini*. Florence 1999.

Bernardi, Walter: Uno scienziato aretino protagonista della modernità. In: Mangani, Lorella/Martini, Giuseppe (eds.): *Francesco Redi Aretino*. Arezzo 1999, pp. 17–36.

Biondi, Albano; Tra corpo ed anima: medicina ed esorcistica nel Seicento (*L'Alexicacon* di Candido Brugnoli). In: Prodi, Paolo/Penuti, Carla: *Disciplina dell'anima, disciplina del corpo e disciplina della società tra medioevo ed età moderna*. Bologna 1994.

Biraben, Jean-Noël: *Les hommes et la peste en France et dans les pays européens et méditerranéens*. Vol. II. Paris/La Haye 1975/1976.

Blair, Ann: Autorship in the Popular "Problemata Aristotelis". In: *Early Science and Medicine* 4/3 (1999), pp. 189–227.

Boccadamo, Giuliana: A Napoli: 'mori negri' fra Cinque e Seicento. In: Salvatore, Gianfranco (ed.): *Il chiaro e lo scuro: gli africani nell'Europa del Rinascimento tra realtà e rappresentazione*. Lecce 2021.

Bonacchi, Gabriella: *Legge e peccato. Anime, corpi e giustizia alla corte dei papi*. Rome/Bari 1995.

Bonazza, Giulia: Slavery in the Mediterrenean. In: Pargas, Damians A./Schiel, Juliane (eds.): *The Palgrave Handbook of Global Slavery throughout History*. Cham 2023, pp. 227–242.

Bono, Salvatore: *Schiavi. Una storia mediterranea (XVI–XIX secolo)*. Bologna 2016.

Bono, Salvatore: *Schiavi musulmani nell'età moderna. Galeotti, vu' cumprà, domestici*. Naples 1999.

Bono, Salvatore: Schiavi musulmani sulle galere e nei bagni d'Italia dal XVI al XIX secolo. In: Ragosta, Rosalba (ed.): *Le genti del mare Mediterraneo*. Vol. II. Naples 1980, pp. 837–875.

Bonora, Elena: *La Controriforma*. Rome/Bari 2001.

Borris, Kenneth/Rousseau, George (eds.): *The Sciences of Homosexuality in Early Modern Europe*. London/New York 2008.

Borris, Kenneth: Sodomizing Science. Cocles, Patricio Tricasso, and the constitutional Morphologies of Renaissance Male Same-Sex Lovers. In: Borris, Kenneth/Rousseau, George (eds.): *The Sciences of Homosexuality in Early Modern Europe*. London/New York 2008.

Borris, Kenneth: The Prehistory of Homosexuality in the Early Modern Sciences. In: Borris, Kenneth/Rousseau, George (eds.): *The Sciences of Homosexuality in Early Modern Europe*. London/New York 2008.

Boswell, John: *Christianity, Social Tolerance and Homosexuality. Gay People in Western Europe from the Beginning of the Christian Era to the Fourteenth Century*. Chicago/London 1980.

Brambilla, Elena: *Corpi invasi e viaggi dell'anima. Santità, possessione, esorcismo dalla teologia brocca alla medicina illuminista*. Rome 2010.

Braudel, Fernand: *Civiltà e imperi del Mediterraneo nell'età di Filippo II*. Vol I.–II. Turin 2010 (trad. Paris^1 1949).

Brendecke, Arndt: Attention and Vigilance as Subjects of Historiography. An Introductory Essay. In: Brendecke, Arndt/Molino, Paola (eds.): *The History and Cultures of Vigilance. Historicizing the Role of Private Attention in Society. Storia della Storiografia* 74/2 (2018), pp. 17–27.

Brendecke, Arndt: Warum Vigilanzkulturen? Grundlagen, Herausforderungen und Ziele eines neuen Forschungsansatzes. In: *Mitteilungen des Sonderforschungsbereichs 1369 'Vigilanzkulturen'* 1 (2020), pp. 11–17.

Brendecke, Arndt/Molino, Paola (eds.): *The History and Cultures of Vigilance. Historicizing the Role of Private Attention in Society. Storia della Storiografia* 74/2 (2018).

Brockliss, Laurence/Jones, Colin: *The Medical World of Early Modern France*. Oxford 1997.

Broggio, Paolo: *Evangelizzare il Mondo. Le missioni della Compagnia di Gesù tra Europa e America (secoli XVI–XVII)*. Rome 2004.

Brogini, Anne: L'esclavage au quotidien à Malte au XVIe Siècle. In: *Cahiers de la Méditerranée* 65 (2002), pp. 137–158.

Bruijn, Iris: *Ship's Surgeons of the Dutch East India Company. Commerce and the Progress of Medicine in the Eighteenth Century*. Leiden 2009.

Brunelli, Giampiero: *Soldati del papa. Politica militare e nobiltà nello Stato della Chiesa (1560–1644)*. Rome 2003.

Buchet, Christian (ed.): *L'Homme, la santé et la mer. Actes du colloque international tenu à l'Institut Catholique de Paris les 5 et 6 décembre 1995*. Paris 1997.

Burg, B. R.: *Pirati e sodomia*. Milano 1994 (trad. New York 1982).

Burke, Peter: *Cultura popolare nell'età moderna*. Milano 1980 (trad. New York 1978).

Burke, Peter: *Scene di vita quotidiana nell'Italia Moderna*. Rome 1988 (trad. Cambridge 1987).

Bynum, William F.; Porter, Roy (eds.): *Companion Encyclopedia of the History of Medicine*. Vol. 1. London/New York 1993.

Bynum, William F./Bynum, Helen: *Dictionary of Medical Biography*. London 2007.

Cadden, Joan: *Nothing Natural is Shameful. Sodomy and Science in Late Medieval Europe*. Philadelphia 2013.

Calafat, Guillaume/Grenet, Mathieu: *Méditerranées. Une histoire des mobilités humaines (1492–1750)*. Paris 2023.

Calafat, Guillaume/Kaiser, Wolfgang: The Economy of Ransoming in the Early Modern Mediterranean. A form of Cross-Cultural Trade Between Southern Europe and the Maghreb (Sixteenth to Eighteenth Centuries). In: Trivellato, Francesca/Halevi, Leor/Antunes, Catia (eds.): *Religion and Trade. Cross-Cultural Exchanges in World History, 1000–1900*. New York 2014, pp. 108–130.

Calcagno, Paolo: "Per sfogare la sua brutale libidine". Pratiche di sodomia a bordo delle galee nel XVIII secolo. In: Ferrando, Francesca/La Rocca, Maria Cristina/Morosini, Giulia (eds.): *Storie di violenza. Genere, pratiche ed emozioni tra Medioevo ed età contemporanea*. Rome 2020, pp. 171–182.

Calisse, Carlo: *Storia di Civitavecchia*. Florence 1936.

Calvi, Giulia: L'oro, il fuoco, le forche. La peste napoletana del 1656. In: *Archivio Storico Italiano* 139/3 (1981), pp. 405–458.

Calzolari, Monica/Di Sivo, Michele/Grantaliano, Elvira: *Giustizia e criminalità nello Stato pontificio. Ne delicta remaneant impunita*. Rome 2002.

Camaioni, Michele: *Il governo dei pulpiti. Predicatori, potere e pubblico nell'Italia della prima età moderna*. Rome 2024.

Campbell, Gwyn/Stanziani, Alessandro (eds.): *Debt and Slavery in the Mediterranean and Atlantic Worlds. Financial History* 22 (2013).
Camporesi, Piero: *Camminare il mondo. Vita e avventure di Leonardo Fioravanti, medico del Cinquecento.* Milano 1997.
Camporesi, Piero: *Lo "Speculum cerretanorum" di Teseo Pini, "Il vagabondo" di Rafaele Frianoro e altri testi di "furfanteria".* Turin 1973.
Candiani, Guido: *Dalla galea alla nave di linea. Le trasformazioni della marina veneziana (1527-1699).* Novi Ligure 2012.
Candiani, Guido: La gestione degli equipaggi nei vascelli veneziani tra Sei e Settecento. In: Dattero, Alessandra, Levanti, Stefano (eds.): *Militari in età moderna. La centralità di un tema di confine. Milano, 20 giugno 2004.* Milano 2006, pp. 171-195.
Canosa, Romano: *Storia di una grande paura. La sodomia a Firenze e a Venezia nel Quattrocento.* Milano 1991.
Capponi, Niccolò: *Lepanto 1571. La lega Santa contro l'Impero Ottomano.* Milano 2010.
Caputo, Sara: Treating, Preventing, Feigning, Concealing. Sickness, Agency and the Medical Culture of the British Naval Seaman at the End of the Long Eighteenth Century. In: *Social History of Medicine* 35/3 (2021), pp. 749-769.
Casagrande, Carla/Crisciani, Chiara/Vecchio, Silvana (eds.): *"Consilium". Teorie e pratiche del consigliare nella cultura medievale.* Tavarnuzze (Impruneta) 2004.
Casanova, Cesarina: *Crimini nascosti. La sanzione penale dei reati "senza vittima" e nelle relazioni private (Bologna, 17. secolo).* Bologna 2007.
Casanova, Cesarina: Meglio non dire che punire. La sanzione penale dei crimini nefandi. In: Lagioia, Vincenzo (ed.): *Storie di invisibili, marginali, ed esclusi.* Bologna 2012, pp. 33-42.
Cassiani, Gennaro: Medici, magistrati e filosofi contro i miasmi della peste. Ricerche in margine ad alcuni documenti sull'epidemia di Roma del 1656-57. In: *Ricerche di storia sociale e religiosa* 46 (1994), pp. 187-215.
Cattaneo, Massimo: "Vitio nefando" e Inquisizione romana. In: Formica, Marina/Postigliola, Alberto (eds.): *Diversità e minoranze nel Settecento. Atti del Seminario di Santa Margherita Ligure, 2-4 giugno 2003.* Rome 2006, pp. 55-77.
Cavallo, Francesca: La fama di santità di un cappuccino nella Livorno di fine Seicento. p. Ginepro da Barga (1630-1709). In: *Nuovi Studi Livornesi* X (2002/2003), pp. 31-58.
Cavallo, Sandra: *Artisans of the Body in Early Modern Italy. Identities, Families and Masculinities.* Manchester/New York 2007.
Cavarzere, Marco: Cosimo I, pater ecclesiae tra eresia, riforma religiosa e ragion di stato. In: *Annali di storia di Firenze* 9 (2014), pp. 77-86.
Cerutti, Simona: *Étrangers. Étude d'une condition d'incertitude dans une société d'Ancien Régime.* Paris 2012.
Cerutti, Simona/Pomata, Gianna (eds.): *Fatti. Storie dell'evidenza empirica. Quaderni storici* 108/3 (2001).
Chizzolini, Benedetta: Enslavement and Female Agency in Early Modern Livorno. In: *Slavery & Abolition* (2025), pp. 1-20.
Chizzolini, Benedetta: Medicare, osservare, controllare. Medici e galeotti a Livorno tra XVI-XVIII secolo. In: Dyble, Jake/Lo Bartolo, Alessandro/Morelli, Elia (eds.): *Il teatro della Turchia. Visioni del Vicino Oriente in età moderna.* Rome 2024, pp. 71-92.

Chizzolini, Benedetta: Navigating Ambiguities. Fluid Identities and Legal Nuances among Galley Rowers (Great Duchy of Tuscany, 16th–18th centuries). In: *Journal of Global Slavery* 10/1 (2025), pp. 53–79.

Chizzolini, Bendetta: The "taglio degli schiavi": Evaluating the Value of Galley Slaves in Early Modern Livorno (17th–18th centuries). In: *Sources for Medicine, Slavery & Race-Making in the Early Modern World*, https://www.mmor.co.uk/blog/the-taglio-degli-schiavi-evaluating-the-value-of-galley-slaves-in-early-modern-livorno-17th-18th-centuries [last accessed: 15.07.2025]

Ciano, Cesare: *La Sanità Marittima nell'Età Medicea*. Pisa 1976.

Cipolla, Carlo M.: La città di fronte alle crisi di mortalità. In: *Società italiana di demografia storica. La demografia storica delle città italiane. Relazioni e comunicazioni presentate al Convegno tenuto ad Assisi nei giorni 27–29 ottobre 1980*. Bologna 1982.

Cipolla, Carlo M.: *Contro un nemico invisibile. Epidemie e strutture sanitarie nell'Italia del Rinascimento*. Bologna 1986.

Cipolla, Carlo M.: *Miasmi e Umori*. Bologna 1989.

Cipolla, Carlo M.: Origini e sviluppo degli uffici di sanità in Italia. In: *Annales Cisalpines d'Histoire Sociale* 4 (1973), pp. 84–101.

Cipolla, Carlo M.: *Il pestifero e contagioso morbo. Combattere la peste nell'Italia del Seicento*. Bologna 1981.

Cipolla, Carlo M.: *Public Health and Medical Profession in the Renaissance*. Cambridge 1976.

Ciuti, Francesco: Il medico e l'ospedale. Il nosocomio di Santa Maria Nuova e le professioni sanitarie a Firenze in età moderna. In: *Medicina & storia. Rivista di storia della medicina e della sanità* XI (2011), pp. 63–88.

Civale, Gianclaudio: *Guerrieri di Cristo. Inquisitori, gesuiti e soldati nella battaglia di Lepanto*. Milano 2009.

Clark, Michael/Crawford, Catherine: *Legal Medicine in History*. Cambridge 1994.

Coccoli, Lorenzo: Perchè il colpo passi la pelle. La Casa di correzione del San Michele nel suo tempo. In: *Giornale di Storia. Dialoghi sul carcere* 38 (2021), pp. 1–20.

Cohn, Samuel: *Cultures of Plague. Medical Thinking at the End of the Renaissance*. Oxford 2009.

Conforti, Maria/De Renzi, Silvia: Sapere anatomico negli ospedali romani. Formazione dei chirurghi e pratiche sperimentali (1620–1720). In: Romano, Antonella (ed.): *Rome et la science moderne. Entre Renaissance et Lumière*. Rome 2008, pp. 433–472.

Colombo, Emanuele: *Quando Dio chiama. I gesuiti e le missioni nelle Indie (1560–1960)*. Bologna 2023.

Conrad, Lawrence I./Neve, Michael/Nutton, Vivian/Porter, Roy/Wear, Andrew: *The Western Medical Tradition. 800 BC to AD 1800*. Cambridge [10]2011.

Cosmacini, Giorgio: *Storia della medicina e della sanità in Italia. Dalla peste nera ai giorni nostri*. Bari 2005.

Coturri, Enrico, Le scuole ospedaliere di chirurgia del Granducato di Toscana (sec. XVII–XIX). In: *Minerva Medica* XLIX (1958), pp. 3–18.

Courouve, Claude: *Les origins de la repression de l'Homosexualité*. Paris 1978.

Crawford, Catherine: Medicine and the Law. In: Bynum William F./Porter Roy (eds.): *Companion Encyclopedia of the History of Medicine*. Vol. 2. London/New York 1993.

Crisciani, Chiara: "Exempla" in Medicina. Epistemologia, insegnamento, retorica (secoli XIII–XV). Una proposta di ricerca. In: Gadebusch Bondio, Mariacarla/Ricklin, Thomas (eds.): *Exempla medicorum. Die Ärzte und ihre Beispiele (XIV–XVIII Jahrhundert)*. Florence 2008, pp. 89–108.

Crisciani, Chiara: Consilia, Responsi, Consulti. I pareri del medico tra insegnamento e professione. In: Casagrande, Carla/Crisciani Chiara/Vecchio Silvana (eds.): *"Consilium". Teorie e pratiche del consigliare nella cultura medievale*. Tavarnuzze (Impruneta) 2004, pp. 259-279.

Criscuolo Vincenzo (ed.): *I Cappuccini. Fonti documentarie e narrative del primo secolo (1525-1619)*. Rome ²2020.

Croce, Benedetto: La vita infernale delle galere. In: Croce, Benedetto: *Varietà di storia letteraria e civile, II*. Bari 1949, pp. 83-92.

Curcio, Giovanna/Zampa, Paolo (eds.): Civitavecchia. In: *Quaderni della ricerca in Architetture sociali nello Stato Pontificio* (1988), pp. 59-71.

Curcio, Giovanna/Zampa, Paolo: Il porto di Civitavecchia dal XV al XVIII secolo. In: Simoncini, Giorgio (ed.): *Sopra i porti di mare, vol. IV. Lo Stato Pontificio*. Florence 1995, pp. 159-232.

D'Angelo, Edoardo (ed.): *Liber gomorrhianus: omosessualità ecclesiastica e riforma della Chiesa San Pier Damiani*. Alessandria 2001.

Daston, Lorraine/Lunbeck, Elizabeth (eds.): *Histories of Scientific Observation*. Chicago 2011.

Davidson, Arnold: *L'emergenza della sessualità. Epistemologia storica e formazione dei concetti*. Macerata 2001 (trad. Harvard 2001).

Dean, Trevor/Lowe, Kate J.P (eds.): *Crime, Society and the Law in Renaissance Italy*. Cambridge 1994.

De Castelsangiovanni, Gabriele: *L'assistenza religiosa ospedaliera dei cappuccini in Italia*. Rome 1967.

De Ceglia, Francesco Paolo (ed.): *The Body of Evidence. Corpses and Proofs in Early Modern European Medicine*. Leiden/Boston 2020.

De Lucia, Lori: The Exceptional History of a Black Saint in Sixteenth Century Palermo and Why It Matters. In: *"Transition"* (2021), pp. 54-67.

De, Monya: Towards Defining Paternalism in Medicine. In: *AMA Journal of Ethics* 6/2 (2004), pp. 69-71.

De Polis Barbara (ed.): *1685-1697. Storie di galee e soldati della squadra navale pontificia nel conflitto di Morea. Oltre un decennio di guerre raccontate dai Padri cappuccini. Società Storica Civitavecchiese* 14 (2013).

De Renzi, Silvia: Per una biografia di Paolo Zacchia. Nuovi documenti e ipotesi di ricerca. In: Pastore, Alessandro/Rossi, Giovanni (eds.): *Paolo Zacchia. Alle origini della medicina legale (1584-1659)*. Milano 2008, pp. 50-73.

De Renzi, Silvia: Medical Competence, Anatomy and the Polity in Seventeenth-Century Rome. In: *Renaissance Studies* 21/4 (2007), pp. 551-567.

De Renzi, Silvia: Medical Expertise, Bodies, and the Law in Early Modern Courts. In: *Isis* 98 (2007), pp. 315-322.

De Renzi, Silvia: La natura in tribunale. Conoscenze e pratiche medicolegali a Roma nel XVII secolo. In: Cerutti, Simona/Pomata, Gianna (eds.): *Fatti. Storie dell'evidenza empirica. Quaderni storici* 108/3 (2001), pp. 799-822.

De Renzi, Silvia/Bresadola, Marco/Conforti, Maria: *Pathology in Practice. Diseases and Dissections in Early Modern Europe*. London 2018.

De Renzi, Silvia: Witnesses of the Body. Medico-legal Cases in Seventeenth Century Rome. In: *Studies in History and Philosophy of Science* 33A (2002), pp. 219-242.

De Vito, Christian/Schiel, Juliane/Van Rossum, Matthias: From Bondage to Precariousness? New Perspectives on Labor and Social History. In: *Journal of Social History* 54/2 (2020), pp. 644-662.

De Vito, Christian/Lucrezio Monticelli, Chiara: Pluralità dei regimi punitivi. Periodizzazioni, circolazioni, modelli cattolici. In: *Meridiana. Regimi Punitivi* 101 (2021), pp. 9-22.

Di Nepi, Serena: *I confini della salvezza. Schiavitù, conversione e libertà nella Roma di età moderna*. Rome 2022.

Di Renzo Villata, Maria Gigliola: Paolo Zacchia, la medicina come sapere globale e la "sfida" al diritto. In: Pastore, Alessandro/Rossi, Giovanni (eds.): *Paolo Zacchia. Alle origini della medicina legale (1584-1659)*. Milano 2008, pp. 9–49.

Di Tommaso, Noemi: The Erudite Practitioner. Francesco Redi's Communication Strategies. In: Giannini, Giulia (ed.): *Visual, Material and Print culture in the Early Stages of the Institutionalisation of Science. The Accademia del Cimento as a Multifaceted Case Study. Physis* 2 (2024), pp. 373–408.

Donato, Maria Pia: *Morti improvvise. Medicina e religione nel settecento*. Rome 2010.

Donato, Maria Pia/Berlivet, Luc/Cabibbo, Sara/Michetti, Raimondo/Nicoud, Marilyn (eds.): *Médecine et religion. Collaboration, compétitions, conflits (XIIe-XXe siècle)*. Rome 2013.

Donato, Maria Pia: Medicina e religione : percorsi di lettura. In: Donato, Maria Pia/Berlivet, Luc/ Cabibbo, Sara/Michetti, Raimondo/Nicoud, Marilyn (eds.): *Médecine et religion. Collaboration, compétitions, conflits (XIIe-XXe siècle)*. Rome 2013.

Duranti, Tommaso: Confidentia tamen de medico debet precedere. La fiducia verso i medici tra pieno medioevo e prima età moderna. In: Malatesta, Maria (ed.): *L'invenzione della fiducia. Medici e pazienti dall'età classica a oggi*. Rome 2021, pp. 59–80.

Elias, Norbert: *Il processo di civilizzazione. La civiltà delle buone maniere*. Bologna 1982 (trad. Basel[1] 1939).

Elliot, Dyan: *The Corrupter of Boys. Sodomy, Scandal, and the Medieval Clergy*. Philadelphia 2020.

Favarò, Valentina: *La modernizzazione militare nella Sicilia di Filippo II*. Palermo 2009.

Feci, Simona: *L'acquetta di Giulia. Mogli avvelenatrici e mariti violenti nella Roma del Seicento*. Rome 2024.

Felici, Lucia (ed.): *La Livornina. Alle origini di Livorno, città cosmopolita in età moderna*. Rome 2024.

Fenicia, Giulio: *Il Regno di Napoli e la difesa del Mediterraneo nell'età di Filippo II (1556-1598). Organizzazione e Finanziamento*. Bari 2003.

Ferragud, Carmel: The Role of Doctors in the Slave Trade during the Fourteenth and Fifteenth Centuries within the Kingdom of Valencia (Crown of Aragon). In: *Bulletin of the History of Medicine* 87 (2013), pp. 143–169.

Ferrara, Santo Davide, *Malpractice and Medical Liability. European State of the art Guidelines*. Berlin 2013.

Filioli Uranio, Fabrizio: I cappellani delle galere pontificie tra Lepanto e l'anno giubilare 1625. In: Greco, Gaetano (ed.): *Il principe, la spada e l'altare*. Pisa 2014, pp. 215–230.

Filioli uranio, Fabrizio: *La squadra navale pontificia nella Repubblica internazionale delle galere. Secoli XVI-XVII*. Rome 2016.

Fischer-Homberger, Esther: *Medizin von Gericht. Gerichtsmedizin von der Renaissance bis zur Aufklärung*. Bern 1983.

Fissel, Mary E.: The Medical Marketplace, the Patient, and the Absence of Medical Ethics in Early Modern Europe and North America. In: Baker, Robert B./McCullough Lawrence (eds.): *The Cambridge World History of Medical Ethics*. Cambridge 2008, pp. 531–539.

Fiume, Giovanna: *Schiavitù Mediterranee. Corsari, rinnegati e santi di età moderna*. Milano 2009.

Fontenay, Michel: L'esclave galérien dans la Méditerranée des Temps modernes. In: Bresc, Henri (ed.): *Figures de l'esclave au Moyen-Age et dans le monde moderne*. Paris 1996, pp. 115–143.

Fontenay, Michel: Esclaves et/ou captifs. Préciser les concepts. In Kaiser, Wolfgang (ed.): *Le commerce des captifs. Les intermédiaires dans l'échange et le rachat des prisonniers en Méditerranée, XVe-XVIIIe siècles*. Rome 2008, pp. 14–24.

Fosi, Irene: *Convertire lo straniero. Forestieri e inquisizione a Rome in età moderna*. Rome 2011.

Fosi, Irene: *La giustizia del papa. Sudditi e tribunali nello Stato Pontificio in età moderna*. Rome/Bari 2007.

Fosi, Irene (ed.): *La Peste a Roma (1656–1657)*. Rome 2006.

Fosi, Irene: Sudditi, tribunali e giudici nella Roma barocca. In: *Roma moderna e contemporanea* I (1997), pp. 19–40.

Foucault, Michel: *Folie et déraison. Histoire de la folie à l'âge Classique*. Paris 1961.

Foucault, Michel: *Histoire de la sexualité I. La volonté de savoir*. Paris 1976.

Foucault, Michel: *L'archeologie du savoir*. Paris 1969.

Foucault, Michel: *Naissance de la clinique. Une archéologie du regard médical*. Paris 1963.

Foucault, Michel: *Surveiller et punir. Naissance de la prison*. Paris 1975.

Frattarelli Fischer, Lucia: Il bagno delle galere in "terra cristiana". Schiavi a Livorno fra Cinque e Seicento. In: *Nuovi Studi Livornesi* VIII (2000), pp. 69–94.

Frattarelli Fischer, Lucia: La Livornina. Alle origini della società Livornese. In: Prosperi, Adriano (ed.): *Livorno 1606–1806. Luogo di incontro tra popoli e culture*. Turin 2009, pp. 43–62.

Frattarelli Fischer, Lucia/Papi, Maria Lia (eds.): *Studi di storia. Livorno dagli archivi alla città*. Livorno 2001.

Frattarelli Fischer, Lucia/Villani, Stefano: "People of Every Mixture". Immigration, Tolerance and Religious Conflicts in Early Modern Livorno. In: Isaacs, Katherine (ed.): *Immigration and Emigration in Historical Perspective*. Pisa 2007, pp. 93–107.

French, Roger Kennet: The Medical Ethics of Gabriele Zerbi. In: Wear, Andrew/French Roger Kennet/Geyer Kordesch, Joanna (eds.): *Doctors and Ethics. The Earlier Historical Setting of Medical Ethics*. Amsterdam 1993, pp. 72–97.

Furfaro, Domenico: *La vita e l'opera di Leonardo Fioravanti*. Bologna 1963.

Fusco, Irene: *La grande epidemia. Potere e corpi sociali di fronte all'emergenza della Napoli spagnola*. Naples 2017.

Gadebusch Bondio, Mariacarla: Avoidable Mistakes – Premodern Medical Fallibility as an Ethical Problem with Epistemological Implications. In: Finucci, Finucci (ed.): *Understanding Medical Humanities*. Berlin 2022.

Gadebusch Bondio, Mariacarla: I denasati e i medici. Discussioni sulla funzione di una protuberanza più o meno necessaria. In: Varanini, Gian Maria (ed.): *Deformità fisica e identità della persona tra Medioevo ed età moderna*. Florence 2015, pp. 159–180.

Gadebusch Bondio, Mariacarla: Erkundungen des Fremden. Über 'korrigierende' Eingriffe an den weiblichen pudenda in der vormodernen Medizin. In: *Medizinhistorisches Journal* 57/1 (2022), pp. 2–35.

Gadebusch Bondio, Mariacarla/Paravicini Bagliani, Agostino (eds.): *Errors and Mistakes. A Cultural History of Fallaibility*. Florence 2012.

Gadebusch Bondio, Mariacarla/Ricklin, Thomas (eds.): *Exempla medicorum. Die Ärzte und ihre Beispiele (XIV–XVIII Jahrhundert)*. Florence 2008.

Gadebusch Bondio, Mariacarla/Kaiser, Christian/Förg, Manuel (eds.): *Menschennatur in Zeiten des Umbruchs. Das Ideal des politischen Arztes in der Frühen Neuzeit*. Berlin/Boston 2020.

Gadebusch Bondio, Mariacarla (ed.): *Medical Ethics. Premodern Negotiations between Medicine and Philosophy*. Stuttgart 2014.

Gadebusch Bondio, Mariacarla: Appelle zum Schutz von Gesundheit und Leben. Wilhelm Fabrys Einsatz zur Responsibilisierung von Ärzteschaft, Obrigkeiten und Betroffenen. In: Gadebusch Bondio, Mariacarla/Hengerer, Mark/Lepsius, Susanne (eds.): *Techniken der Responsibilisierung. Historische und gegenwartsbezogene Studien.* Hannover 2023, pp. 79–126.

Gadebusch Bondio, Mariacarla/Förg, Katharina-Luise: Der vigilante Chirurg und seine Embleme. Wilhelm Fabrys multimedialer Einsatz. In: *Tierische Symbole und Embleme ärztlicher Vigilanz. Working Paper des SFB 1369 'Vigilanzkulturen'* 1 (2021), pp. 14–20.

Gadebusch Bondio, Mariacarla: Verità e menzogna nel dialogo fra medico e paziente (XV–XVII sec.). In: Gadebusch Bondio, Mariacarla (ed.): *La conversazione. Un tema fra storia, arte e filosofia dal Medioevo al Settecento.* Padova, 2011. *I Castelli di Yale. Quaderni di Filosofia* (2012), pp. 71–85.

Gardiner, Robert/Morrison, John: *The Age of The Galley. Mediterranean Oared Vessels Since Pre-Classical Times.* London 1995.

Gelfand, Toby: The History of the Medical Profession. In: Bynum, William F./Porter, Roy (eds.): *Companion Encyclopedia of the History of Medicine.* Vol. 1. London/New York 1993, pp. 1119–1150.

Geltner, Guy: *The Medieval Prison. A Social History.* Princeton/Oxford 2008.

Gentilcore, David: *Medical Charlatanism in Early Modern Italy.* New York 2006.

Gentilcore, David: I Protomedicati come organismi professionali in Italia durante la prima età moderna. In: Betri, Maria Luisa/Pastore, Alessandro (eds.): *Avvocati, medici, ingegneri. Alle origini delle professioni moderne, secoli XVI–XIX.* Bologna 1997, pp. 93–105.

Gentilcore, David: Negoziare rimedi in tempo di peste: alchimisti, ciarlatani, protomedici. In: Fosi, Irene (ed.): *La Peste a Roma (1656–1657).* Rome 2006.

Geremek, Bronisław: *Inutiles au monde: truands et misérables dans l'Europe moderne (1350–1600).* Paris 1980.

Geremek, Bronisław: *Mendicanti e miserabili nell'Europa moderna (1350–1600).* Rome/Bari 1999 (trad. Paris 1980).

Geremek, Bronisław: *Poverty. A History.* Oxford 1994.

Geremek, Bronisław: Il pauperismo nell'età preindustriale (secoli XIV–XVIII). In: Romano Ruggiero/Vivanti, Corrado (eds.): *Storia d'Italia.* Vol 5: *I documenti.* Turin 1973, pp. 667–698.

Gilly, Carlos: *Theodor Zwinger e la crisi culturale della seconda metà del Cinquecento.* Florence 2012.

Ginzburg, Carlo: Folklore, magia, religione. In: *Storia d'Italia.* Vol. I: *I caratteri originali.* Turin 1972, pp. 604–676.

Goldtwhwaite, Richard: *The Economy of Renaissance Florence.* Baltimore 2008.

Goldtwhwaite, Richard/Mandich, Giulio: *Studi sulla moneta Fiorentina: (secoli 13-16).* Florence 1994.

Grassi, Umberto: *L'Offizio sopra la Sanità. Il controllo della sodomia nella Lucca del Cinquecento.* Milano/Udine 2014.

Grassi, Umberto: *Sodoma. Persecuzioni, affetti, pratiche sociali (secoli V–XVIII).* Rome 2019.

Grassi, Umberto/Marcocci, Giuseppe (eds.): *Le trasgressioni della carne. Il desiderio omosessuale nel mondo islamico e cristiano, secc. XII–XX.* Rome 2015.

Grassi, Umberto/Marcocci, Giuseppe: L'intreccio dei desideri, la tolleranza della carne: per una nuova storia delle relazioni tra musulmani e cristiani. In: Grassi, Umberto/Marcocci, Giuseppe (eds*.): Le trasgressioni della carne. Il desiderio omosessuale nel mondo islamico e cristiano, secc. XII–XX.* Rome 2015, pp. 7–22.

Grassi, Umberto/Lagioia, Vincenzo/Romagnani, Gian Paolo (eds.): *Tribadi, sodomiti, invertite e invertiti, pederasti, femminelle, ermafroditi.... Per una storia dell'omosessualità, della bisessualità e delle trasgressioni di genere in Italia*. Pisa 2017.

Groebner, Valentin: *Defaced. The Visual Culture of Violence in the Late Middles Ages*. New York 2008 (trad. Munich 2003).

Groebner, Valentin: *Storia dell'identità personale e della sua certificazione. Scheda segnaletica, documento di identità e controllo nell'Europa moderna*. Bellinzona 2008 (trad. Munich 2004).

Guarnieri, Gino: *I cavalieri di Santo Stefano nella storia della marina italiana (1562-1859)*. Pisa 1960.

Guillén, Fabienne P./Trablesi, Salah (eds.): *Les Esclaves en Méditerranée. Espaces et dynamiques économiques*. Madrid 2012.

Halperin, David: *How to do the history of homosexuality*. Chicago 2002.

Harrison, Mark: *Contagion. How Commerce Had Spread Disease*. Yale 2012.

Härter, Karl: Disciplinamento sociale e ordinanze di polizia nella prima età moderna. In: Prodi, Paolo/Penuti, Carla (eds.): *Disciplina dell'anima, disciplina del corpo e disciplina della società tra medioevo ed età moderna*. Bologna 1994, pp. 635-658.

Henderson, John: *The Renaissance Hospital. Healing the Body and Healing the Soul*. New Haven/London 2006.

Heng, Geraldine: *The Invention of Race in the European Middle Ages*. Cambridge 2018.

Hengerer, Mark Sven/Demichel, Sébastien: *Vigilance and the Plague. France Confronted with the Epidemic Scourge during the 17th and 18th Centuries*. Berlin/Boston 2024.

Hershenzon, Daniel: *The Captive Sea. Slavery, Communication, and Commerce in Early Modern Spain and the Mediterranean*. Philadelphia 2018.

Herzig, Tamar: Enslavement, Religion, and Cultural Commemoration in Livorno. In: *Religions* 14/5 (2023), p. 607.

Herzig, Tamar: Slavery and Interethnic Sexual Violence. A Multiple Perpetrator Rape in Seventeenth-Century Livorno. In: *The American Historical Review* 127/1 (2022), pp. 194-222.

Holenstein, André/Schläppi, Daniel: *Empowering Interactions: Political Cultures and the Emergence of the State in Europe 1300-1900*. London/New York 2007.

Hudson, Geoffrey L.: *British Military and Naval Medicine, 1600-1830*. Amsterdam/New York 2007.

Inì, Marina: Architecture and Plague Prevention: Lazzaretti in the Eighteenth-Century Mediterranean. In: Gharipour, Mohammad (ed.): *Public Health in the Early Modern City in Europe*. Cham 2023, pp. 83-123.

Johnson, Walter: On Agency. In: *Journal of Social History* 37/1 (2003), pp. 113-124.

Jones, Colin: The Plague and its metaphors in early Modern France. In: *Representations* 53 (1996), pp. 97-127.

Jones, Colin/Porter, Roy (eds.): *Reassessing Foucault: Power, Medicine and the Body*. London/New York 1994.

Jütte, Robert: *Poverty and Deviance in Early Modern Europe*. Cambridge 1994.

Kaba, Riyaz/Sooriakumaran, Prassana: The evolution of the doctor-patient relationship. In: *International Journal of Surgery* 5/1 (2007), pp. 57-65.

Kefler, Mathew: *The Boswell's Thesis. Essays on Christianity, Social Tolerance, and Homosexuality*. Chicago 2005.

Koslofsky, Craig/Zaugg, Robert (eds.): *A German Barber-Surgeon in the Atlantic Slave Trade. The Seventeenth-Century Journal of Johann Peter Oettinger*. Charlottesville 2020.

Kuru, Selim S.: Il genere del desiderio. L'amore per i bei ragazzi nella letteratura ottomana della prima età moderna. In: Grassi, Umberto/Marcocci, Giuseppe (eds.): *Le trasgressioni della carne. Il desiderio omosessuale nel mondo islamico e cristiano, secc. XII–XX*. Rome 2015, pp. 81–102.

Lagioia, Vincenzo: "Agitato da passione amorosa". Tracce omoerotiche a processo (secc. XVII–XVIII). In: Scaramella, Tommaso (ed.): *Alla prova delle passioni. Sessualità non conforme e soggettività fra età moderna e contemporanea*. Pisa 2024, pp. 85–102.

Lagioia, Vincenzo (ed.): *Storie di invisibili, marginali ed esclusi*. Bologna 2012.

Laquer, Thomas: *Making Sex. Body and Gender from the Greeks to Freud*. Cambridge 1990.

Laugier, Laurence (ed.): *Paul Zacchias Questions Médico-Légales. Des fautes médicales sanctionnées par la loi*. Aix-en-Provence 2006.

Lavenia, Vincenzo: Tra Cristo e Mare. Disciplina e catechesi del soldato cristiano in età moderna. In: Brizzi, Gian Paolo/Olmi, Giuseppe (eds.): *Dai cantieri della storia. Liber amicorum per Paolo Prodi*. Bologna 2008, pp. 1–18.

Lavenia, Vincenzo: I diavoli di Carpi e il Sant'Uffizio (1636–1639). In: Rosa, Mario (eds.): *Eretici, esuli e indemoniati nell'età moderna*. Florence 1998, pp. 77–139.

Lavenia, Vincenzo: *Dio in uniforme. Cappellani, catechesi cattolica e soldati in età moderna*. Bologna 2017.

Lavenia, Vincenzo: Tra eresia e crimine contro natura. Sessualità, islamofobia e inquisizioni nell'Europa moderna. In: Grassi, Umberto/Marcocci, Giuseppe (eds.): *Le trasgressioni della carne. Il desiderio omosessuale nel mondo islamico e cristiano, secc. XII–XX*. Rome 2015, pp. 103–131.

Lavenia, Vincenzo: *Un' eresia indicibile. Inquisizione e crimini contro natura in età moderna*. Bologna 2015.

Lavenia, Vincenzo: La medicina dei diavoli. Il caso italiano, secoli XVI–XVII. In: Donato, Maria Pia/Berlivet, Luc/Cabibbo, Sara/Michetti, Raimondo/Nicoud, Marilyn (eds.): *Médecine et religion. Collaboration, compétitions, conflits (XIIe–XXe siècle)*. Rome 2013, pp. 163–194.

Lavenia, Vincenzo: Schiavi. In: Petrolini, Chiara/Lavenia, Vincenzo/Pavone, Sabina (eds.): *Sacre metamorfosi. Racconti di conversione tra Roma e il mondo in età moderna*. Rome 2022, pp. 3–42.

Lawrence, Susan: Medical Education. In: Bynum, William F./Porter, Roy (eds.): *Companion Encyclopedia of the History of Medicine*. Vol. 2. London/New York 1993, pp. 1151–1179.

Homosexuality. New York 1981.

Le Roy Ladurie, Emmanuel: Un concept. L'unification microbienne du monde (XVIe–XVIIe siècles). In: *Revue Suisse d'Histoire* 23/4 (1973), pp. 627–696.

Lever, Maurice: *Les bûchers de Sodome*. Paris 1985.

Liboni, Gionata: Humanist Post-Mortem. Philology and Therapy. In: De Renzi, Silvia/Bresadola, Marco/Conforti, Maria (eds.): *Pathology in Practice. Diseases and Dissections in Early Modern Europe*. London 2018, pp. 20–38.

Licata, Salvatore J./Petersen, Robert P. (eds.): *Historical Perspectives on*

Linden, David: Gabriele Zerbi's De cautelis medicorum and the Tradition of Medical Prudence. In: *Bullettin of the History of Medicine* 73 (1999), pp. 19–37.

Linden, David: The perfect physician. 16th century perspectives from the Iberian Peninsula. In: *Sudhoffs Archiv* 84/2 (2000), pp. 22–31.

Linte, Guillaume: *Hygiène navale et médicine des colonies en France XVIe–XVIIIe siècle*. Paris 2023.

Lomas, Manuel: *Governing the galleys. Jurisdiction, Justice and Trade in the Squadrons of the Hispanic Monarchy (Sixteenth–Seventeenth Centuries)*. Leiden/Boston 2020.

Lo Basso, Luca: *A vela e a remi. Navigazione, guerra e schiavitú nel Mediterraneo (sec. XVI–XVIII)*. Ventimiglia 2004.
Lo Basso, Luca: Condannati alla galera nell'Italia dell'età moderna. Gli esempi di Venezia e Genova. In: Antonelli, Livio (ed.): *La polizia in Italia e in Europa. Punto sugli studi e prospettive di ricerca*. Soveria Mannelli 2006, pp. 117–144.
Lo Basso, Luca: *Uomini da remo. Galee e galeotti del Mediterraneo in età moderna*. Milano 2003.
Lucrezio Monticelli, Chiara: Trastevere come spazio della reclusione tra XVIII e XIX secolo. Il carcere femminile di S. Michele a Ripa. In: Ermini Pani, Letizia/Travaglini, Carlo (eds.): *Trasformazioni urbane. Il caso del rione Trastevere*. Rome 2010, pp. 397–420.
Lucrezio Monticelli, Chiara: Prigioni e rappresentazioni. Il modello romano di carcere in prospettiva storica. In: *Giornale di Storia, Dialoghi sul carcere* 38 (2021), pp. 1–9.
Lyon, David: *Surveillance Studies: An Overview*. Cambridge 2007.
MacKinney, Loren C.: Medical Ethics and Etiquette in the Early Middle Ages. The persistence of Hippocratic Ideals. In: *Bulletin of the History of Medicine* 26/1 (1952), pp. 1–31.
Maclean, Ian: *Logic, Signs and Nature in the Renaissance. The Case of Learned Medicine*. Cambridge 2007.
Malatesta, Maria (ed.): *L'invenzione della fiducia: medici e pazienti dall'età classica a oggi*. Rome 2021.
Malcolm, Noel: *Forbidden Desires in Early Modern Europe. Male-Male Sexual Relations, 1400–1750*. Oxford 2024.
Malcolm, Noel: Forbidden Love in Istanbul: Patterns of Male-Male Sexual Relations in the Early-Modern Mediterranean World. In: *Past & Present* (2022), pp. 55–88.
Mallinckrodt, Rebekka von/Lentz, Sarah/Köstlbauer, Josef (eds.): *Beyond Exceptionalism. Traces of Slavery and the Slave Trade in Early Modern Germany, 1650–1850*. Berlin/Boston 2021.
Mandressi, Rafael: Medicus politicus. Notas sobra una historia política de la medicina. In: *Claves. Revista de Historia* 6/10 (2020), pp. 273–283.
Marchisello, Andrea: "Culpa habet sociam poenam". La responsabilità del medico nelle Quaestiones medico-legales di Paolo Zacchia. In: Pastore, Alessandro/Rossi, Giovanni (eds.): *Paolo Zacchia. Alle origini della medicina legale (1584–1659)*. Milano 2008, pp. 221–248.
Marotta, Achille: The Muslim Friend. Cross-Confessional Male Intimacy in Eighteenth-Century Italy. In: *Journal of Early Modern History* 28/3 (2024), pp. 230–252.
Martin, Meredith/Weiss, Gillian: *The Sun King at Sea. Maritime Art and Galley Slavery in Louis XIV's France*. Los Angeles 2022.
Martini, Gabriele: *Il "vitio nefando" nella Venezia del Seicento. Aspetti sociali e repressione di giustizia*. Rome 1988.
Massad, Joseph A.: *Desiring Arabs*. Chicago/London 2007.
Maxwell, Stewart P: G: Representations of Same-Sex Love in Early Modern Astrology. In: Borris, Kenneth/Rousseau, George (eds.): *The Sciences of Homosexuality in Early Modern Europe*. London/New York 2008, pp. 165–182.
McClintock, Aglaia: Dal Servus Poenae Alla "Servitù Penale". Storia Di Una Nozione. In: *Koinonia* 44 (2020), pp. 1072–1085.
Meillassoux, Claude: *Anthropologie de l'esclavage. Le ventre de fer et d'argent*. Paris 1986.
Mendelsohn, Andrew/Kinzelbach, Annemarie/Schilling, Ruth (eds.): *Civic Medicine. Physicians, Polity, and Pen in Early Modern Europe*. London 2020.
Mibillion, Jean: *Riflessioni sulle prigioni degli ordini religiosi*. Lecce 2020.
Minois, Georges: *Il prete e il medico. Fra religione, scienza e coscienza*. Bari 2016 (trad. Paris 2015).

Montecón Movellán, Tomas A.: Oltre la repressione. Relazioni omosessuali tra musulmani e cristiani nella Spagna del Cinque e Seicento. In: Grassi, Umberto/Marcocci, Giuseppe (eds.): *Le trasgressioni della carne. Il desiderio omosessuale nel mondo islamico e cristiano, secc. XII–XX.* Rome 2015, pp. 133–154.

Moulinier-Brogi, Laurence: *L'uroscopie au Moyen Âge. "lire dans un verre la nature de l'homme".* Paris 2012.

Mustakeem, Sowande' M.: *Slavery at Sea. Terror, Sex, and Sickness in the Middle Passage.* Champaign 2016.

Murphy, Hannah: Re-writing Race in Early Modern European Medicine. In: *History Compass* 19/11 (2021), pp. 1–12.

Münster, Ladislao: Studi e ricerche su Gabriele Zerbi. Nota 1. Nuovi Contributi biografici, la sua figura morale. In: *Rivista di storia delle scienze mediche e naturali* 41 (1950), pp. 64–83.

Münster, Ladislao: In tema di deontologia medica. Il "De cautelis medicorum" di Gabriele Zerbi. In: *Rivista di storia delle scienze mediche e naturali* 46 (1956), pp. 60–83.

Nardi, G. M.: Statuti e documenti riflettenti la dissezione anatomica umana e la nomina di alcuni lettori di Medicina nell'antico "Studium Generale Fiorentino". In: *Rivista di Storia delle Scienze Mediche e Naturali* 47 (1956), pp. 245–248.

Neal, Derek: Disorder of Body, Mind or Soul. Male Sexual Deviance in Jacques Despars's Commentary on Avicenna. In: Borris, Kenneth/Rousseau, George (eds.): *The Sciences of Homosexuality in Early Modern Europe.* London/New York 2008, pp. 43–56.

Nicolson, Michael: The Art of Diagnosis. Medicine and the Five Senses. In: Bynum, William F./ Porter, Roy (eds.): *Companion Encyclopedia of the History of Medicine.* Vol. 2. London/New York 1993, pp. 801–825.

Nutton, Vivian: Beyond the Hippocratic Oath. In: Wear, Andrew/French, Roger Kennet/Geyer-Kordesch, Joanna (eds.): *Doctors and Ethics. The Earlier Historical Setting of Medical Ethics.* Amsterdam 1993, pp. 10–37.

Nutton, Vivian: The Reception of Fracastoro's Theory of Contagion. The Seed that Fell Among Thorns. In: *Osiris, Renaissance Medical Learning. Evolution of a Tradition* 6 (1990), pp. 196–234.

Nutton, Vivian: Medicine in Late Antiquity and the Early Middle Ages. In: Conrad, Lawrence I./ Neve, Michael/Nutton, Vivian/Porter, Roy/Wear, Andrew: *The Western Medical Tradition. 800 BC to AD 1800.* Cambridge ^{10}2011.

Nutton, Vivian: Medicine in the Greek World, 800-50 BC. In: Conrad, Lawrence I./Neve, Michael/ Nutton, Vivian/Porter, Roy/Wear, Andrew: *The Western Medical Tradition. 800 BC to AD 1800.* Cambridge ^{10}2011, pp. 11–38.

Nutton, Vivian: Roman Medicine, 250 BC to AD 200. In: Conrad, Lawrence I./Neve, Michael/Nutton, Vivian/Porter, Roy/Wear, Andrew: *The Western Medical Tradition. 800 BC to AD 1800.* Cambridge ^{10}2011, pp. 39–70.

Nutton, Vivian: Medicine in Medieval Western Europe, 1000-1500. In: Conrad, Lawrence I./Neve, Michael/Nutton, Vivian/Porter, Roy/Wear, Andrew: *The Western Medical Tradition. 800 BC to AD 1800.* Cambridge ^{10}2011, pp. 139–206.

Nutton, Vivian: The Seeds of Disease. An Explanation of Contagion and Infection from Greeks to the Renaissance. In: *Medical History* 27/1 (1983), pp. 1–32.

Paglia, Vincenzo: *La Pietà dei carcerati. Confraternite e società a Roma nei secoli XVI–XVIII.* Rome 1980.

Park, Katharine: The Criminal and the Saintly Body. Autopsy and Dissection in Renaissance Italy In: *Renaissance Quarterly* 47/1 (1994), pp. 1–33.

Park, Katharine: *Doctors and Medicine in Early Renaissance Florence*. Princeton 1985.
Park, Katharine: The Myth of the "One sex" Body. In: *Isis* 114/1 (2023), pp. 150–175.
Parker, Charles H.: *Global Interactions in the Early Modern Age*. Cambridge 2010.
Partner, Peter: *Corsari e crociati. Volti e avventure del Mediterraneo*. Turin 2003.
Pastore, Alessandro: Errori e peccati dei medici nei manuali dei confessori (secoli XV–XVI). In: *Archivio Storico Italiano* 666 (2020), pp. 775–797.
Pastore, Alessandro: Maladies vraies et maladies simulées. Les opinions des juristes et des médecins (XVIe–XVIIe siècles). In: Barras, Vincent/Porret, Michel (eds.): *Homo Criminalis. Pratiques et théories médico-légales (XVIe–XXe siècles). Equinoxe. Revue de sciences humaines* 22 (1999), pp. 11–26.
Pastore, Alessandro: *Il medico in tribunale. La perizia medica nella procedura penale d'antico regime (secoli XVI–XVIII)*. Bellinzona ²1998.
Pastore, Alessandro: *Le regole dei corpi. Medicina e disciplina nell'Italia moderna*. Bologna 2006.
Pastore, Alessandro/Rossi, Giovanni: *Paolo Zacchia. Alle origini della medicina legale (1584–1659)*. Milano 2008.
Pastura Ruggiero, Maria Grazia: *La Reverenda Camera Apostolica e i suoi archivi: secoli XV–XVIII*. Rome 1984.
Pavone, Sabina: *I gesuiti dalle origini alla soppressione. 1540–1773*. Rome/Bari 2004.
Pennuto, Concetta: La natura dei contagi in Fracastoro. In: Pastore, Alessandro/Peruzzi, Enrico (eds.): *Girolamo Fracastoro fra medicina, filosofia e scienze della natura. Atti del Convegno internazionale di studi in occasione del 450. anniversario della morte Verona–Padova, 9–11 ottobre 2003*. Florence 2006, 57–71.
Pennuto, Concetta: *Simpatia, fantasia e contagio. Il pensiero medico e il pensiero filosofico di Girolamo Fracastoro*. Rome 2008.
Piaia, Gregorio: *Pietro D'Abano. Filosofo, medico e astrologo europeo*. Milano 2020.
Piccialuti, Maura (ed.): *La sanità a Roma in età moderna, Roma moderna e Contemporanea* XIII/1 (2005).
Pizzolato, Nicola: "Lo diavolo mi ingannao". La sodomia nella campagna siciliane (1572–1664). In: *Quaderni Storici* CCXII/22 (2006), pp. 449–480.
Pomara Saverino, Bruno: *Rifugiati. I moriscos e l'Italia*. Florence 2017.
Pomata, Gianna/Siraisi, Nancy G.: *Historia. Empiricism and Erudition in Early Modern Europe*. Cambridge 2005.
Pomata, Gianna: Praxis Historialis. The Uses of Historia in Early Modern Medicine. In: Pomata, Gianna/Siraisi, Nancy G. (eds.): *Historia: Empiricism and Erudition in Early Modern Europe*. Cambridge 2005, pp. 105–146.
Pomata, Gianna: Sharing Cases. The Observationes in Early Modern Medicine. In: *Early Science and Medicine* 15/3 (2010), pp. 193–236.
Pomata, Gianna: *La promessa di guarigione, malati e curatori in antico regime*. Rome 1994.
Povolo, Claudio: La vittima nello scenario del processo penale. Dai crimini senza vittime all'irruzione della vittima nel dibattito sociale e politico. In: *Acta Histriae* 12/1 (2004), pp. I–XIV.
Preto, Paolo: *Peste e Società a Venezia nel 1576*. Venice 1984.
Prodi, Paolo/Penuti, Carla: *Disciplina dell'anima, disciplina del corpo e disciplina della società tra medioevo ed età moderna*. Bologna 1994.
Prodi, Paolo: *Il sovrano pontefice. Un corpo e due anime: la monarchia Papale nella prima età moderna*. Bologna 1982.
Prodi, Paolo: *Una storia della giustizia*. Bologna 2000.

Prosperi, Adriano: *Dare l'anima. Storia di un infanticidio.* Turin 2005.
Prosperi, Adriano/Lavenia, Vincenzo/Tedeschi, John (eds.): *Dizionario Storico dell'Inquisizione.* Pisa 2010.
Prosperi, Adriano (ed.): *Livorno 1606-1806. Luogo di incontro tra popoli e culture.* Turin 2009.
Prosperi, Adriano: *Missionari. Dalle Indie remote alle Indie interne.* Rome/Bari 2024.
Prosperi, Adriano: "Otras Indias". Missionari della Controriforma tra contadini e selvaggi. In: *Scienze, credenze occulte, livelli di cultura. Convegno internazionale di studi (Firenze, 26-30 giugno 1980).* Florence 1982, pp. 205-234.
Prosperi, Adriano: *Tribunali della Coscienza. Inquisitori, confessori, missionari.* Turin 1996.
Prosperi, Adriano: *La vocazione. Storie di gesuiti tra Cinquecento e Seicento.* Turin 2016.
Puccini, Clemente: *Il Methodus Testificandi di Giovan Battista Codronchi.* Sala Bolognese 1987.
Reinhard, Wolfgang: Konfession und Konfessionalisierung in Europa. In: Reinhard, Wolfgang (ed.): *Bekennins und Geschichte. Die Confessio Augustana im historischen Zusammenhang.* Munich 1981, pp. 165-189.
Ricci, Giovanni: *I turchi alle porte.* Bologna 2008.
Rigato, Daniela: Medico divino e razionale, carnifex e amicus. A chi dare fiducia? In: Malatesta, Maria (ed.): *L'invenzione della fiducia. Medici e pazienti dall'età classica a oggi.* Rome 2021, pp. 37-57.
Rissel, Magnus: Hamburg and the Lower Elbe in the Atlantic Slave Trade of the Early Modern Period. In: *Werkstatt Geschichte* 66/67 (2014), pp. 75-96.
Robison, Kira: *Healers in the Making. Students, Physicians, and Medical Education in Medieval Bologna (1250-1550).* Leiden/Boston 2021.
Rocciolo, Daniele: Il tribunale del cardinal Vicario e la città. Brevi note tratte dall'opera di Nicolò Antonio Cuggiò. In: *Roma Moderna e Contemporanea* V (1997), pp. 175-184.
Rocke, Michael J./Grendi, Nicola: Il controllo dell'omosessualità a Firenze nel XV secolo. Gli Officiali di Notte. In: *Fonti criminali e storia sociale, Quaderni Storici. Nuova Serie* 22/66 (3) (1987), pp. 701-723.
Rocke, Michael J.: *Forbidden Friendships. Homosexuality and Male Culture in Renaissance Florence.* New York/Oxford 1996.
Roest, Bert/Mixon, James D. (eds.): *A Companion to Observant Reform in the Late Middle Ages and Beyond.* Leiden/Boston 2015.
Ronzoni, Daniele/Piaia, Gregorio (eds.): *Pietro D'Abano il Conciliatore. Crocevia di culture.* Padova 2021.
Rousseau, George: Policing the Anus. Struprum and Sodomy According to Paolo Zacchia's Forensic Medicine. In: Borris, Kenneth/Rousseau, George (eds.): *The Sciences of Homosexuality in Early Modern Europe.* London/New York 2008, pp. 75-91.
Rowson, Everett K.: The Effeminates of Early Modern Medina. In: *Journal of the American Oriental Society* 111 (1991), pp. 671-693.
Rowson, Everett K.: General Irregularity as Entertainment. Instituzionalized Transvestitism and the Caliphal Court in Medieval Baghdad. In: Farmer, Sharon/Braun Pasternack, Carol (eds.): *Gender and Difference in the Middle Ages.* Minneapolis 2003, pp. 45-72.
Rowson, Everett K.: Omoerotismo ed élite mamelucca tra Egitto e Siria nel tardo medioevo. In: Grassi, Umberto/Marcocci, Giuseppe (eds.): *Le trasgressioni della carne. Il desiderio omosessuale nel mondo islamico e cristiano, secc. XII-XX.* Rome 2015, pp. 23-52.

Ruggiero, Guido: The Cooperation of Physicians and the State in the Control of Violence in Renaissance Venice. In: *Journal of the history of medicine and allied sciences* 33/2 (1978), pp. 156–166.

Russell, Andrew (ed.): *The Town and the State Physician in Europe from the Middle Ages to the Enlightenment*. Wolfenbüttel 1981.

Salvadorini, Vittorio: Traffici e schiavi fra Livorno e Algeria nella prima decade del '600. In: *Bollettino storico pisano* 51 (1982), pp. 67–104.

Sandrini, Enrico: *La professione medica nella dottrina del diritto comune. Secoli XIII-XVI*. Padova 2008.

Sani, Roberto/Zurlini, Fabiola (eds.): *La formazione del medico in età moderna (secc. XVI-XVIII)*. Macerata 2012.

Santus, Cesare: *Il turco a Livorno. Incontri con l'Islam nella Toscana del Seicento*. Milano 2019.

Savoia, Paolo: Skills, Knowledge, and Status. The Career of an Early Modern Italian Surgeon. In: *Bulletin of the History of Medicine* 93/1 (2019), pp. 27–54.

Sbriccoli, Mario: *Crimen laesae maiestatis. Il problema del reato politico alle soglie della scienza penalistica moderna*. Milano 1974.

Sbriccoli, Mario: Fonti giudiziarie e fonti giuridiche. Riflessioni sulla fase attuale degli studi di storia del crimine e della giustizia criminale. In: *Studi storici* 29 (1988), pp. 491–501.

Scaramella, Tommaso: *Un doge infame. Sodomia e nonconformismo sessuale a Venezia nel Settecento*. Venice 2021.

Scaramella, Tommaso: La storia dell'omosessualità nell'Italia moderna. Un bilancio. In: *Storicamente* 12 (2017), pp. 1–21.

Scaramella, Tommaso: "Ti ho sospettato di non avermi aperto i cuore". Amori nascosti nelle lettere di un erudito: Agostino Vivorio (1743–1822). In: Scaramella, Tommaso (ed.): *Alla prova delle passioni. Sessualità non conforme e soggettività fra età moderna e contemporanea*. Pisa 2024, pp. 103–118.

Schiel, Juliane: Slaves' Religious Choice in Renaissance Venice. Applying Insights from Missionary Narratives to Slave Baptism Records. In: *Archivio Veneto* 9/146 (2015), pp. 23–45.

Schiera, Pierangelo/Prodi, Paolo: "Disciplinamento" nella prima età moderna. In: *Quaderni Storici* 17/49 (1) (1982), pp. 349–351.

Schilling, Heinz: *Konfessionskonflikt und Staatsbildung*. Gütersloh 1981.

Schleiner, Winfried: *Medical Ethics in the Renaissance*. Washington 1995.

Schlosser, Hans: *Tre secoli di condannati bavaresi alle galere*. Venice 1984.

Sewell, Willia/Walkowitz, Judith/Eley, Geoff/Zimmerman Angela/Tejada,Vivien: The Agency Dilemma: A Forum. In: *The American Historical Review* 128 (2023), pp. 883–937.

Shemesh, Abraham O.: 'All Men Have Been Considered Equal by Me'. The Attitude of Amatus Lusitanus towards Treating Gentiles according to his Physician's Oath. In: *HTS Teologiese Studies/Theological Studies* 75/3 (2019).

Siena, Kevin: The Strange Medical Silence on Same-Sex Transmission of the Pox, c. 1660–c. 1760. In: Borris, Kenneth/Rousseau, George (eds.): *The Sciences of Homosexuality in Early Modern Europe*. London/New York 2008, pp. 115–34.

Siraisi, Nancy G.: *History, Medicine, and the Traditions of Renaissance Learning*. Ann Arbor 2007.

Siraisi, Nancy G.: *Medieval and Early Renaissance Medicine. An Introduction to Knowledge and Practice*. Chicago 1990.

Siraisi, Nancy G./Bresadola, Marco: Segni evidenti, teoria e testimonianza nelle narrazioni di autopsie del rinascimento. In: Cerutti, Simona/Pomata, Gianna (eds.): *Fatti. Storie dell'evidenza empirica, Quaderni storici* 108/3 (2001), pp. 719–744.

Sisk, Bryan/Frankel, Richard/Kodish, Eric/Harry Isaacson, J.: The Truth about Truth-Telling in American Medicine. A Brief History. In: *The Permanente Journal* 20/3 (2016), pp. 74–77.
Smallwood, Stephanie E.: *A Middle Passage from Africa to American Diaspora*. Cambridge 2007.
Smallwood, E.: *Saltwater slavery. A middle passage from Africa to American diaspora*. Cambridge 2007.
Snyder, Jon R.: *Dissimulation and the Culture of Secrecy in Early Modern Europe*. Berkeley 2009.
Stanziani, Alessandro: *Le metamorfosi del lavoro coatto. Una storia globale, XVIII–XIX secolo*. Bologna 2022.
Stevens Crawshaw, Jane: *Plague Hospitals. Public Health for the City in Early Modern Venice*. Farnham 2012.
Stolberg, Michael: *Learned Physicians and Everyday Medical Practice in the Renaissance*. Berlin/Boston 2022.
Thompson, Edward: *The Making of the English Working Class*. London 1963.
Tognotti, Eugenia: *L'altra faccia di Venere. La sifilide dalla prima età moderna all'avvento dell'Aids (XV–XX sec.)*. Milano 2006.
Tomassetti, Stefano: Dentro e Fuori l'Ospedale Di Età Moderna. Idee, Pratiche, Contesti. In: *Storica* 74 (2019), pp. 91–127.
Trivellato, Francesca: *The Familiarity of Strangers. The Sephardic Diaspora, Livorno, and Cross-Cultural Trade in the Early Modern Period*. Yale 2012.
Tröhler, Ulrich: *"To Improve the Evidence of Medicine". The 18th century British Origins of a Critical Approach*. Edinburgh 2000.
Vanzan Marchini, Nelli Elena: Medici ebrei e assistenza cristiana nella Venezia del '500. In: *La Rassegna Mensile di Israel* 45/4–5 (1979), pp. 132–161.
Valensi, Lucette: *Ces étrangers familiers. Musulmans en Europe (XVIe–XVIIIe siècles)*. Paris 2012.
Veyne, Paul: *Les Grecs ont-ils cru à leurs mythes*. Paris 1983.
Viafora, Corrado/Furlan, Enrico/Tusino, Silvia (eds.): *Questioni di vita. Un'introduzione alla bioetica*. Milano 2019.
Viaro, Andrea: La pena della galera. La condizione dei condannati a bordo delle galere veneziane. In: Cozzi, Gaetano (ed.): *Stato, società e giustizia nella repubblica veneta. Secoli XV–XVIII*. Rome 1980.
Vitalini Sacconi, Vittorio: *Gente, personaggi e tradizioni a Civitavecchia dal Seicento all'Ottocento*. Vol. I. Rome 1982.
Walsham, Alexandra: "The Social History of the Archive. Record-Keeping in Early Modern Europe". In: *Past & Present* 230/11 (2016), pp. 9–48.
Watson, Katherine D.: *Forensic Medicine in Western Society. A History*. London 2010.
Wear, Andrew: Medicine in Early Modern Europe, 1500-1700. In: Conrad, Lawrence I./Neve, Michael/Nutton, Vivian/Porter, Roy/Wear, Andrew: *The Western Medical Tradition. 800 BC to AD 1800*. Cambridge [10]2011, pp. 215–362.
Wear, Andrew/French, Roger Kennet/Geyer-Kordesch, Joanna (eds.): *Doctors and Ethics. The Earlier Historical Setting of Medical Ethics*. Amsterdam 1993.
Winnebeck Julia/Sutter, Ove/Hermann, Adrian/Antweiler, Christoph/Conermann, Stephan: The Analytical Concept of Asymmetrical Dependency. In: *Journal of global slavery* 8/1 (2023), pp. 1–59.
Zagorin, Perez: *Ways of lying. Dissimulation, Persecution, and Conformity in Early Modern Europe*. Cambridge 1990.
Zeeden, Ernst Walther: *Die Entstehung der Konfessionen. Grundlagen und Formen der Konfessionsbildung im Zeitalter der Glaubenskämpfe*. Munich 1965.

Zemon Davies, Natalie: *La doppia vita di Leone l'Africano*. Rome/Bari 2008 (trad. New York[1] 2006).
Zemon Davies, Natalie: *Fiction in the Archives. Pardon Tales and Their Tellers in Sixteenth-Century France*. Stanford 1997.
Ziegler, Hannes: The Preventive Idea of Coastal Policing. Vigilance and Enforcement in the Eighteenth-Century British Customs. In: Brendecke, Arndt/Molino, Paola (eds.): *The History and Cultures of Vigilance. Historicizing the Role of Private Attention in Society. Storia della Storiografia* 74/2 (2018), pp. 75–98.
Zuccolin, Gabriella: The Hermaphroditic Hyena. In: Hopwood, Nick/Flamming, Rebecca/Kassell, Lauren (eds.): *Reproduction. Antiquity to the Present Day*. Cambridge 2018, p. 672.
Zysberg, André: *Les galériens. Vies et destins de 60 000 forçats sur les galères de France, 1680–1748*. Paris 1987.

Index of Persons

d'Abano, Pietro 36, 239–240
Adam, Guillaume 200, 221
Aristotle 19, 23, 45, 145, 237–239
– Pseudo-Aristotle 238
Augustine 157, 200
Avicenna 18

Baldigara, Battista 96
Barbaro, Antonio 86
de Beaumanoir, Philippe 204
Beccaria, Cesare 90f.
Bernard of Cluny 199
Bernardi, Filippo 76, 80, 90–92, 104, 108, 110, 134, 138, 150–152, 159, 208f.
Bindi, Giovan Battista 146
Bonomo, Giovanni Cosimo 133f., 138f.
Botallo, Leonardo 22, 28f., 32, 43
Brugnoli, Candido 165

Cardano, Girolamo 51
of Carinthia, Herman 222
Casella, Francesco 146f.
Castelli, Pietro 32 32
Castiglione, Baldesar 40
de Castro, Rodrigo 31, 39f., 44f., 47, 55, 64f., 167, 173f., 244
Celsus 41f., 45
Claro, Giulio 85
of Cluny, Peter 222
Codronchi, Giovan Battista 62f., 164f.
Cortesi, Giovanni Battista 41
Cospi, Antonio Maria 58–60
Della Croce, Giovanni Andrea 41
Cremani, Luigi 86

Damascenus 45
Davies, William 98f.
Deciani, Tiberius 85
Dioscorides 155

Falcucci, Niccolò 36, 42
Farinacci, Prospero 54, 225
Fedeli, Federico 61, 63

Fei, Giovanni 180
Fioravanti, Leonardo 125
Fracastoro, Girolamo 142
Fragoso, Juan 174, 176

Galen 15–16, 20–21, 33, 36, 41, 22, 28, 141, 164
of Gandersheim, Hroswitha 220
Gastaldi, Girolamo 145f.
Grimaldi, Domenico 135, 147
Guglielmotti, Alberto 79, 81, 113, 115f., 118, 126, 149

Hippocrates 16, 21, 23–24, 36, 44, 48, 51, 141, 143, 164

Ingrassia, Giovanni Filippo 33, 61f., 66

of Ketton, Peter 222

Lancisi, Giovanni Maria 53, 134, 155
de Lille, Alain 220
Locke, John 93
Lusitanus, Amatus 226

de Marcy, Henry 220
Marteilhe, Jean 95

Oresme, Nicole 237

Pantera, Pantero 72, 82f., 94, 126f., 135, 149, 168, 172
Peter the Venerable 222
Pini, Teseo 98
Pliny 199
of Poitiers, Peter 222
Pomponazzi, Pietro 164
Priori, Lorenzo 52, 54, 86, 225

Ramazzini, Bernardo 137, 141
Redi, Francesco 133–135, 138f.

Savelli, Marcantonio 54, 86, 225

Scetti, Aurelio   95–97, 150, 169
Sforza Pallavicino, Francesco Maria   28, 145
Sicco, Giovanni Antonio 32, 42–43   32, 38, 42 f.

of Tarsus, Paul 94, 198
Thomas Aquinas   184, 196, 201 f.
of Toledo, Peter   222
Trincavelli, Vittore   32

degli Ubaldi, Baldo   57
Ulpianus   88

da Villanova, Arnaldo   25, 47
Von Friemar, Heinrich   227

Zacchia, Paolo 20, 30, 45, 49 f., 56, 62, 64–68, 165, 225 f.
de Zancari, Alberto   34, 36
Zerbi, Gabriele   1, 10, 27, 32, 34–38, 42 f., 46, 157

Index of Subjects

Auditore delle galere   111–113, 123
Autopsy   23f., 51–53, 60f., 181, 187f.

Bagno de' forzati/ degli schiavi   75f., 80, 88, 91f., 100, 103–112, 129, 133, 137f., 150f., 156, 159–163, 208–212

Capitano del Bagno   105, 108, 111, 208
Captivitas   75, 98f., 152, 158, 246
Captivitas   84
Capuchins   72–75, 91, 109–111, 114, 122, 151, 153, 157–162, 209
Cautela/ae   10, 15, 32-34, 37, 43–48, 157
Coggia   159
Confessionalization   9f., 72
Consilia   25f., 50
Corsairing   4f., 103, 115
Cyprus War   69, 122, 220

Discipline   2–4, 9–13, 15, 18, 21, 38, 41, 55, 57, 62, 65, 68–73, 75, 87f., 92f., 97, 105f., 108, 113f., 117, 119–122, 179, 192, 195f., 207f., 216, 233, 237, 241f., 244f., 249f.

Exorcism   160–165

Forensic Medicine   11, 48–52, 59, 61f., 65f., 179, 225, 236, 242

Galea polmonare   147f.
Galea polmonare/Galee polmonari   147–149
Galley doctor   2, 7, 13, 98, 120, 126, 128–130, 132–135, 144, 146f., 153f., 168, 177f., 186, 188, 230f., 246f., 250
Governor of Civitavecchia   119, 123f., 154, 185, 224
Governor of Livorno   181

Hippocratic Oath   1, 26, 33, 67, 248
Hospital   2, 13, 17, 72–74, 125, 127–133, 135, 141, 145, 147–159, 161f., 164–167, 170f., 188, 206, 212, 227, 230, 233, 247f.
– Hospital of San Giovanni di Dio   145

– Hospital of Santa Barbara   149, 153, 157f., 166, 185
– Hospital of Santa Chiara   129
– Hospital of Santa Maria Nuova   127, 129f.
– Hospital of Sant'Antonio   127
– Hospital of Santo Spirito   150
– Hospital of the Bagno   133, 150f., 159, 160–164, 171
– Ospedale del Ceppo   130
– Ospedale Nuovo Pisa   130
Humoral Theory   15, 49, 63, 134, 141, 236

Inquisition   65, 102, 109f., 112, 162, 202, 214f., 219, 238
Iron sock   162, 166, 181, 209

Jesuits   73f., 122, 154f.

Lepanto (Battle of)   5f., 69, 72, 95, 115, 121f., 135, 148, 157, 220, 247

Medical Ethic   1, 10, 33–36, 38, 46, 67, 248
Medical examination   3f, 8, 34–37, 46, 67f., 87, 168, 169, 173ff., 179, 187, 195, 225f.
Methodus Testificandi   12, 60–62, 64, 67, 167, 225
Metical Etiquette   37–40
Mission   73–75, 96, 154f., 245

Observationes   26
Optimus Medicus   1, 11f., 15, 27, 32, 34, 61f., 64

Papal States   4, 13, 64, 81f., 89, 102, 113–117, 145, 149, 155, 169, 171, 205, 213, 249
Papasso   158
Plague   7, 11, 24, 28, 53, 55, 95, 116, 133, 136f., 139–147, 151, 153–155, 157f., 187f.
Prison   2, 75, 88f., 103–105, 111, 116f., 149, 161f., 169–171, 181, 183, 205, 209

Ransom   4, 75, 83, 173, 176–178

Simulation 45–47
– Disease Simulation 48, 63, 65, 167
Sodomy 3, 13, 54f., 69, 102, 107, 109f., 119, 121, 191–198, 200–207, 209–211, 213–219, 221–226, 228–231, 233–242, 248f.
Surveillance 2, 8, 10f., 40, 70, 72, 100, 106, 121, 231, 243, 245
– Surveillance studies 8, 10

Taglio degli schiavi 173

Turks 4–6, 69, 83, 96, 98, 100, 103, 109, 113, 134, 152, 172, 216, 219, 221
"Turks" 80, 83f., 158f.

Vigilance 5–8, 10–15, 21, 35f., 38, 41, 43, 46–48, 50, 54, 68–71, 99, 106, 110, 118, 121f., 141, 160, 191, 206, 209, 223f., 243–245, 247
– Medical vigilance 1, 3, 6, 8, 10, 12f., 15, 48f., 54, 68, 77, 139, 243–245, 248, 250
– Vigilance Studies 10, 243
Vigilanti di Maria 110, 151, 209f., 213

www.ingramcontent.com/pod-product-compliance
Lightning Source LLC
Chambersburg PA
CBHW031724230426
43669CB00007B/232